TS
155.8
.M32
1992

Making manufacturing
cells work.

$60.00

DATE			

BAKER & TAYLOR BOOKS

Making Manufacturing Cells Work

Lee R. Nyman
Editor

INGERSOLL ENGINEERS

Richard Perich
Publications Administrator

Published by

Society of Manufacturing Engineers
in cooperation with the
Computer and Automated Systems
Association of SME
One SME Drive
P.O. Box 930
Dearborn, Michigan 48121

McGraw-Hill, Inc.
New York   San Francisco   Washington, D.C.   Auckland
Bogotá   Caracas   Lisbon   London   Madrid
Mexico City   Milan   Montreal   New Delhi   San Juan
Singapore   Sydney   Tokyo   Toronto

Making Manufacturing Cells Work

Copyright © 1992
Society of Manufacturing Engineers
Dearborn, Michigan 48121

First Edition

Third Printing

Library of Congress Catalog Number: 92-80972

International Standard Book Number: 0-07-047977-1

Manufactured in the United States of America

ACKNOWLEDGEMENTS

Writing a book is *not* one of the easiest things for a group of manufacturing consultants to do, believe me. This effort has been one of the most painful activities I have ever seen us attempt at Ingersoll Engineers. As a consulting firm, we have a lot of know-how when it comes to conceptualizing, developing, detailing, installing, coddling, and debugging cellular manufacturing. But trying to get us to put that knowledge into coherent, black-and-white wording—what an experience!

Fortunately, we are blessed with some writing and editorial talent that was patient and persevering with all the consulting professionals who contributed to this wonderful effort at *finally* documenting what we have been preaching for so long. Our thanks to the people at the Society of Manufacturing Engineers for their forbearance while we went through our struggles, notably Reference Publications Division Director Tom Drozda, Publications Development Manager Bob King, and Publications Administrator Richard Perich.

Each chapter of the book was originally drafted by an Ingersoll Engineers staff member experienced in the subject matter (hence the varying formats) and reviewed by his colleagues. This was a deliberate effort on our part to solicit the ultimate mix of approaches, views, and experience in the variety of subjects presented throughout the 21 chapters of the book. We must acknowledge the contributions, direct and more subtle, of the following people:

James B. Ayers	James J. McCarthy
George J. Ballantyne	J. Stanton McGroarty
Robert L. Callahan	Ronald F. Miller
D. Gregory Cummins	James E. Murphy
William A. Dvorak	Richard W. Poseley
Marc J. Fisher	Thomas P. Reif
Robert K. Hill	Steven E. Sadler
James M. Hornik	Joseph A. Smolar
Mark W. Huffman	Owen L. Spannaus
Kenneth W. Lang	Edward Thomas
Herbert A. Marsh	

Special appreciation must go to a few like Stan, Herb, Bob H., Ken, Marc F., and Ed, who actually *volunteered* for more than their share, and to my reviewer, Ingersoll Engineers President Bob Callahan. Editing was done by Heidi Tennant and Polly Kahler, graphic arts by Mike Phelps and Lee Ann Gustafson. The text coordinator was Peggy Johnson, and Debbie Palmer served as proofreader.

Finally, we were all very pleased with the enthusiastic responses of our clients in allowing us to detail some of their experiences with cellular applications. For their commitment to the advancement of manufacturing, we recognize:

Automatic Switch Company Aiken, South Carolina	Mr. Larry Kimball, director of operations Mr. Lee Taylor, manufacturing engineer
Ingersoll-Rand Company Centrifugal Compressor Division Mayfield, Kentucky	Mr. Charles G. Blair, project manager Ms. Judy A. Hendrick, controller
Ingersoll-Rand Company Engineered Pump Division Phillipsburg, New Jersey	Mr. Edward G. Adamusinski, manufacturing engineering planning supervisor Mr. Curt Sittler, manager of manufacturing engineering
John Deere Harvester Works A Factory of Deere and Company East Moline, Illinois	Mr. Richard G. Kliene, general manager Mr. Daryl E. Schloz, combine manufacturing manager Mr. Roger L. Boostrom, Focus '90s project manager
Metalsa, S.A. de C.V. Monterrey, N.L., Mexico	Mr. Antonio R. Zarate, chief executive officer Mr. Hernan Garcia, director of Monterrey operations
Union Special Corporation Huntley, Illinois	Mr. Joseph E. Mercure, executive vice president of operations worldwide Mr. Leland J. Brasile, director of manufacturing engineering

We'd enjoy hearing from readers of this book, with thoughts, concerns, or suggestions for a possible revision in a few years. This will assist us and SME in learning what you are facing in your day-to-day quest to "make things." Please contact me at:

Ingersoll Engineers
1021 N. Mulford Road
Rockford, IL 61107
(815) 395-6502

Lee Nyman

TABLE OF CONTENTS

INTRODUCTION

This book is based on more than 10 years of hands-on cellular manufacturing experience with hundreds of businesses, installing several thousand flexible manufacturing cells—in North and South America, Europe, Australia, and Japan. These manufacturing businesses produce a broad range of products and parts in varying volumes, shapes, materials, and sizes. It is on the success of these installations, as well as the failures and those that were simply frustrating, that this book is based.

Without exception, the first lesson that experience teaches is that no two flexible manufacturing cells (FMCs) are the same. For countless reasons, each business environment and the inherent conditions within that environment require a different approach and yield a unique end result. Even companies producing essentially the same products in about the same quantities reach different FMC solutions. However, enough has been learned to provide a solid body of knowledge about this subject; generic approaches now exist to act as a guide for those who are involved with FMCs, especially those who have no prior experience with them.

Successful FMC implementation is a complex process. One of the reasons for this is the erroneous way the FMC concept was introduced into manufacturing companies in the Western world. It's important for FMC practitioners to understand this introduction along with numerous other aspects of the concepts underlying a financially successful FMC.

THE BEGINNING

Flexible manufacturing cells began their rise to popularity in Western manufacturing companies in the early 1980s. As products manufactured offshore with FMC techniques made their way into Western markets—markets once completely dominated by Western manufacturers—the new production processes also began moving west. In response to this invasion, thousands of curious Western manufacturing executives and engineers visited the plants of these offshore competitors to learn how they were able to outperform Western manufacturers in price, quality, and service.

These visitors noticed numerous and significant differences in the way most of the offshore competitors operated their businesses. Many returned home and began experimenting with the innovations they had witnessed. One of these differences is the subject of this book—flexible manufacturing cells. Other differences, incidently, included things such as quality circles (which today

i

Western manufacturers refer to as employee empowerment or involvement), Just In Time delivery methods, and others, some with foreign sounding names such as Kanban. All of this was new, a little strange, and undoubtedly confusing, but many Western manufacturers concluded that they had little choice but to imitate these offshore competitors.

WE BEGAN TO COPY

With respect to FMCs, what the Western visitors saw were factories producing low- to medium-volume part quantities with plant layouts that were distinctly different. Rather than having factories laid out by process—with similar machines grouped together in the traditional method—they saw low- to medium-volume product factories with *dissimilar* machines grouped together. These groups of machines were performing dissimilar processes such as cutting, grinding, and drilling on the same part, or parts that looked alike. The economic advantages of this cellular method of arranging machines over the traditional Western method was not readily apparent. Nevertheless, many Western manufacturers began experimenting with arranging dissimilar machines close together and dedicating these machines to making just one part or group of parts.

After much experimentation, cells were perceived as moneymakers for their users, even though the financial benefits were often difficult to quantify using traditional financial measurement systems. Increasing numbers of manufacturers began to experiment. In every case they learned that numerous other parts of the business required changes once they began implementation. In the cell environment, traditional setup, scheduling, maintenance, and supervisory practices were inappropriate. In this case, major and significant errors occurred—cells were introduced into environments unsuited to them. Trying to mount a fuel injection pump onto a 1932 Model A engine will produce the same results. The technology is great but the environment is all wrong. And as we entered the 1990s, *this* was still happening with FMC installations.

AN EXPENSIVE ERROR

The offshore originators of cells didn't develop them this way. Rather, cells were a natural outgrowth of a grander strategy to dominate markets. They were a natural byproduct of an entirely new way to run a manufacturing business that has proven to be lean, quick, and simple. For instance, in the offshore manufacturing environment, chopping setup times to a fraction of their historical requirement preceded cell development, as did the establishment of new maintenance, scheduling, and supervisory practices. The eminently successful and practical Kanban scheduling practices were in place well before cells were installed. So were Just In Time practices. Likewise, an environment in which cells could work was established through the use of quality circles involving the people in the shop. Similarly, measurement systems, rather than "variances," were aimed at measuring important cost areas such as quality, throughput time, and work-in-process inventories—all of which are positively impacted by cells.

The point of all of this discussion is that the major problem cells face in Western manufacturing companies is the lack of preparation for their use. When cells are placed in a traditionally run manufacturing business, it becomes evident fairly quickly that the peripheral changes must also occur if the cells are to produce their inherent economic benefits. Some Western companies are quick to recognize the necessity to make such changes; most are not. Frequently, years of little positive activity and great confusion follow a cell installation while the peripheral issues are half-heartedly, haphazardly addressed.

Cells simply don't work well, if at all, when they are not part of an *overall* strategy of change undertaken by their users. Cells standing alone are worthless. They are isolated islands remote from the rest of the world. These cells don't fit into the traditional scheduling, product costing, or operational parameters established for the rest of the facility. The *combination* of cells and overall change to peripherals provides the users with the significant competitive advantage enjoyed by the originators of cells.

To compound the error, some businesses became enamored with high technology. It was the thing to do. In most cases during the early 1980s, neither the culture of the people nor the environment where the technology was to be used was appropriate. Technology may be the right thing to do; however, *there is a right time to do it*. Given these conditions, perhaps Roger Smith, former chairman of General Motors, said it best: ". . .Technology only gives us the opportunity to make scrap faster." Cellular manufacturing is complex! It's best to start simply, using as much existing machinery and as many existing processes as possible, and taking care of the peripheral issues before advancing into technology.

LESSONS LEARNED

To help avoid such errors, Chapters One through Four of this book deal with the up-front strategy, thinking, and planning that is mandatory if the full benefit of cells is to be realized. (We note that most cellular installations in place today were not accompanied by such up-front work.) The following pages describe the learning process that large numbers of manufacturers went through as the cellular concept was adapted and developed in Western manufacturing businesses. For example, shortly after their first cell became operational, most users discovered the following:

- Machine utilization percentages decreased in the new way of grouping machines, because the cycle time of these dissimilar machines varied widely.
- Machine operators had to be trained to run a number of different kinds of machines because there was not enough work on each machine in the group to keep them busy at it.
- The new method of arranging machines reduced a number of costs, most

iii

notably in overhead accounts such as material handling and quality.

- The new method also significantly reduced the amount of time the parts had to spend in the shop, because parts did not leave the area and processes could be quickly accomplished.
- This reduction in the time that parts spent in the shop resulted in less work-in-process inventory, in turn requiring less floor space.
- The use of loading and unloading robots became more feasible because they could service more than one machine when the machines were grouped closely together.
- On the other hand, the new arrangement appeared to be messy, with dissimilar machines placed, for example, in a series of circles compared to similar machines placed in straight, attractive rows.
- Furthermore, moving machines to the new configuration could be expensive, especially if large machines with foundations had to be moved.
- Finally, costly production interruptions could be encountered in making the required changes in plant layout.

It took several years for Western manufacturers to put these advantages and disadvantages into perspective. By the mid-1980s, however, it became clear that the advantages far outweighed the disadvantages.

As enough people became involved in installations of groups of dissimilar machines and exchanged views in seminars, media articles, and on the job, disciplined approaches to this new way of arranging machines began to develop. At the same time, as knowledge of this new machine grouping methodology grew, a host of peripheral issues surfaced. For example, grouping machines this new way required:

- Changes in the scheduling of parts into the shop, because the existing scheduling systems (usually MRP) were found to be unsuitable;
- Changes in data collection for accounting and the subsequent reporting;
- New approaches in machine setup, maintenance, and tool supply, because machines grouped in this new way could not be shut down or delayed much or the financial benefits would quickly disappear;
- New methods of supervision and operator training, because operators took on added responsibilities due to multitasking requirements within cells;
- More flexible and modular fixturing, because two or more parts with different dimensions could be moved through the same machine group (in some cases including the fixture).

As these types of change resulted from the new, better method of grouping machines, confusion and, worse, dissension occurred. People simply did not want to take on the additional work created by the new production process. Regardless, the economic advantages of grouping machines this way prevailed, and more companies attacked these peripheral issues. These too were tackled often enough so that disciplined approaches began to develop.

Detailed attention to minimizing the time to perform setup tasks was a critical step in making it possible to process parts in small quantities on dissimilar standard machines. Setup times were reduced by staggering amounts—reductions of over 90% were not uncommon. Once these reductions were achieved, a standard, normally low-volume machine could be made to run almost continuously. When grouped with other machines also with short setup times, the whole group of machines could run virtually nonstop.

Machine tool builders then recognized the value of keeping setup times minimal, and they began to supply standard machines designed for minimum setup. One example is in punch presses, where traditionally, press setup was a laborious, time-consuming procedure. Today, quick die change is a growing concept. Another example involves numerical control of metalcutting machines. NC permits huge reductions in setup times on this kind of equipment.

CELLULAR CONCEPT MIGRATION

Today, cells are no longer restricted to only metalworking manufacturing environments. Cellular principles have migrated into numerous other types of manufacturing, including castings, electronics, food, and plastics.

In just the last few years, FMC principles have also made their way into office functions or paperwork processing areas. As in manufacturing, where the machines performing the work steps are linked together, the work steps involved in paper flow are linked by placing the people doing the work together so that delays are minimized.

The principle underlying this new method of accomplishing successive work steps is that, where a series of successive work steps must be carried out to finish an overall task, directly coupling or linking these tasks is much more cost-effective than doing the work steps in separate locations. This is true whether the tasks are performed by machines or people.

Initially, most cells were machining cells. Today, of course, flexible manufacturing cells are commonly referred to in a variety of ways, depending on who is referring to them. For example, they are called *manufacturing cells* instead of *machining cells* because work other than machining is increasingly found in cells, processes such as assembly, fabrication, heat treat, washing, packaging, and shipping. The word *flexible* in the name became appropriate as growing numbers of techniques were developed to send a wider variety of part dimensions and shapes through a group of dissimilar machines, and to adjust the cell configuration and output as market and customer needs dictated.

SOME OTHER THOUGHTS

While the word "new" describes FMCs throughout this introductory chapter, this method of arranging machines and equipment in manufacturing dates back to the Industrial Revolution. What is new, however, is the FMC's adaptation of the production line concept to lower-volume, higher-variety operations.

Furthermore, most FMCs (in Western companies) were *not* constructed using the sequence outlined in this book. Manufacturers used (and still use) a variety of approaches when implementing cells. Some, for example, simply moved all the machines required to complete the work steps on a specific part to one location. Then, through trial and error, they slowly developed an environment in which FMCs could be successful.

Today, many such installations are producing good results despite their undisciplined origins. Furthermore, those manufacturers who have had several years of experience with FMCs are finding that once they have installed operationally and financially sound FMCs, they must continuously refine them to keep them that way. This need for continuous refinement is understandable if an FMC is likened to a transfer line; that is, any change in size, shape, volume, or process can have a profound effect on the output of an entire transfer line. These changes must be sorted out as they occur. Similarly, as these changes occur in an FMC, the methods for accommodating them must be developed if the FMC output is to remain optimal. The difference between a transfer line and an FMC when these inevitable part changes occur is:

- In a transfer line, it usually takes engineers to redesign the mechanisms in the line and machine tool builders to apply them to the complex changes required.
- In an FMC, the operators usually can be relied upon to design and apply most of the methods for accommodating the change.

TECHNOLOGY ADDED TO THE FMC CONCEPT

Experienced users of FMCs have found that, once they mastered the basics of cellular manufacturing, they could apply technological innovations to allow FMCs to perform many of the work steps with mechanisms instead of people.

- Computer-enhanced machine tools (using NC, CNC, or DNC instructions) with computer-driven tool changers could machine parts.
- Automated guided vehicles (AGVs) could be used to move parts, especially large or heavy parts, between machines. They could also take the parts to the next manufacturing operation.
- Robots could remove parts from AGVs, place them on machines, and afterward return the parts to the AGVs.

A part of the continuous technological refinement attempted by a few experienced FMC users was dubbed *CIM*—computer-integrated manufacturing. The push for CIM was backed by the belief that AGVs, robots, and computer-instructed machines could be integrated with other computer systems in a manufacturing company—for example, with databases, production and inventory control systems, and accounting systems. This, however, proved to be too much of a technological reach for virtually all manufacturers; the overwhelming majority of them were simply not ready for such sophisticated technology. Even today, for most manufacturers, CIM is an automation goal more than a reality.

THE CONTENTS OF THIS BOOK

In short, there are few, if any, hard and fast rules for successful FMC development. There is, however, a generic approach to FMC development for the beginning user of FMCs. We have restricted this book to describing a general approach to FMCs in a factory environment only. The material is presented for those starting from scratch with FMCs, beginning with understanding the strategic goals of the business and the market (the customers of the parts coming off an FMC) and ending with an operational, financially sound FMC installation.

The interests of the people reading this book will vary widely. Some parts or chapters will be of great or little interest depending on whether the reader is associated with a company whose managers:

- Are well-versed in the FMC concept, or have yet to install their first one;
- Plan to use FMCs as a strategic competitive weapon, or are relatively free of market pressures and simply want to experiment to determine if FMCs will work in their environment;
- Are comfortable with proposing, paving the way for, and promoting changes in the way work is done and the business is run, or are hesitant to do so.

The functional background of the reader also will dictate where his or her interest lies. Some manufacturing engineers, for example, might think about skipping the first four chapters because the subject matter seems to have little to do with their functional role in the company. But FMCs are, by definition, cross-functional, and if their financial benefits are to be fully exploited, the engineer must understand the influences other functions will have on what he or she is doing, and how to deal with the effects cells will have on these other functions. Successful cell implementation also requires complete top-down, bottom-up support within the company. So our engineer must understand what the concerns of top management will be, and how the hourly work force will react to planned changes.

The 21 chapters of this book fall roughly into five parts, as follows.

Chapters One Through Four: Up-Front Planning For FMCs

FMCs can have a huge impact on the customers (internal or external) for whom the products or parts are produced. It is logical, therefore, to find out what these customers want or need *before* setting out to design an FMC (Chapter One).

It's been proven time and again that the benefits, and sometimes the costs of FMCs can be substantial. It makes great sense to do considerable initial homework in the financial area (Chapter Two) *before* beginning.

Eventually, if they're going to succeed, FMCs will have impact throughout the organization (Chapter Three). As a result, it pays (and it's good politics, too) to spend time up-front considering these impacts and how to involve, early on in some form or another, the people who will be affected; this often means the whole company, and it may mean varying degrees of training and education. In addition, there is a strong trend today, led by more sophisticated and venturesome manufacturers, to make FMC development a part of a larger business improvement program, perhaps called "restructuring" or "world-class manufacturing." The first three chapters also examine how cells fit into these programs, and how they're affected by them.

Finally, FMC design and implementation are *projects* that must be carried out, usually while trying to maintain production. Projects require a different management style than running production. Specific project management techniques must be employed if the FMC project is not to drift on for several years (Chapter Four).

Most of the early designers of FMCs were engrossed in the newness of the FMC concept and spent their time learning how to overcome the engineering and technical problems they encountered. They weren't able to plan. Years can now be chopped from the FMC schedule with such planning, and the information in the first chapters of this book is critical to getting the job done in a reasonable amount of time.

Chapters Five Through Eight: Conceptual FMC Development

The first two chapters of this section (Chapters Five and Six) describe the transition from the up-front planning to the factory floor activities. Discussed are definitions and descriptions of the various types of FMCs, including group technology, product, and process. These chapters take the user through the methodologies of transferring goals and objectives into a useable FMC program and into further detail for individual cell development. The reader will note that FMCs should be developed at a conceptual level first. Detailed engineering is not required at this early stage; it comes later, presuming the go-ahead is granted by upper management.

All programs of this magnitude must include the evaluation of numerous alternatives. Chapter Seven provides a detailed methodology for this evaluation so that all pertinent factors are considered, and the appropriate decisions made.

Chapter Eight is aimed at the all-important "selling to management" step in any project that involves change and has significant financial impact. This process is especially critical if the changes involve a number of different functional organization units or departments—which they invariably do.

In more enlightened companies, such selling and convincing of management may not be as important as in companies where management has to be educated and *then* convinced. Furthermore, if the development process includes on-going management involvement and commitment (as it should), the selling process will be automatic and only a formality.

Chapters Nine Through 13: Special FMC Considerations

Developing an FMC involves much more than just moving machines into new groupings. Most major programs of this kind do require a certain amount of additional capital and equipment. In Chapter Nine, the reader will obtain an understanding of using this capital judiciously.

This section also delves into some supporting systems that must be considered to help FMCs operate more efficiently, such as material handling (Chapter 10) and robotics (Chapter 11).

The impact that FMCs will have on quality is covered in Chapter 12; reducing inventories and throughput time demands a level of quality and precision unheard of 10 years ago. Chapter 13 addresses the need for change within the manufacturing control systems. With people working on various parts or orders simultaneously, traditional scheduling programs become inadequate.

Chapters 14 Through 18: Implementing FMCs

The first four chapters take the reader through the detail planning and the actual implementation of an FMC. An important part of any implementation program is the measurement of results; Chapter 18 identifies the need to establish a base line from which results can be measured, and the various ways the user can measure the FMC's impact on the bottom line.

Chapters 19 And 20: Integration And Trends

In Chapters 19 and 20, a number of post-FMC installation observations are made, concluding with a series of predictions on the future of the FMCs. Chapter 21 contains case studies on some manufacturers who are breaking new ground with cells.

SUMMARY

Although this book is written primarily for the reader who may have little experience with the FMC concept, those with experience should be able to benefit from it as well. For instance, the economic justification of installing FMCs is a complex subject mastered by few to date. Similarly, the eventual impact of FMCs on the entire organization is still being evaluated.

One thing is certain—the FMC concept is remaking the factories and offices of Western business enterprises. The concept is exciting, and a successful installation is most gratifying.

The overall difficulty with the FMC concept is that the concept means "change," and on a wholesale level! While some relish such change, most do not; this is the principal reason that this book spends considerable time on the careful planning, handling, and selling of change issues. Unwilling, uninvolved, unsold people are a chief cause for the failure of FMCs and they need not be.

AN FMC STRATEGY BEGINS WITH THE CUSTOMER

Flexible manufacturing cells (FMCs) are transforming manufacturing. Companies now enjoying the advantages of cells find that these gains come in two forms: higher profits on the bottom line from reduced costs, and better performance in the marketplace through improved service and greater flexibility.

This chapter and the next describe why and how companies might start their own process of FMC development. Chapter One describes how cells can better satisfy customers. Chapter Two discusses ways cells can improve financial performance. The remainder of the book details specific aspects of cell design. While technically oriented people may wish to skip to that point, the company-wide involvement required in a cellular project makes it important that *everyone* involved in the effort understand all of the aspects of such a significant undertaking.

WHY CELLS?

Many companies will never enjoy the benefits of flexible manufacturing cells. Some will never consider them at all. Or cellular manufacturing may be tried but, because of a lack of success, the effort will be abandoned. These situations—never trying or failed attempts—can usually be traced to fundamental omissions in the planning effort. Seldom are such dead-end efforts the result of some technical problem.

Most planning shortfalls are failures to take into account the needs of customers and the financial environment. When cell implementation is not linked to the customer service and financial strategic drivers, the program becomes simply a functional manufacturing effort. Lacking broad, top-level support, cells fail. Cells should be part of an "enterprise" solution for better customer service, market position, and financial performance.

The Enterprise Perspective

Cells have tremendous potential for transforming relationships throughout the enterprise. This includes not only the touch labor of direct manufacturing operations, but also support organizations like scheduling, engineering, procurement, upstream suppliers, and downstream customers.

Figure 1-1 is a useful model for describing this enterprise perspective. At the top of the triangle is the enterprise. It is the whole of the organization— encompassing manufacturing, design, sales and marketing, distribution, suppliers, and the support departments a company needs to operate. Within the enterprise is the focused factory. This "business within a business" represents a natural division along customer, product, or process lines. A focused factory is specialized in some way to achieve an advantage in the marketplace for the goods it produces.

The "cell" is the work group, with associated process equipment and tools to produce the goods and services of the focused factory. The cell becomes the base of the enterprise. Because this enterprise architecture provides one of the most effective ways to compete, Ingersoll Engineers describes it as the Agile Enterprise[SM]. For the company seeking to improve its competitive position, using cells as building blocks to focused factories is the best way to start on the road to becoming an agile enterprise.

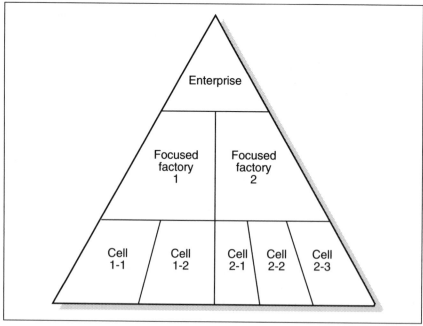

Figure 1-1. *The cell and cell team form the base of the agile enterprise.*

The Manufacturing Enterprise—Past, Present, And Future

A company considering cells often needs motivation to proceed. After all, a commitment to cellular manufacturing is a major one. Successful implementation will change the company forever.

In late 1991, an industry-led consortium, under the auspices of the Office of the Secretary of Defense and the Iacocca Institute at Lehigh University, published a report which provided a vision of the agile manufacturing enterprise of the year 2006. The authors quote a study that traced the evolution of manufacturing since 1800; *Figure 1-2*[1] is an excerpt from it, showing how manufacturing systems have evolved, and the productivity increase associated with each transformation. Each new system produced a substantial improvement in productivity over the system it displaced, with gains of three to one and more being the norm.

The report also states that industry is in the midst of a transition to agile manufacturing, with an expected threefold increase in performance. The Lehigh report lists 18 characteristics of the agile enterprise, as shown in *Figure 1-3*.[2] The figure also indicates where cellular manufacturing will help an organization achieve agility; while not a total answer in any category, cells support at least two-thirds of the characteristics. The measures described in this book are an important first step in preparing the organization to be a 21st-century survivor.

Short-Term Reasons For Cells

But many companies are more concerned with short-term survival. To them the year 2006 is a long way off—they only want to survive and thrive until next year. For those with more immediate objectives and obstacles to overcome, there are many good short-term reasons for incorporating cells into a strategy:

- *Lack of performance and accountability.* The company cannot meet its quality, delivery, or financial goals. Cells are seen as a means to identify and systematically address production problems.
- *Time-based management.* The system for designing and delivering goods and services is expected to be fast and responsive to customer needs. Beating the competition to market with a new product has a dramatic impact on that product's profitability. Keeping customers for existing products requires continuous improvement in making and supporting the product through its life cycle.
- *Saving money.* Cost reduction is often a necessary, but not sufficient, requirement for implementing cells. Those controlling the approval process frequently need financial justification before supporting a plan. Unfortunately, such approval processes often assume that competitors are standing still. Because some benefits from cellular manufacturing are hard to identify (see Chapter Two), the projects do not move forward.
- *Job enrichment.* Redesign of manufacturing-related processes will make the workplace more attractive to employees. Cells broaden responsibility, bring suppliers closer to customers, and instill a spirit of teamwork. The

3

EVOLUTION OF MANUFACTURING

	English system (1800-1850)	American system (1850-1900)	Taylor's scientific management (1900-1950)	Dynamic world (1950-)	NC era (1970-)	CIM (1980-)	Agile Enterprise (2000-)
Typical number of machines	3	50	150	150	50	30	Count systems, not machines
Minimum number of people for efficiency	40	150	300	300	100	30	Count people on team
Staff/line ratio	0:40	20:130	60:240	100:200	50:50	20:10	40:10?
Productivity increases over previous epoch	4:1	3:1	3:1	3:2	3:1	3:1	3:1?
Rework as fraction of total work	.8	.5	.25	.08	.02	.005	.002?
Engineering ethos	Mechanical	Manufacturing	Industrial	Quality	Systems	Knowledge	Systems understanding
Process focus	Accuracy	Repeatability	Reproducibility	Stability	Adaptability	Versatility	Agility

Figure 1-2. *Adapted from R. Jaikumar, "From Filing and Fitting to Flexible Manufacturing: A Study in the Evolution of Process Control," Harvard Business School, 1988, as it appeared in* **21st Century Manufacturing Enterprise Strategy**, *Volume 2, November 1991. The column "Agile Enterprise" was added in the second publication. Small arms manufacturing served as the basis for the historical study.*

Agile Enterprise Characteristics	Impact Of Cellular Manufacturing As An Enabling Subsystem
Concurrency in design and manufacturing	Substantial
Continuous education	
Customer responsive	Substantial
Dynamic multi-venturing	Substantial
Employees valued	
Empowered employees in teams	Substantial
Environmentally benign	
Flexible (re-)configuration	Substantial
Information accessible and used	Moderate
Knowledgeable employees	Substantial
Open architecture	
Optimum first-time design	
Quality over product life	Moderate
Short cycle time	Substantial
Technology leadership	Substantial
Technology sensitive	
Total enterprise integration	Substantial
Vision-based management	Moderate

Figure 1-3. *Cellular manufacturing contributes to achieving many of the characteristics of an agile enterprise.*

payoff is reduced turnover, retention of scarce resources, and improved customer satisfaction.

- *Customer/competitor pressure.* Companies, particularly those who are "followers," want to avoid being bypassed by the competition. So they pursue cellular manufacturing, not out of conviction, but to catch up to others.
- *Widespread dissatisfaction.* Often financial performance is satisfactory, but management knows it could be better. Changes to the current system, however, bring only small improvements (less than 10%) in terms of speed or cost. Cells become a way to achieve real breakthroughs in performance.
- *Growth/contraction.* Rapid changes in business level—up or down—produce a need to restructure. Cells become a logical alternative to the current functional organization.
- *Overhead reduction.* Often overlooked in financial justification for cells is their impact on overhead. Well-designed cells will reduce support functions like production control, engineering support, inventory-related costs, expediting, and accounting, and will increase the capacity of fixed assets.

In summary, in the cell environment, high-quality products should flow through the factory faster than ever before, as should information. Inventories will be substantially reduced, as will scrap, rework, and material obsolescence. In addition, companies will likely enjoy increased market share. Even in declining markets, the company that has better delivery and reliability gains market share—frequently a significant amount over a short period of time.

MARKET STRATEGY PLANNING

Cells often are overlooked when manufacturers plan new products or seek to improve the performance of existing products. But this is a mistake; cells have a role throughout the product life cycle. Teams from marketing, design, finance, and manufacturing will increasingly participate in joint planning that will include flexible manufacturing cells. These teams must view product lines for opportunities for FMCs and develop a strategy for FMC support of the product line.

Product Concepts

The product concept is an important tool in planning flexible manufacturing cells; it is the nucleus around which all company activities should be oriented. The definition of a product is "any tangible physical object which satisfies wants or needs that may be presented to the market for acquisition, usage, and/or consumption." But the idea of a product concept extends far beyond the physical product itself.

To understand the product concept from an enterprise perspective, it is useful to examine the varying levels of a product, as shown in *Figure 1-4*. This figure shows that not only is a product a basic tangible object, but it also includes such

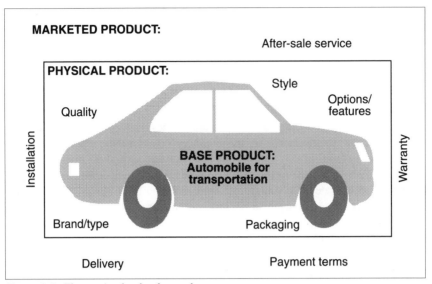

Figure 1-4. *The varying levels of a product.*

things as packaging, brand name, quality, style, and features. Augmentation factors such as installation, delivery, credit, after-sale service, and warranty complete the product definition.

Companies such as IBM and Caterpillar sell more than the basic product; they sell service, reliability, and aftermarket support as strongly as the actual product and its features. Who doesn't know that a Mercedes-Benz automobile represents the ultimate in engineering, or that one is buying quality when purchasing a John Deere tractor?

The product concept has significance for the cell designer. Too often, cell design is viewed as an isolated shop floor activity; the process is then limited to the physical arrangement of machines, rewriting work instructions, engineering the flow of material, and arranging construction of electrical and piping systems. While these are important activities, other considerations may be just as important: order flows from customer to the cell, support requirements for spare parts, and flexibility to accommodate design changes. If the designer only considers basics like direct labor productivity, the cell design may fall short in terms of strategy implementation. A discussion later in this chapter covers "quality function deployment" (QFD), a methodology for capturing these issues during cell design.

Product Classification

There are many ways to classify products. Experienced marketing and manufacturing professionals will know the various methods for performing these analyses. All planning begins with logical groupings of product lines that are examined for their importance both today and in the future. To do this, one

7

should begin with available sales and profit data, the history of each product line, its competitive challenges over time, and its prospects for the future. This information helps identify the best product line candidates for cells, decide which areas to cellularize first, and justify the cost of the effort.

Segmentation

Industrial companies often define their opportunities in terms of customer segments. A cell designer should be aware of these segments, since each may have different requirements in terms of cell performance parameters. An example is a plumbing components company serving several different segments. *Figure 1-5* shows a three-step segmentation of the plumbing components market.

One of the strategic advantages of cells is the ability to tailor the cell to the needs of each customer segment. In this way, the producer avoids the "one size fits all" type of factory; this factory is vulnerable to competitors who focus on the needs of each segment and design their facilities accordingly.

Position Within The Product Life Cycle

The position of the product relative to its life cycle creates dynamic forces in the way the product is accepted in the marketplace, and in the way the company must carry out activities to support it. Awareness of the product life cycle also is important to the cell designer—the life cycle says a lot about the future potential of a product line and how cells can contribute to a product's success. Here are some ways cells can complement a product strategy during the different life cycle phases:

- *Development.* Since there is no production, the role of the cell planner is to support product introduction. He or she should be designing the production environment and ensuring that forecasted market demand can

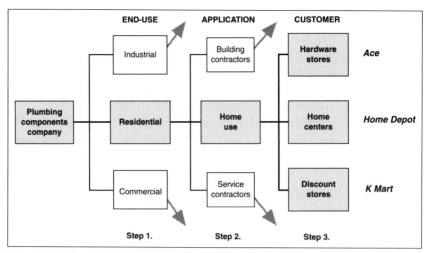

Figure 1-5. *Three-step segmentation of customer markets.*

8

be met. A process called quality function deployment can improve the product's chances of success. Companies successful in reducing their time to market for new products view this early manufacturing involvement as key to their success.

- *Introduction.* During this period of fast growth and market uncertainty, cells can offer flexibility in responding to demand as it develops. Cell design should provide for agility in terms of capacity, lead times, and quality, to ensure that the company locks in a leadership role in the new market. A company can use cells to stabilize the production process, defining the best levels of automation, identifying key suppliers, and organizing for future high-volume production.
- *Growth.* As volumes and profits grow, cells provide "modules" for expansion. A refined cell design can be replicated in new locations in response to the increasing level of sales. This will avoid the loss of market share from an inability to fill orders or poor performance in terms of quality and cost.
- *Maturity and decline.* Cells are often part of a plan to squeeze cost out of a mature product line. If the company is shifting from a traditional functional manufacturing structure to cellular, the move may boost sales by providing better service, quality, or cost. A shift to cells often brings increased market share as competitors clinging to outmoded manufacturing structures are forced to exit the market.

Cells can make a contribution to any product line strategy at any point in the product life cycle. More and more companies will capitalize on this link between the life cycle and cellular technology in their product planning.

MAKING IT HAPPEN

This concept is one of the most important. Plans that are not well-structured and implemented don't add much value to the enterprise. Here is a description of how strategic and market plans can be formed and converted to action through cellular manufacturing. This deployment should include three steps, which can be presented in the form of questions:

- "Where are we now?"
- "Where do we want to go?"
- "How do we get there?"

Step 1: Where Are We Now?

Good planning requires a realistic assessment of the current state of the company. That state should include an understanding of the company's competitive position, the marketplace's view of the company's products and services, and the potential of the business's product lines. This can be a difficult task, as managers must expose and agree upon the company's shortcomings as well as strengths. Management must also articulate the goals and objectives (strategic drivers) that define success for any plan, and make a realistic assessment of the internal potential to reach them. Customer input must be sought through feedback in such critical areas as quality, cost/price, delivery, flexibility, innovation, and service.

Describing the environment. Typically, a company's objectives include measurements such as growth per year, profitability to maintain or grow, establishing or maintaining a leadership position in the marketplace or in the use of technology, or minimizing risk by diversification. Plans for how these strategic drivers are to be accomplished are necessary for the factory planning process. Some basic questions that must be answered include:

- Which customers is the company trying to satisfy?
- What are their needs?
- What products represent future growth opportunities?
- What products are declining or becoming obsolete with new developments?
- How long is the life cycle of the various products in the current product line?
- What technologies are important to this business?

Also useful in this planning are descriptions of the background, history, and planning assumptions of the company—topics might include the following:

- The history of the organization, including major events affecting production facilities and product development;
- Future business and industry environment assumptions;
- The preferences and behavior patterns of current managers and owners, including hurdles for approving change;
- Available financial and technical resources, and
- The current understanding of the company's distinctive competencies.

If a planning team is being utilized, coming to agreement on these basics can save much future discussion and dissention. Some firms hold off-site management retreats to devote adequate quality time to these issues. After answering these questions and gathering this information, a company should identify the competencies on which it will compete.

Methodology for finding distinctive competencies. Strategies are built by focusing company strengths on opportunities in the marketplace. Companies may spend much time and effort in studies to define these strategies; resources may include strategic planning departments, consultants, research organizations, and many hours put in by line managers. Often this effort is required if data is missing concerning markets, technology, internal capability, and competitors.

A SWOT analysis, an examination of the firm's *strengths*, *weaknesses*, *opportunities*, and *threats*, is an effective tool to use in the process, and it can be done in a very short time:

- *Strengths* are resources or capacities that can be used effectively to achieve objectives within the company.
- *Weaknesses* are limitations, faults, or defects within the organization, perhaps as compared to the competition or industry in general.

10

- *Opportunities* are defined as any favorable situation in the environment affecting the company. These can be developing trends, changes in the applicable laws affecting the industry, emerging technology, or an unmet need existing in the market.
- *Threats* are any unfavorable situations that could be potentially dangerous to the organization.

Included in the analysis would be information about the company's market, financial resources, manufacturing/technology capabilities, inventories, marketing and distribution systems, competition, research and development, environmental impacts, historical detail, and projected image.

An understanding of opportunities and threats helps define market areas in which the company can be most effective. Examining the organization's strengths and weaknesses helps to determine the distinctive competencies of a company. This can be the basis for the strategic development that follows. For broad input, the analyses can be done independently and the results compared and discussed in a group. If a strategic direction has already been set for a company, this method is an effective test to understand the validity of this strategy or provide support for changes or iterations. An effective strategy is one which pursues opportunities that utilize the strengths or distinctive competencies of a company, and simultaneously guards against threats by minimizing, fixing, or avoiding its weaknesses.

An example of a completed SWOT analysis appears in *Figure 1-6*. In the scenario created in this SWOT analysis, FMCs could support the strategy in several areas. For example, an opportunity exists to increase sales if delivery was faster and more reliable. Well-designed cells provide fast turnaround on orders. This thrust supports overcoming a perceived weakness, which is long lead times.

The threats indicate that the company has had difficulty maintaining market share, particularly in fast-growing markets. Cells could not only quicken lead time, but also reduce inventory, freeing capital to invest in expansion and product development.

Step 2: Where Do We Want To Go?

Providing a vision for the future is a key task for management. In surveys of manufacturing companies, Ingersoll Engineers has found *lack of vision* to be a key obstacle in improving operations.[3] Too many executives are content to react to the present rather than to manage for the future. An example of this is U.S. manufacturers' overwhelming lack of planning for compliance with ISO 9000 standards, as discussed in *Figure 1-7*.[4] With a good understanding of where a company is, a tool that can speed the answering of the question, "Where do we want to go?" is benchmarking.

Benchmarking. At the heart of the benchmarking activity is identifying a company that is the very best at an activity, function, or capability that is important to one's own success. After an exchange of data with that company regarding the function under consideration, a strategy is developed to incorpo-

S	
	Strong image for 25 years
	One of the leaders
STRENGTHS	International company
	Broad product line
	Large distribution
	Direct sales

W	
	Long leadtimes
	Premium price image
WEAKNESSES	Late on new R & D projects, long development cycle
	Higher costs than many competitors
	Some quality issues

O	
	Automation trend requires more of this product
	45% of competitors' distribution would carry product if given line
OPPORTUNITIES	Many target accounts surveyed unaware of product line
	Many potential customers identified delivery as only factor preventing purchase

T	
	Many competitors with same or larger market share
	Market has been growing faster than our company
THREATS	Some industry segments with fastest growth require long specification acceptance period
	Internal resistance to change

Figure 1-6. *Performing a SWOT analysis helps a company to define markets in which it can be effective using its distinctive competencies.*

rate those aspects of the model company's approach that could be effective, and to better the other company if possible.

WHAT IS ISO 9000?[4]

More than 26 countries, including the United States, have supported the International Standards Organization's (ISO) 9000 series quality guidelines, which are quickly being adopted as the national standards in countries within the European Community and beyond.

It is highly probable that certification of operations under ISO 9000 will be mandatory in certain countries and market segments in the near future; the standard merits serious consideration in determining a company's future vision. Unfortunately, less than 5 percent of U.S. manufacturers that market products overseas have completed the ISO 9000 certification process, which takes at least six months. This lack of a quality program registration under the standard could become a nontariff trade barrier, blocking American firms from NATO contracts and other European business.

Five standards make up the 9000 series:

- ISO 9000 is the road map, providing guidelines for selection and use of the remaining standards.

- ISO 9001 covers quality system requirements primarily for preventing nonconformity at all stages, from design through servicing.

- ISO 9002 covers supplier requirements for preventing or detecting nonconformance during production and installation.

- ISO 9003 covers final product inspection and test on the supplied product.

- ISO 9004 provides guidelines to users in the process of developing in-house quality systems.

In the United States, ANSI will serve as the official U.S. member body to the ISO; the ANSI/American Society for Quality Control Series Q90 is identical to the ISO 9000 series standards.

Figure 1-7. ISO 9000 standards are one of the many influences manufacturers should be considering in defining their companies' future.

Although this sounds simple, locating and exchanging the data is becoming a competitive process in itself. The unique benefit of this process, however, is that it focuses management effort toward *external* best practices—aimed squarely at the marketplace and the company's competition.

Other companies are seeking the advice of the Malcolm Baldrige National Quality Award winners. Xerox and IBM-Rochester, for example, have established formal programs to expose and even train industry managers in the techniques and strategies that helped them win the award. Others use outside resources like manufacturing consultants to bring the necessary outside perspective.

Matching market strategy to cells. The market is the primary consideration for any company strategy. Knowing the company's business, who it serves, and how it serves them provides the framework for a manufacturing strategy that may include FMCs. The unique demands of customers, products, or management are the inputs needed for the detail design.

A decision to install cells, unfortunately, is usually not made as part of answering the "Where do we want to go?" question at the enterprise level; it is usually the result of the manufacturing people alone trying to reduce inventory, improve throughput, improve quality, reduce costs, or maybe even respond to customer demands. All could be good reasons for cells, but their connection to anything higher than plant goals is usually obscure. Making ties between the market strategy and the manufacturing strategy will ensure that cells are applied in the most profitable manner.

Serving product strategies. Product strategies vary in answering the "Where do we want to go?" question, from making commodity fasteners by the millions and selling to anyone who will buy at a fraction of a cent below the competition, to building one-of-a-kind, multimillion dollar turbines for power generation, where price is less important than contract performance. High-volume, repetitive manufacturing, by definition, does not require flexibility. One-piece manufacturing requires high flexibility, but repetition must be found in families of parts.

While these extreme situations could have applications for cells, the most common scenario is more likely to be midway between them. Cells are not the answer in every case; traditional functional manufacturing may serve the need well. In fact, zealous cell designers, in their quest for part families in extremely low-volume environments, can actually encourage inventory building by trying to load cells with similar parts. With this warning in mind, here are some examples of products where cells have competently supported a product strategy:

- *Power hand tools*. Product innovation is important and some customers require delivery from stock. There is variety in the designs and some seasonality in demand. Responding to new designs, volume shifts, and repair parts is a differentiation for the market leader.
- *Aftermarket tailpipes and mufflers*. Huge variety exists in the final shape of the product, with moderate volumes. Raw material is nearly uniform, but it must be bent to final shape just in time to avoid excessive storage and shipping costs.
- *Automotive switches*. Hundreds of new products are required annually. The development cycle requires close interaction between engineering, sales, and manufacturing. Cells provide a way to produce a variety of items and

14

respond to frequent changes in design.

- *Industrial bearings*. Volumes ranging from several thousand to fewer than 100. Sizes vary, but families are well defined and geometries are very similar. Outer races, inner races, cages, and rollers make up every bearing. Customers are distributors, OEMs, and service parts. Designs are mature, but variations happen frequently. Cells provide flexibility for size and volume, as well as redundancy of capacity.

In all cell installations, the opportunity for quality confirmation at each operation increases. Discipline is required to realize improvements, and operators must be trained to use gages, read instructions, communicate, and sometimes to read and write. If quality is part of a product strategy and the volume and mix is right, FMCs usually fill the need; nonconforming quality measurements often drop by 50% or more when cells are implemented. The closeness of operations creates valuable, immediate, and continuous feedback to front-line operations. In this way, the pace of organizational learning accelerates, enabling the manufacturing company to create high-quality products at minimum cost.

Step 3: How Do We Get There?

In reality, the move to cellular manufacturing is rarely a single decision or part of an overall plan to support higher strategies. Instead, it tends to be an incremental learning process. Some people get discouraged by the anemic results from their first attempts, but as the next one gets better, the learning process is alive and growing. Planners learn that, if a cell can function the way one machine used to, then maybe several cells can replace an entire department. If a department can function like a cell, maybe a focused factory is possible, and so on toward the agile enterprise. QFD is one technique that can add some discipline to the process.

Quality function deployment (QFD). Benchmarking can be especially valuable as input into the technique of quality function deployment. QFD requires some concrete objectives for product and process design, plus a realistic understanding of where a company stands with respect to its competitors. It is an approach for designing or redesigning a product or service, and involves teams in developing the actual product or service, plus the necessary support structure to deliver it. It was first practiced in Japan in the early 1970s, and is now quickly gaining popularity throughout the world.

QFD makes use of a series of matrices—often called "houses" because of their shape—as shown in *Figure 1-8*. Each house is a matrix, with requirements along the left and the means to fulfill them along the top. These houses, linked together, translate customer requirements into a definition of the production environment.

The process begins on the left of the figure, with the "house of quality." This matrix translates customer requirements (up the left) into product engineering features (along the top). The matrix shows which customer requirements are met by which product feature, shown in the completed matrix with notations at the

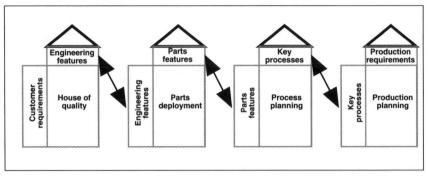

Figure 1-8. *Full deployment requires completing all of the houses.*

intersection of the requirement and the feature. When a requirement is not covered by a feature, the QFD team must decide whether to add a product feature that will meet the customers' requirements.

The output of the first matrix is the input of the second, and so on. After the components of the product and service are defined, the key processes—including the use of manufacturing cells—are defined in the last two matrices.

QFD enables a multidisciplined team to track production processes all the way back to basic customer requirements. What if a customer requirement cannot be fulfilled with the current production system? The customer expectation will have to be changed, or a new production system—often with FMCs—will have to be planned.

Users of QFD are finding many benefits. The discipline of the process requires a complete definition of product requirements. This speeds product development, and it also forces early manufacturing involvement in the design process, reducing delays when the transition from engineering to manufacturing occurs.

The technique overcomes interdepartmental barriers that exist in many companies. To complete the matrices, a team from marketing, customer service, engineering, procurement, and manufacturing must meet frequently and communicate continuously. This shortens decision-making by avoiding traditional chains of command.

The technique is not always easy to use, however. Traditional organizations present barriers to the effectiveness of multifunctional teams. In addition, many companies don't have the discipline to gather and process all of the information required by QFD. Also, particularly at the beginning of a new development project, there is often sparse and incomplete information about what customers want or even who the customers are for a new product.

Since QFD, done properly, will define the production environment, it will identify where deployment of cells will help satisfy customer requirements. In many cases, customer requirements for quality, lead time, and cost can only be satisfied with cellular manufacturing. In other cases, cells may provide the competitive edge that the development team has been seeking.

The use of QFD will grow rapidly and become an important application in FMC planning. This will be particularly true with the growth of the focused factory, where off-the-floor functions like engineering and procurement become part of manufacturing teams within cells.

This bottom-up pressure caused by innovations on the shop floor like cellular manufacturing and the multifunctional teaming of QFD keeps the competitive game interesting. Marketing strategists must understand the implications of these innovations or risk failure in the marketplace. No strategic planner would have factored six-sigma quality performance into a plan in the mid-1980s. No planner would have demanded five-day lead times instead of five weeks. No planner would have told customer service that repair parts can be shipped overnight. All this has been made possible by progress made from the bottom up. The methods described in this book are an important part of staying ahead in the competitive race.

ENDNOTES

1. Nagel, Roger, and Dove, Rick (Editors), *21st Century Manufacturing Enterprise Strategy: An Industry-Led View (Volume 2)*, Bethlehem, PA: Iacocca Institute-Lehigh University, 1991, pp. 95-96.
2. Ibid (*Volume 1*), p. 38.
3. *Change: The Good, The Bad, And The Visionary*, Warwickshire, Great Britain: Ingersoll Engineers, November 1991.
4. Coleman, John R., "ISO 9000 Or Isolation?" *Manufacturing Engineering*, March 1992, p. 4.

IIIII 2 IIIII

MAKING FMCs PAY THEIR WAY

Chapter One described how FMC technology can support customer service and market strategies. But what about the other important constituencies of the company—parent companies, boards of directors, shareholders, banks, and others with a stake in the proverbial "bottom line" (including the company's own financial people)? While an FMC strategy should begin with the customer, a company cannot ignore financial goals and constraints; the influence that these constituents have over the company's decision-makers can be paralyzing.

The financial environment has a number of crosscurrents. How profitable is our market? Is it growing? Shrinking? What kind of financial muscle does our company have? Strong? Weak? How do our competitors stand on the same issues? The answers to these questions will affect the willingness to start an FMC effort, the priorities for the program, and the pace at which it proceeds.

Chapter One's discussion of how to ensure that cellular technology will meet customers' needs considers the external elements influencing the company. This chapter addresses the obstacles to cellular manufacturing inside the organization and some measures to overcome resistance to change and move forward; it is aimed primarily at nonfinancial people who will have to win the support of financial people and others for the FMC project. (Chapter Three discusses the "softer" organizational impacts of change.) This chapter will not tell nonfinancial people how to prepare a cost justification for cells. It will introduce them to ways they can influence the thinking of people who are concerned with such justification.

FMCs must pay their way. If they don't, they may be successes to those outside the company but failures to those within. (Chapter Eight describes ways to "package" this information for selling to top management.) *Figure 2-1* illustrates the situation faced by many companies when considering cellular manufacturing—this "force field analysis" is from a technique developed by Kurt Lewin. The force field shows driving forces for change and restraining

19

FORCES FOR CHANGE			FORCES RESISTING CHANGE
External:			**All Internal:**
Rising "world class" standards for manufacturing performance	$\Rightarrow$	$\Leftarrow$	Requires cultural changes in the way business is done
Immediate competitive threats	$\Rightarrow$	$\Leftarrow$	Unfamiliarity with concept
Internal:		$\Leftarrow$	Organization changes required
Belief in continuous improvement culture	$\Rightarrow$	$\Leftarrow$	Discipline necessary
Better visibility, easier problem-solving	$\Rightarrow$	$\Leftarrow$	Investment in control and scheduling systems required
Increased capacity, ROA	$\Rightarrow$	$\Leftarrow$	Worker inflexibility
Improved working environment	$\Rightarrow$	$\Leftarrow$	Current justification process
Cost reductions: —Direct labor —Support/overhead —Material —Quality	$\Rightarrow$	$\Leftarrow$	"If it ain't broke, don't fix it" culture
		$\Leftarrow$	Emphasis on higher machine utilization

Figure 2-1. *"Force field" analysis for cellular manufacturing. The factors for change must exceed those opposing it to make change begin.*

forces opposing change. When the two are in balance, little movement one way or the other takes place. When forces *for* change exceed those opposing it, change starts to happen.

While the driving forces for change on the left are a mix of internal and external factors, the resisting forces are usually all internal. They constitute the cultural, organizational, and procedural inertia to be overcome before an organization moves ahead with cellular manufacturing. This chapter is in three parts, divided according to the life cycle of a cellular manufacturing project:

- The first part describes overcoming obstacles in the justification process prior to implementation;
- The second addresses the implementation process itself, with attention to ensuring that an effort, once begun, continues;

- The third describes ways to look at what changes were accomplished, using information on what other experienced cell implementers have found (based on the results of a survey by Ingersoll Engineers of companies who have had cells in place long enough to have evaluated the benefits).

JUSTIFYING FMCs

Before a manufacturer commits to its first cell, several events must have happened to cause the forces for change to push back the forces resisting change. These almost always include the following:

- A combination of external market/competitive forces makes change necessary. Chapter One describes some of these forces.
- Management must have "bought in" to the need for what will probably be radical change to existing ways of doing business. This has required some level of education and awareness about cellular manufacturing.
- Finally, a proposal for cellular manufacturing will have faced the hurdles of an approval process. These are usually associated with capital budgeting and, since cells may require a capital expenditure, proposals for cells often face the capital appropriation approval cycle and its many associated issues.

The Approval Cycle—Roadblock To Improvement

To understand the hurdles faced in justification, current management practices with respect to approving proposals for change must be examined. At most United States manufacturing companies, the approval cycle for proposals like cellular manufacturing is a financially oriented one. Cost of implementation is weighed against benefits—financial only—to determine if the project offers a return on investment.

A proposal document, called by different names in different companies, describes the project, the costs, and the benefits. By communicating in financial terms, approvers have a common base for understanding. Hard-to-measure, nonmonetary benefits (often labelled "intangible") may be described in some detail. The writers of these proposals usually rely on the existing cost accounting system for their data, which in turn, rely on a direct labor base over which indirect costs are allocated.

The proposal is passed from approver to approver until authorized or rejected. It's not uncommon to have over a dozen signatures required for major requests. The approved project becomes a part of the business's capital and operating plan, which often has severe advanced timing hurdles.

Interestingly, Japanese manufacturers don't rely on financially oriented approval processes. A study[1] reported in *The Wall Street Journal* indicated that Japanese companies use a far more subjective process. The study concludes that the Japanese are under less short-term financial pressure; their concern is with the human elements, including organization and motivation.

21

But some good projects—particularly those related to cells—do not clear the hurdles in this type of approval process. Reasons can include the lack of financial return or fear that the inherent risks of the project are too great. The following paragraphs describe some methods used with increasing frequency to justify projects like FMCs. They should be extremely helpful in selling sound cellular projects.

Most Approval Processes Assume A Static Competitor

One problem with the process just described is that it is internally focused. That is, if the return on investment from the project's cash savings to the company does not justify the outlays, the project fails.

But such a position assumes that the competitive environment is static—that is, competitors are going to stand still and not undertake any program to increase their market share. One certainty in today's world is that the "quality threshold" for all products is constantly rising. *Figure 2-2* illustrates the situation, showing the quality threshold. This threshold is the product package (see *Figure 1-4*) required to be a player in any market; it encompasses product features, workmanship, and skill in delivering those features consistently and reliably. Any competitor that stands still risks falling below the threshold and being forced from the business. This is illustrated by the position of two competitors, A and B. Company A initially has a larger market share, as represented by the size of its bubble. B has a smaller market share but rates higher than A in quality. Furthermore, A begins right on the threshold but shows no improvement. In time it slips below the threshold, and its market share shrinks. B, on the other hand, improves, raising the threshold and gaining market share.

A trap in the traditional approval process is the assumption of static market shares, a static quality threshold, and static market position. Competitive markets are, in fact, quite dynamic; cells proposals should reflect this reality.

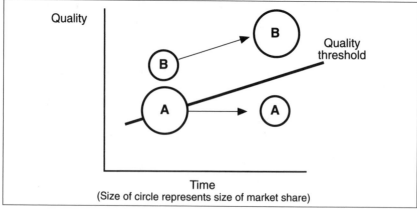

Figure 2-2. *Company A assumed a static market, and as the quality threshold rose, customers switched to Company B's product, with its improved quality.*

22

Chapter One described techniques like benchmarking to determine where a product stands with respect to world-class standards. This information will help overcome the tendency of approvers to think that their current competitive position is static.

Use Function- Or Task-Based Costing

Figure 2-1 showed that a motivator for cellular manufacturing is cost savings. Savings sources can include direct labor, overhead, inventory, and materials. Yet, with the exception of direct labor, these savings are not very visible to most approvers.

Direct labor is measured in detail in most manufacturing operations as part of standard cost accounting systems. The "conventional view" in *Figure 2-3* represents this perspective. In the example, direct labor is $1; overhead is $2. In most accounting systems, overhead in this case would be allocated over processes and products at a rate of 2 to 1.

The consequence is that overhead is often a "given," and projects must be justified on direct savings alone. Because it is not accounted for in detail, overhead cannot be part of the savings benefits.

FUNCTION-BASED COSTING		
	Conventional view	Function-based view
Direct labor	$1.00	$1.00
Overhead	2.00	x
Engineering	x	0.25
Manufacturing support	x	1.25
Sales/Marketing	x	0.50
Material:	4.00	x
—Services		0.50
—Subcontracted		2.00
—Commodity		1.50
TOTAL	**$7.00**	**$7.00**

Figure 2-3. *Costing methods that better link costs to their sources will make it easier to show the effects of cellular manufacturing.*

But cellular manufacturing has financial benefits far beyond direct labor categories. In fact, direct labor in many companies is only 5 to 10% of total cost. Examples of other high-cost categories include support areas or functions like material handling, maintaining inventories, floor space, production scheduling, quality control, production materials, and engineering support.

The function-based costing view, shown very simplistically on the right in *Figure 2-3*, highlights additional areas where savings can occur. If such an approach is taken, the benefits obtained *off* the shop floor are usually several times what is achieved on the shop floor. So a project that fails the justification test in the conventional view might pass with flying colors if a function-based view is applied.

Let The Balance Sheet Finance The Project

The balance sheet also offers opportunities for financially justifying cellular manufacturing. *Figure 2-4* is an example of the asset side of a typical balance sheet. Of the items listed, companies have found cellular benefits in the accounts receivable, inventories, and fixed-asset categories.

Accounts receivable. Accounts receivable can benefit if cellular manufacturing can increase the timeliness, throughput time, and accuracy of deliveries. For example, cells can help shift the company from a "build to stock" to a "build to order" customer service philosophy. This means the company can ship a higher percentage of complete orders, reducing problems associated with shortages, which make getting paid more problematical.

Many customers are implementing "supplier partnership" arrangements. For their most reliable vendors, payment becomes automatic and paperless, with delivery of the complete order.

Inventories. Inventories are the most common balance sheet item targeted by cellular manufacturing. Improvements in quality in cells will increase first-pass yields, reducing the need for contingency raw material stock. Cells also reduce production lead time by 70% and more, having a dramatic impact on work-in-process inventory.

BALANCE SHEET Assets ($ in millions)	
Cash and marketable securities	$50
Accounts receivable	100
Inventories	200
Fixed assets	250
TOTAL	$600

Figure 2-4. *Implementing cellular manufacturing affects most balance sheet asset categories.*

In a build-to-order environment, finished goods inventory is greatly diminished; only contingency stock need be carried. Some companies have achieved 90% reductions in the amount of finished goods. Inventory turn ratios in excess of 30 are easily realistic.

Fixed assets. Fixed asset savings are another target of cellular manufacturing. Cells usually require less factory and warehouse space. Equipment needs may increase if some redundant capacity is needed, but this is not always the case. Most companies implement cells without significant capital investments.

Cells will also decrease the investment needed in systems infrastructure. Control of shop floor production becomes a local activity; large computer systems are used for longer range material planning and master scheduling only.

In a growing company, cells will help avoid additions to physical facilities and improve the yield on existing investments. Many times, growing companies over-expand to match burgeoning demand, but pay later when diminished demand catches them with overcapacity. Deploying cells early in the product life cycle will reduce this risk later on.

Recognize The Cost Of Quality

As the quality movement spreads, more and more companies recognize that a significant share of their costs are nonvalue-adding in terms of customer needs. That is, if the activity causing the cost ceased, the customer would never know the difference. Examples include material movement in the factory, inspection, rework, and expediting material. When companies add up all these nonvalue-adding activities, they find that these costs represent 20 to 50% of their cost of goods sold. Categories used in calculating the cost of quality include:

- *Prevention*—activity undertaken to avoid quality problems. Many organizations are trying to spend more on this category because it is more cost-effective than spending money to correct problems after they occur.
- *Detection*—inspection and auditing activity to catch defects after they occur. This is a nonvalue-adding activity often accounted for as overhead.
- *Correction*—tasks in the factory and field to fix problems. This includes rework on the shop floor, field service, and warranty expense. This is the most expensive activity of all.
- *Customer goodwill*—some companies are trying to capture this cost. Defects have an inevitable impact on the product's position in the marketplace due to inconveniences imposed on customers and their loss of faith in the product.

Many companies have found that cells reduce the cost of quality by causing:

- Less movement and storage of material.
- Fewer defects in fabrication and assembly, because feedback in and from cells is much faster than in a traditionally organized factory.
- Reduced inspection and oversight activity, by making cell operators responsible for quality.
- Improved design-to-build teaming efforts, with earlier debugging and efficient, quality designs.

- Improved goodwill, by allowing direct access between cell workers and customers.
- Removal of production planning, control, and expediting functions, because of simplified layout, reduced inventory, and simplified process design.

Engineers preparing a feasibility study for cellular manufacturing should prepare a cost-of-quality estimate for the affected processes, determine how each component might be affected, and use this data as part of the justification procedure. Cost of quality is a different perspective on the cost of operations. Unfortunately, this perspective is difficult to glean from conventional reporting systems, which bury these costs in overhead.

PROPAGATING FMCs

Often, the initial commitment may be for only one cell; managers want to try the concept, validate it in their operations, measure benefits, and make sure the cell does not upset production. Perhaps there is a longer range plan for more cells—perhaps not. A cellular program needs to be restarted, however, even after initial attempts have been highly successful. There are several ways to ensure that a cellular program, once begun, will continue:

- Maximizing the benefits from the initial effort to ensure continued support.
- Using a cash-sensitive restructuring approach so the organization's exposure to negative cash flow is minimized.
- Building enthusiasm through a ''brushfire'' approach to gain organization-wide enthusiasm.

Use The High-Leverage Project To Maximize Initial Benefits

Any company reviewing cells for implementation will probably have a choice of where to begin. A macro plan covering the whole production system is recommended (see Chapter Five); that plan may call for many cells. But *where* the effort begins will have a major effect on the success of the first cell and how much of the plan actually gets implemented.

Assuming a company has alternatives for starting, each cell will have a distinctive impact on operations, customer service, and financial performance. By picking the alternative with the highest leverage on these factors, a company can get the most out of its initial effort. *Figure 2-5* illustrates the concept; different projects require different levels of effort to produce different levels of benefit. The top graphic shows the relative effort and benefit for Cell A—a large benefit for relatively little effort. Some examples of reasons for the higher benefits are shown below the graphic. Note that many go beyond the immediate impact on shop floor operations to issues covered in Chapter One.

Cell B requires a higher level of effort but produces a lower relative benefit. Its impact does not extend beyond the shop floor. Direct labor or floor space

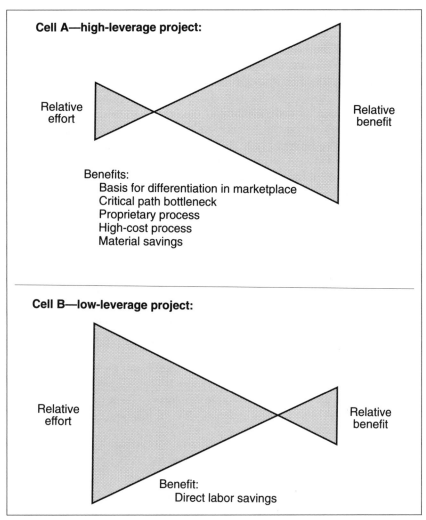

Figure 2-5. *As an initial effort, cell A will result in the greatest benefits for the least effort. This type of "success story" goes far in paving the way for future cells.*

savings may be the only benefit. This does not mean the cell is not a good project. It is just that its scope is limited. Successful implementation of Cell A is more likely to build a broader base of support throughout the organization.

A successful cell that provides benefits beyond the shop floor will reinforce the initiative for additional cells. It will also support the growing view that, for many industries, market success will depend on operating effectiveness. In master planning where multiple cells are to be implemented, the contribution of each to strategic position should be evaluated. Contributions to quality, delivery, and cost from a customer's perspective should be determined for each cell project. The alternatives should then be ranked to help decide where to start.

Plan A Cash-Sensitive Restructuring

As mentioned, cells can be justified by reductions in operating expense and improvements in the balance sheet. A cellular manufacturing master plan can also be structured around the need to conserve cash. In this case, projects may not be sequenced based on strategic impact as discussed, but on the basis of the cash impact.

This will require an understanding of the cost of each cell for capital and expenses, and impact on the balance sheet and income statement accounts. Implementation can be timed to keep net cash flow positive. *Figure 2-6* shows how implementation can be planned to reduce capital exposure. For Plan 1, investment, shown in the shaded area, is far greater than in Plan 2. In Plan 2, initial implementation has focused on lower cost cells with higher cash flow benefits. The capital exposure is far less, and since cash flow is often a primary consideration in managing a company, developing a cash-sensitive restructuring plan can make the changes associated with cellular manufacturing painless from a cash perspective.

Another caution to consider in a cash-sensitive restructuring is that cash flow benefits from cells are often nonlinear. That is, benefits grow proportionately larger as more cells are implemented. This results because a single cell may have only a small impact, cutting a material handler here, or an inspector there. Fractional reductions do not achieve much visibility. *Figure 2-7* illustrates the concept.

As more cells are implemented, total return accelerates. Early cells are islands in a traditional factory structure. As the factory becomes more cellular, the total company reflects the new leaner structure. Top management should be educated to this phenomena before the program begins. In this way, expectations will match likely patterns of benefit flow.

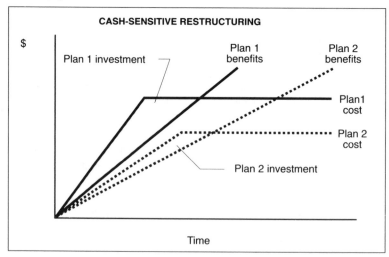

Figure 2-6. *Plan 2 is more sensitive to cash flow and requires less capital exposure.*

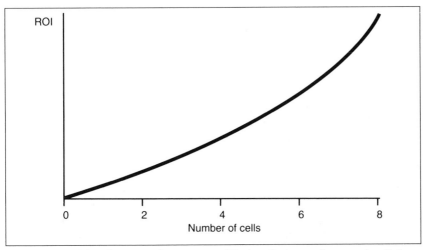

Figure 2-7. *Cell benefits are nonlinear—most of the benefits come after enough cells have been implemented to affect a major portion of manufacturing.*

Build Enthusiasm With Brushfires

Most of the justification and propagation approaches described here have focused on winning top management approval for FMC implementation. But cells can have a major positive impact on worker satisfaction and enthusiasm for the job. Cells can be designed to broaden the scope of a job and the autonomy and the control over how the job is done by the individual. This movement toward enrichment of the task is very much complementary to the movement toward total quality and employee empowerment.

Propagation can occur through brushfires, where enthusiasm from the grass roots spurs further effort throughout the plant; at some point, the brushfires turn into a conflagration. In fact, all the quantitative benefits in the world will not matter if the people who will work in the restructured plant reject the new structure.

RESULTS FROM EXPERIENCED USERS

As with any change, there are always leaders and followers. Followers are particularly interested in the results achieved by the leaders. In describing the experience, of a sample of leading cellular manufacturing implementers, data from a 1990 Ingersoll Engineers' survey of manufacturers is used[2]. The data will demonstrate the extent to which others have realized the benefits of cellular manufacturing. Many of the benefit areas described can serve as a checklist, along with other suggestions in this book, for the company considering cells.

Why Companies Started Cells

The survey shows that most companies have pursued cells only recently. Fewer than 10% had experimented with cells before 1980. Market demands for better delivery were the prominent motivator for shifting to cellular manufacturing. Next, users sought to reduce work-in-process inventory, freeing valuable capital for other purposes. A few cited a better workplace and job satisfaction as a main objective. *Figure 2-8* is a summary of the main objectives for introducing cell manufacturing.

The idea to employ the concept of cellular manufacturing originated most often within the operations function. The idea was also introduced through formation of a business strategy, by an internal champion, or through an external recommendation.

Results Achieved

Improving delivery response and reducing inventory were two primary motivators for FMCs; a majority of companies achieved their goals. *Figure 2-9* shows the percentage of companies reporting shorter lead times (an indicator of more reliable delivery and improved response) and reduced inventory. Other indicators of improved operating effectiveness include those that are shown in *Figure 2-10*.

Implementation Cost

A majority of companies rated their overall investment as *small* or *none*. But in the 30% of cases where significant investment was reported, the allocation

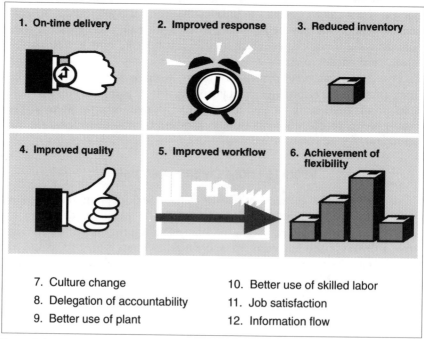

Figure 2-8. *Main objectives of cell introduction. The survey revealed that market pressures are the main drivers.*

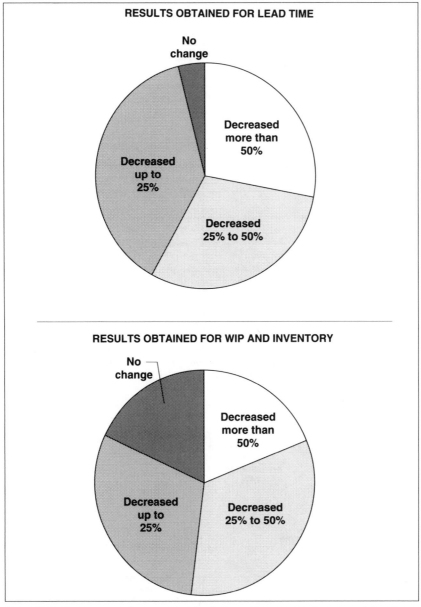

Figure 2-9. *For most people surveyed, introducing cells seemed to be a zero-risk option with many benefits.*

was shown as in *Figure 2-11*. Most of the investment was made by larger companies with sales in excess of $150 million. In most companies, cellular manufacturing was achieved with the rearrangement of existing production equipment.

Indicator:	Companies Reporting Improvements (Percent):
Distance work travels	90
On-time delivery	90
Number of handlings	87
Flexibility	82
Productivity	80
Morale	76
Quality	76
Changeover times	76
Accuracy of information	68
Timeliness of information	65
Documentation	52
Sales	38

Figure 2-10. *Results obtained by the survey respondents, expressed through other common indicators.*

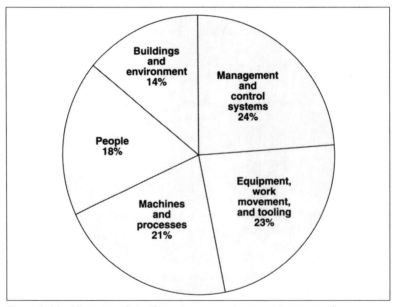

Figure 2-11. *Allocation of significant investment in companies surveyed.*

For manufacturers of products suitable for cellular manufacturing in terms of manufacturing processes and market environment, cells are a low-risk alternative. In fact, the survey showed that only 1% of those who had tried cells planned to abandon the effort. For many others, cells had become a "way of life," as one manager expressed it.

Cells are quietly revolutionizing manufacturing around the world. At some point they will no longer have to be "justified" they will become the natural model of choice for companies organizing to be agile, competitive enterprises.

ENDNOTES

1. Powell, Scott, "Japanese Execs: Financial Wizards, They're Not," *The Wall Street Journal*, April 17, 1989.
2. *Competitive Manufacturing: The Quiet Revolution*, Warwickshire, Great Britain: Ingersoll Engineers, 1990.

IIII 3 IIII

ASSESSING THE
ORGANIZATION IMPACTS

Many manufacturers today think FMCs are only a new way to process parts through a series of work steps by rearranging factory machines and equipment. While this is true, it is only part of the story. The idea behind the FMC concept implies an entirely new way to run a manufacturing business. Underscoring this conclusion are a handful of businesses today that have entirely restructured their companies in response to fierce offshore competition. One of the reasons they did so was their initial experiments with FMCs. This chapter describes how and why FMCs have organizational impact in both the near and long term.

IN THE NEAR TERM

Once FMCs are in place, full benefits cannot be realized until changes, problem solving, and opportunities in other departments, functions, or areas of the company have been pursued. If these other departments do not change the way they operate, the FMC benefits are watered down, and sometimes even completely destroyed. The greatest benefits are realized quickly in companies that include all affected functions from the beginning of the FMC project. The purpose of identifying and discussing these areas of potential impact on FMCs is so that they can be anticipated and planned for, and the appropriate people from these functions can be involved in developing the FMCs. An FMC development organization plan is pictured in *Figure 3-1*.

If these other functions become involved too late, delays can be expected while they make the necessary adjustments in the way they conduct their business to accommodate FMCs. The enthusiasm for the new way of doing work that could have been achieved by involving the people up-front will also never be attained. FMCs are not exclusively a shop technique. They demand extensive changes in nonshop areas. It is far more productive to anticipate and begin to

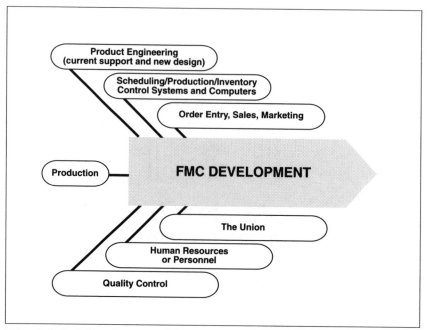

Figure 3-1. *Including all aspects of the organization in FMC planning results in a plan with the greatest chance of success.*

handle these areas from the outset. It is important to understand how each of these peripheral departments can be impacted and, if changes don't occur in them, how the FMC can be negatively impacted.

Product Engineering

It is not unusual, when planning an FMC installation, to find within the factory an overly complex product line, a myriad of parts with little difference between them, an infinite variety of sizes (of holes, for instance), etc. Such conditions are costly in any shop environment. In a traditional nonFMC environment, such product design practices are frequently justified or tolerated because of two factors:

- The continued acceptance of the "we've always done it this way" attitude, so that such varieties in the products are never challenged;
- The shuffling of parts between WIP inventories that dampens many of the costly visible effects.

An FMC environment is much less forgiving, and such product design and engineering practices must be challenged. If part proliferation is high, the involvement of the product engineering staff members will need to be high; in fact, a part rationalization and/or standardization exercise should probably be required. The product engineering people must also have a good understanding of FMC restrictions and capabilities. The aim is to move parts through the FMC

36

at a high rate of speed. The faster the parts move, the more the financial benefits accrue. Part variety in shape, size, type, and dimension all depend on setup, fixturing, and tool variety, and this slows the work accomplished in an FMC. Even if the products being manufactured are simple and the number of parts is under control, input from the product engineering people could allow the FMC planners to consider possible future production needs. Also, with a working knowledge of the FMC's capabilities, the design speed for future new products could be accelerated.

Scheduling/Production/Inventory Control/Computer Systems

These systems areas are almost always heavily impacted by the introduction of FMCs into a traditional environment. After all, these systems must "mirror" the activity going on in the shop; if the shop activity is changed radically, the systems must be altered to reflect it.

If we examine what changes occur in the shop when the switch is made from the traditional approach to an FMC approach, we can appreciate why significant systems changes must occur:

- Six machines are grouped into an FMC; the six machines are now one unit to be scheduled, rather than six. Or 10 FMCs exist in a shop that previously consisted of 66 machines. Fewer monitoring points mean fewer computer transactions, less work, and the generation of less paper. (The "paperless factory" becomes a possibility.)
- Often, the operators of FMCs may decide on the best way to sequence parts through the FMC to minimize setup and tool changing. They assume responsibility for some of the scheduling previously done by the computer department or the production control department. Also, smaller shop scheduling windows must frequently be enforced to avoid bunching of orders or "easy pickings" scheduling by operators.
- As a result of FMCs, there is often no inventory between individual machines. The FMCs may eventually be linked directly to assembly operations, reducing or eliminating preassembly inventory stores. Under certain conditions, FMCs reduce throughput time dramatically enough to eliminate an entire finished goods inventory level in the distribution chain. As a result, inventory control is much easier to do in an FMC shop; in fact, the inventory control function or department is frequently an early casualty in converting from a traditional to an FMC environment. Its potential elimination is a solid piece of the justification for converting to the FMC concept.

All of the changes mentioned above in the function of MRP systems and in the computer department occur because of FMCs. Clearly, then, if the affected systems departments do not conduct business differently, chaos will result and the FMC benefits can be lost. So, it makes little sense to launch an FMC project without involving systems people, seeking initially their understanding and assistance, and later a wholesale change in the way these systems operate.

Order Entry, Sales, Marketing

It is not unusual to see an FMC environment reduce manufacturing lead times from two months to two days. As a result, the company might be able to do some exciting things with its products in the marketplace. Often, however, the time to enter an order is two weeks; the total throughput time (from order entry to finished part) is, therefore, 16 days. While this is preferably better than the two-and-a half months it took previously, applying FMC principles to the order entry work steps (linking them) could perhaps reduce the order entry time too, to two days, for a total throughput time of just four days.

How would such outstanding throughput performance impact customers? The company's sales and marketing people would probably be ecstatic about a consistent four-day response to customer demand; they should be willing to give input on its potential effect on market share. Increased market share represents another justification for FMCs.

The Union

There is little difference in the success rate of FMCs when a union is involved. Union management can see as clearly as company managers the need for companies to remain competitive. The union being made part of the development process of FMCs is mandatory. If a traditional contract is in place, it will have to be changed if the FMCs are ever to achieve their full competitive and financial potential, in terms of the following:

Job assignment restrictions. A traditional union contract will spell out a series of job classifications—sometimes over one hundred of them in one factory! Persons in certain classifications perform certain tasks and no others. The ideal FMC environment argues for just one classification—operator—because FMCs require that people move from job to job to keep parts flowing. Part flow is the essence of a successful FMC. Optimum flow simply cannot occur if the parts have to be handed off to different people because their classifications are different. Furthermore, with part flow as the new priority, the cell operator may have to take on maintenance tasks or tasks involving tools and fixtures previously handled by people with "maintenance" or "toolroom" job classifications.

When such classifications are eliminated, a new method of rewarding people for increasing their skill base or for outstanding performance must be found and negotiated. This topic is discussed later in this chapter. In many union factories where FMCs have been installed, the new operators have expressed great satisfaction with the diverse tasks, need for decision-making, and sense of teamwork in their new positions.

Seniority. A principal objective of most union contracts is the protection of long-term employees. Often, this results in "bumping" practices, especially in times of changes in business climate. Taken to an extreme, bumping could mean that "everyone" had to change jobs to protect the rights of long-term, high-seniority people every time there was a change in one person's job status.

38

These extreme practices are disruptive and expensive in a traditional shop environment. In an FMC environment, however, this can be destructive to the point of making FMCs impossible to maintain, considering it takes about a year for a team of cell operators to make the most of production flow in their specific area of responsibility.

One company incurred this practice during the middle of implementing a large cellular project. The union contract allowed for bidding on new jobs according to internal plant-wide seniority. The mistake was re-posting the cell jobs each time a new cell was started up, about every two months. (It took about that long to choose, move, and retrain the operators.) The bumping that occurred when the next cell would come on-line affected the operator makeup of most of the previously implemented cells, keeping the entire plant in continuous turmoil as everyone jumped from cell to cell until they found one they liked best. This need not occur; negotiations that protect long-term employees as well as the benefits gained employing the FMC concept have been common throughout the 1980s.

Individual incentive systems. Not many companies entered the 1990s retaining individual incentive systems, but a few such systems still exist. Most, about to adopt FMCs, will plan the elimination of individual incentive systems as an early step. FMCs will never produce the desired results while such systems are in use; the individual must be rewarded for acting as part of a team of people. Strong believers in rewarding people for performance can still pay the team a bonus of some kind. This, however, is frightening to some workers; for the first time, the quality of their performance depends on the performance of others.

Making changes in the areas of job classifications, seniority, and incentives in union contracts need not be viewed by the union as threatening. The idea behind the FMC concept is to make the company stronger, and more competitive. Once union representatives understand the sincerity of management's motives and the enriched job positions that result from FMCs, they can enter the negotiations positively.

Human Resources or Personnel

It should be obvious that FMCs have a huge impact on people. This includes machine and equipment operators, supervisors, and people in departments and functions peripheral to the shop as well. Many early FMC practitioners, absorbed in the technical aspects of FMCs, brushed by the people issues. It's been proven that ignoring people does not produce good results. Some technically sound FMCs have been outright failures because the needs of people were overlooked. As a consequence, time and money in quantity should be reserved for communicating plans, soliciting input, training, retraining, team building, managing change, and problem solving. Only people make FMCs pay off! Once employees thoroughly understand and master the concepts, they will solve the everyday problems encountered in any shop and produce the benefits underlying the FMC concept.

Quality Control

A traditional method for achieving respectable quality levels has been having a quality control department, with inspectors reviewing the work being done by the people in the shop. This method never did work well. It was certainly rendered obsolete beginning in the 1970s, when offshore manufacturers set new, higher standards of quality. These new standards were so high they quickly became embarrassing to U. S. manufacturers producing in the traditional way.

One reason for these new higher standards was the use of the FMC concept by these offshore producers. The FMC concept automatically makes an operator (or a team of operators) responsible for performing all of the work steps on a part or series of parts. The next logical step is to also make them responsible for the quality of those parts. The tools and equipment for checking the finished parts must be provided, along with the appropriate training. If poor quality is discovered, the operators will know it instantly and be able to take corrective action. This cannot be done in a traditional environment. Frequently, poor quality work is not even noticed until days or weeks later. By that time, thousands of inferior parts may have been made.

Some companies today have discovered that having FMC operators sign their names to the parts or products made is a meaningful way to instill not only production responsibility to the people in the shop, but quality responsibility as well. People do respond to such implicit "pride in workmanship."

Cell operators become involved in the maintenance and setup of their equipment. The focus on quality in an FMC rapidly changes from using methods of only recording and reporting the amount of deviation, to solving the causes of slippage in the processes. The use of tools such as SPC (statistical process control), which are designed to identify when a step in the work process is slipping from expectations, will be minimized in mature FMC installations.

An FMC environment permits operators to produce incredible improvements in quality levels, totally unachievable in a traditional environment. The FMC environment, coupled with training and proper motivation and reward schemes, produces genuine pride in workmanship, and "perfect" quality can be expected. This can only occur with people who care about and control their environment.

Other Functions

These peripheral departments or functions are among those most commonly impacted by FMCs. Numerous others can be affected as well, however:

- Accounting—If a labor-based accounting system is in place before the introduction of FMCs, another method will have to be found to collect cost information, because "labor" is no longer assigned to single machines (operations) in an FMC.
- Purchasing/material control—Some experienced FMC users find that the operators are clamoring for direct contact with vendors to ensure that materials arrive Just in Time, in certain containers, and at point of use, to make the FMC operate more efficiently.

- Distribution—A factory that can respond to customers quickly may impact the distribution chain, particularly for those companies with a multitiered chain such as those producing consumer products. This chain should be examined to provide some justification for an FMC project; it is amazing how much inventory may reside beyond the factory doors that can be eliminated.

A Conclusion About the Near-Term Organizational Impacts of FMCs

There are two approaches to FMC development and usage. The first, which is the historical approach, is to accomplish all the technical work related to FMCs, get them running, and deal with the numerous impacts on the rest of the organization. Increasingly, a second approach is being adopted, which involves a number of peripheral (nonshop) departments and functions up-front. The second approach is far more pragmatic. It helps ensure that the conversion from traditional concepts to FMCs will occur in a reasonable time period with fully achieved financial benefits.

THE LONG-TERM FMC IMPACT ON ORGANIZATIONS

Once people have mastered the FMC concept, they have a shop which, probably for the first time in the company's history, is quick and responsive to its customers. Once this improved response becomes clear, company executives begin considering applying cellular concepts to other parts of the organization, as shown in *Figure 3-2*:

- After improving order processing throughput from two weeks to two days, a logical question might well be, "Why don't we move the order entry people to the FMCs on the shop floor?" People from both groups would be working along side each other; they could discuss, in real time, any problems a very large customer order might create, or possible methods for filling a rush order. Problems with an order would be instantly known and discussed with customers.
- Historically, product engineering passed part prints and manufacturing instructions "over the transom" to the manufacturing department. The development of such information was not discussed with the people engaged in making the product. But the quick response time of FMCs cannot wait for intercompany mail, or for engineers to fly to the plant from corporate headquarters when problems arise. Instead, companies are implementing "simultaneous" or "concurrent" engineering. This technique links the engineers to the people managing manufacturing and operating the FMCs. Putting engineering and manufacturing personnel together physically and organizationally has resulted in shorter product development and enhancement lead time, fewer prototyping cycles, valuable design improvement and simplification suggestions, shorter learning curves, and fewer problems and changes during production startup.

41

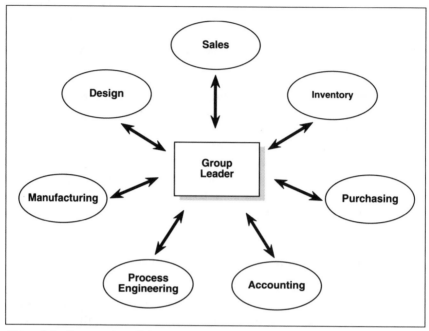

Figure 3-2. *Concept of a cellular organization structure.*

- In companies producing distinctly different product lines or serving different markets, the FMC concept may pave the way to breaking up the entire business into separate organizational pieces (strategic business units, or SBUs, focused factories, or profit centers). In a traditional environment, companies tend to mix different products for different markets together to maximize machine or equipment utilization; this method inhibits routinely meeting quality, cost, and throughput goals. When the shop is converted to FMCs, parts or product groups are processed on machines and equipment dedicated to them; this permits the shop to be broken up into "minifactories." The decision to organize the rest of the company into matching minibusinesses is a logical next step.

These examples of the potential longer term organizational impacts of FMCs are not the end of the story—companies are toying with much more radical ideas about organization realignment. This is exemplified in the following quote from an executive of Corning Glass Works (*Industry Week*, January 7, 1991):

> "In five years, if I had my way, we'd have no sales force—
> just direct plant-to-plant interaction from us to our customers."

Altering an organization (particularly a large alteration), however, requires a new set of skills for many manufacturing companies. The training and education required to succeed in this undertaking starts with teaming.

TEAM ENVIRONMENTS

Operating Teams In Cellular Manufacturing

A nonFMC environment demands few teaming skills, because individual performance dominates. Why are FMC teams more effective than individuals in functional organizations? There are a number of factors involved, all of which are interpersonal in nature. These include things like synergy, motivation, pride of ownership, and shared responsibilities, rewards, and knowledge. In addition to the interpersonal factors, teams generally create higher quality in their work than individuals and are more effective.

One of the most important aspects of teams is their synergistic nature. Momentum, ideas, and energy build on themselves and seem contagious as teams get together and progress starts. Although intangible, synergy is real; it can be seen in sporting events and businesses all over the world, especially in environments where a team is competing against another team or is creating ideas in brainstorming sessions. Many corporate "think tanks" have been developed to stimulate free thinking and product idea generation. The 3M Company is well known for providing positive environments for these idea-generating teams.

Teams naturally create a sharing of responsibilities by the team members. Since everyone is responsible for or at least needs to know about everything, people take a more active interest in all the activities of the team. Outputs of team efforts become the product and property of the group, not just the individual—this includes successes as well as failures, but certainly more of the former than in individual efforts. Training also occurs almost without people noticing it, as they share their knowledge with other team members.

Teaming is especially useful in an FMC project, because factory improvement projects can create a fear of the unknown for many individuals. ("Will I still have a job when this is over? What will it consist of? Will I like it? Can I do it? Who'll be my boss? Why can't things stay the way they are?") Through the use of teams, people in an organization will become involved in the planning and the process of change, and a greater understanding and communication can take place. The result is that individual fears can be dispelled and receptiveness to change increased.

Employee Involvement and Empowerment

Employee involvement through the use of teams also should mean empowerment from management to make decisions. If teams do not have the authority to make decisions and take actions, all of the advantages of having them will be lost.

The empowerment of a team can be a very difficult decision and process in many organizations, especially those that are autocratic in nature. Giving responsibility and power to teams means that someone must also relinquish the responsibility, authority, and power that he or she had in the past. This can be difficult for managers who have been raised with the idea that the broader one's span of control in the organization, the more successful a person is in a career. Getting a manager to refocus on the needs of a business as a whole and be comfortable is a difficult process; it requires careful planning, creative measurements, rewards that promote team thinking, and good communication.

Empowerment of lower levels or teams in an organization is not a substitute for management. The role of managers in any organization is to coordinate, guide, and make strategic decisions regarding the business. The important issue that managers need to resolve is at which level of the organization decisions should be made. In extreme cases, general managers and vice presidents will sit in on daily production meetings to determine what parts will run, on what machines, and in what order. This is a good example of managers not giving responsibility and authority to subordinates who should have it.

The other extreme is the manager who is virtually oblivious to daily operations within the organization. This type of manager is perhaps spending time in community service activities, on boards of other companies, or involved in professional organizations, or in community clubs/societies. Consequently, decisions are not made at all, or they're made by people in the organization who should not be making them.

The proper method is for managers to empower subordinates, through the use of teams, with the responsibility and authority that will achieve the goals of the company, promote teamwork, and keep all members of the organization involved at the correct level. This means having enough understanding of the operations of an organization to make decisions that are required from the management level while getting the people to accept responsibility and make decisions. It sounds easy, but it really is a fine line in a gray area of business management.

Roles of Supervisors in Teams and Cells

The traditional role of the supervisor has been to assign work, keep people working, expedite parts, and report machine efficiency, absenteeism, units produced, and earned hours. With these heavy workloads, it is not surprising that many supervisors in industry take a "hard nosed" approach and become dictatorial leaders. This approach is appropriate in some cases. There are many situations where authoritative supervisors may be necessary and desirable.

However, in the future it will be the role of management to take a harder look at each supervisory situation to determine if new techniques are required. Teams and highly skilled, motivated individuals will need to be guided in a much more personal, participative fashion. Supervisors will need to learn and implement a new set of skills that they may not have today:

- Interpersonal and communication skills;
- Facilitator and coaching skills; and
- Leadership skills.

Interpersonal and communication skills are probably the hardest for an individual to learn. They require a person to move beyond the black and white world of measurable results and into the "process" skills arena—dealing with

44

people one on one. One of the first skills that must be learned is active listening. From the day people are born, they are taught to speak in order to get their way and to be recognized. But in the 1990s and beyond, effective supervisors must do just the opposite of their upbringing—start to really listen to the viewpoints, ideas, and feelings of others.

Projections show that a large part of the workforce will need higher education and will work in jobs that have broader responsibilities and requirements. Many studies also show that literacy problems will continue to grow, creating a larger gap between the educated and skilled workers and the uneducated, general labor workers. Effective supervision of both of these types of people will require diplomacy, tact, and the ability of supervisors to stimulate employee involvement and the sharing of ideas.

EDUCATION AND TRAINING

There is an important distinction to be made between education and training. Education is most easily characterized as an effort to help an individual learn "why" something happens or is important (the conceptual). Training, on the other hand, can be characterized as learning "how" something is done, typically with an improvement in abilities (skills development).

Cellular manufacturing represents a different way of structuring a factory and, ultimately, a business. To effectively install the tools required to support such a new concept, key people in an organization must gain an understanding of the concept through education. Topics that need to be understood by participants include:
- JIT and flow-through manufacturing;
- Quality as a competitive weapon;
- Managing change (especially in an evolving business);
- The workforce of the 1990s; and
- Work teams and their implications.

The importance of continuing education in cellular manufacturing stems back to college courses in classical industrial engineering that focused on the efficiency of direct labor in manufacturing. The combination of efficiency and standard cost systems helped build the departmentalized factories of the 1950s, 1960s, and 1970s. Now that the concept is changing, there is an overpowering need to "unlearn" the concepts of old and relearn new ones. Old ideas and practices can be hard to replace. It is necessary, however, to move on and continue to be competitive in progressive, global markets.

Developing an Education and Training Master Plan

An education and training master plan is an important part of any major improvement project. It must be well thought out and structured to tie the technical factory efforts together with the people efforts, continually building a

45

stronger "people" base. A basic methodology for developing a solid plan for education and training includes:

- Start with the vision and the detailed plans of where the organization and business are going (see Chapter One). This should have a minimum three- to five-year horizon.
- Conduct a company-wide skills assessment to establish baselines for employees.
- Based on the vision, decide what skills and abilities will be needed now and in the future. Define the education and training needs and group people accordingly.
- Develop an education and training curriculum for each of the groupings.
- Determine the starting point for each individual within each curriculum.
- Develop, acquire, or adapt existing courses to the curriculum design.
- Determine education/training schedules that fit in with the factory implementation program. That way, people are trained just before changes start in their areas.
- Schedule classes, arrange facilities, and conduct classes as planned.
- Have students fill out class evaluation sheets; review them and adjust classes as required (usefulness, quality of instruction, difficulty and thoroughness of material).

Every member of the organization in all areas of the business needs to be considered in a master education and training program. This includes the lowest paid janitor or material handler all the way up to and especially the president and CEO. Curriculums in a complete training program should start with basic skills and progress through the complex and technical requirements of the company for the appropriate individuals. Curriculums that should be considered include:

- Basic literacy (reading and writing) and math skills;
- Core manufacturing skills that include topics like quality methods, blueprint reading, machine operations, and basic shop practices;
- Introduction to work teams and problem solving;
- Detailed skills courses that could include sheet metal, machining, fabrication, assembly, materials, clerical, and maintenance;
- Executive management development programs to enhance management skills and keep executives up to date on the latest business approaches and technical innovations;
- Middle management and supervisory development programs designed to improve supervisory, interpersonal, and teamwork skills; and
- Office and nontechnical salaried programs to improve technical areas of operation and promote career development.

The resources required for a major education and training program can be substantial. Companies supporting an FMC program may allocate 20 to 30 hours of training per year for each person. This is in addition to the time required for diagnostics, program development, and instructor, facility, and materials preparation. The price for a less tangible commodity such as training may seem high, but management must recognize that the price for training is relatively small compared to its importance in the success of a major project.

REWARDS

Rewards are an important part of work life. After all, if it weren't for needing paychecks, most people would probably find something better to do! But rewards go beyond a paycheck for most people.

The increase in team activities that is a part of cellular manufacturing will require organizations to develop new, team-oriented reward systems with the appropriate measurement guidelines, including systems that:

- Encourage group success. This can be done by measuring and rewarding the individual's degree of involvement in team efforts, and giving equally shared rewards for the team's obtaining quantifiable measures of quality, results, and output.
- Focus on results that affect the whole organization. If results are successful and have been influenced by individual teams and members, the team rewards should be reflective of this success.
- Encourage continuous improvement. This should always be a factor that is rewarded so that teams are constantly working on new programs to improve the product or service, and complacency is discouraged.

Measuring team efforts in the areas of team interaction and involvement can be more subjective than objective. Managers need to be more observant of, and sensitive to, the team's workings. Those responsible for individual performance reviews may need to sit in on project meetings and gain perspective on a person's progress from other managers and team members.

Gainsharing and other programs designed as group incentives should not be viewed as a total solution to team rewards. Managers should view these preplanned systems in the same way as any other "canned" package; they need to be adapted and tailored to the organization, and supplemented with other systems and procedures. A total understanding of what the organization is trying to accomplish is essential to effectively use such systems.

SUMMARY

FMCs can have an impact on organizations that reaches far beyond the factory floor, whether the cells involve only shop floor teams or the whole organization. The organization will never be the same once FMCs have proven themselves. Initially, these organizational impacts often cannot be predicted, but they become exciting team opportunities as the project progresses.

FMC PROJECT ORGANIZATION

The most important element of a successful manufacturing cell program is a strong project team. This team oversees program activities and serves as the driving force, keeping efforts focused on the program goals. Many FMC implementations require a break from tradition that opens a new path for manufacturing operations. This path becomes the foundation for directing, managing, and assisting the massive change that is required for realization of the full benefits of a manufacturing cell.

In the context of this book, *project team* refers to the group responsible for gathering information, evaluating data, and planning the specific details of the FMC implementation. The team's primary objective is *full* goal achievement *within* the program time frame—these are two most crucial components of the program.

Depending on the size and magnitude of the project to be accomplished, the project team might consist of one full-time person with a few part-time members up to a full-scale project team with dozens of full-time team members and specialists. To define the project team's composition, each project must be evaluated for content, level of sophistication, experience required (previous exposure to cellular manufacturing and new concepts), and resource availability. In the end, the team may be composed of people from diverse areas: in-house engineers, supervisors, outside consultants, corporate specialists, or the hourly plant operators.

In this chapter, the key elements of a project team—overall team organization, composition, dynamics, roles, and operational detail—are described. It is critical that the team have a thorough understanding of each of these elements since it will use them to drive the total program. While operational detail is discussed in the last section of the chapter, it should not be seen as the least important issue. Actually, it provides the most practical information for the project manager faced with getting an FMC project underway.

GETTING STARTED

The first question will probably be "What is the project we're going to 'team' around and where did it come from?" The impetus to pursue FMC or cell development is usually exposure to claims that cells are a good way to improve manufacturing operations. Management will probably accept cells as a potential opportunity for improvement and will approve an FMC project development team. If the firm has previous cell experience, this project team might be charged with identifying the next logical application. In the context of this book, it has been assumed that management has accepted the FMC development project as a method of continuous improvement in the enterprise's desire for greater competitiveness.

Prior to establishing an FMC project team, however, a "steering committee" for overall special project responsibility should be established. As depicted in *Figure 4-1*, the steering committee will reinforce the project's direction and commitments. To ensure an orientation which displays shared leadership, the steering committee is usually composed of seven to 10 senior management directors. The group is expected to:

- Guide company-wide continuous improvement activities such as the FMC projects;
- Evaluate proposed opportunities detailed by the project teams;
- Provide leadership direction to the project, demonstrating management support where appropriate;
- Meet regularly with the project team, at least once every eight weeks;
- Develop and improve systems that allow team members to bring about change;
- Determine and provide the resources required to complete chosen activities and achieve the project objectives;
- Assist in communicating the vision to all levels of the organization, including executive management.

The steering committee members are normally selected from the plant or general manager's senior staff. The steering committee is the entity which will initially confirm the necessity or feasibility of the FMC project team and select the team leader or project champion.

Once the team is accepted as the medium for action, its primary purpose is to provide daily program management and guidance. The team players become the recognized source of information regarding the cell, and this central core of people is charged with motivating and guiding the program efforts. In addition to directing human resources, the team's responsibilities also include managing all of the remaining resources (money, machines, floor space, etc.) critical to the program's success.

Managing change is one of the key elements of an FMC program; almost every process and discipline will be affected to achieve the optimum benefits. The project team must accurately assess the rate of the company's ability to accept and digest change and then maintain the level of change appropriate for

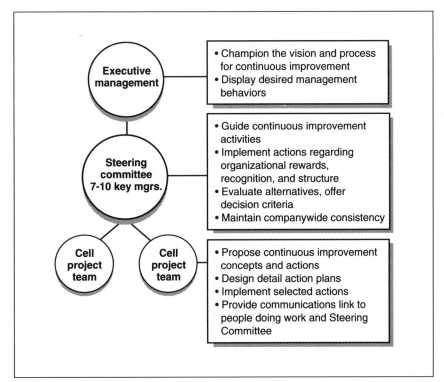

Figure 4-1. *Recommended overall project management structure. This multitiered approach ensures shared leadership.*

the company. The team also determines the actual program baselines. In initial discussions about the FMC project, many goals and objectives may have been reviewed. The team must quantify the current situation, as well as develop the preliminary goals into finite, well-defined program goals. In addition to goal clarification, the program guidelines and operational assumptions must be documented by the team.

Initially, the team organizes the efforts needed for the entire program and continues to manage these efforts as the program matures. After the baselines have been set, the emphasis switches to daily program management responsibility. A project schedule is created and continually updated as cell development progress is measured; program interruptions are handled as they emerge, always concentrating on the targeted goals. Misdirected efforts are re-focused at the program targets. As the program matures, the team will evolve and refine the efforts of program resources to successfully implement the FMC.

TEAM COMPOSITION

Just as each manufacturing cell is designed to satisfy a specific goal or purpose, the exact composition of the core project team is also specific to the cell. In all cases, however, each team member must be committed 100% to the manufacturing cell program and to successful goal achievement. Since the requirements of the team vary with the specific project, and even during a project, a typical team would be made up of the primary manufacturing and support skills needed to complete the program.

Leadership

The project "champion" is one of the most critical facets of team composition. The champion will become the center of program team activities and is the primary voice in the communication of progress and accomplishments. Of the team members, the champion is the driving force that gets things done. The champion should be "hand picked" by the steering committee to lead the FMC program, and he or she is usually a natural leader who is well-respected within the company. In many FMC programs, the champion is also the future FMC manager.

As the team leader, the champion performs critical project duties and must possess a variety of skills; this person must achieve a high level of individual performance while motivating the other team members and communicating the potential and progress of the program to the entire organization. Simply put, the champion must be:

- A highly effective communicator;
- Visionary;
- Able to team with, motivate, and empower others;
- Well-respected for his or her technical abilities and personal traits;
- A strong catalyst and trainer;
- Able to delegate while maintaining control, and
- An experienced project manager with strong analytical skills.

As team leader, the champion is responsible for soliciting, selecting, and managing the project team members. This process is complicated by the high level of performance required of team members, changing demands and availability of human resources, and the ever-present political pressures.

Membership

After the required technical skills have been defined by the steering committee and the champion, the preliminary candidates for team membership should be identified. Each person must fully understand and believe in the changes and requirements of the FMC program. It is the champion's responsibility to educate, motivate, and, ultimately, select the core project team. The easiest selections are those people who have the necessary skills, understand the value of the program, and are eager to participate. Identifying other team members requires an in-depth interviewing/negotiating process between the champion and potential participants.

Since skill requirements will vary over the life of the project, there will be several layers of team representation. For example, facility planning skills will not be required until it is more clear where the FMC will be located. Tooling expertise might also come later in the timing. However, financial assistance might be required early in the project and concentrated again near the latter stages of the concept proposal to management. As seen in *Figure 4-2*, there are some basic needs that should be available within the "core" project team and other specialist roles can be accessed on either a part- or full-time basis.

There are numerous specialist functions available through outside sources, particularly for advanced engineering or technical, analytical, or unique skills. One might look to corporate and/or divisional people for some of these talents or cautiously interview outside consultants who fit the needs.

Part of the project team's responsibility for maintaining internal communication can be satisfied through the periodic steering committee meetings. The necessity of also soliciting advisory assistance from some key operating functions in the company should not be overlooked. The advice and assistance of product development, marketing and sales, and office administration can be particularly useful. There is usually some very valuable experience that is overlooked in retired employees, production training departments, or personnel suggestion system coordinators.

Attributes of a Team Member

The champion must select the members of the core cell team. In addition to the technical skills possessed by this team, many nontechnical abilities are needed. Each member should be an effective communicator since he or she will be seen as an FMC "expert" and must be able to talk intelligently about the project. In most cases, the ability to grasp an abstract concept and translate it into reality is also a desired trait of team members. This ability is essential as the team works to translate the existing operation into the future FMC. Well-developed problem-solving skills are also critical, as barriers must be overcome.

Strong negotiating skills also are a plus for selling ideas in team meetings as well as convincing "non-believers" of the value and necessity for changing to cellular manufacturing. As with all major change programs, convincing others of the program's value will be an ongoing process. Not only must the program be sold to the hourly work force, but serious questions are usually asked by top management when investment is requested.

As the program matures, it is also the team's primary responsibility to uphold and enforce the program implementation schedule. Since the team is normally the author of this plan, the content is best understood by team members. The project team will become policemen, watching over activities. In addition to keeping the program on schedule, key milestone responsibilities will be assigned to each team member to guarantee that the critical program path is maintained. Once the critical path is no longer critical and the whole program is on schedule, the team will shift its focus to the much more pleasant task of tracking and measuring the benefits.

Area/Function	Core Team	Role — Specialists			Advisory
		Part-Time	Full-Time	External	
General management					S
Engineering:					S
— Product design		✓			
— Product development	✓				✓
— Tooling			✓		
— Product research					✓
— Process	✓				
— Manufacturing	✓				
— Industrial		✓			
— Analyst				✓	
Marketing:					S
— Sales					✓
— Service					✓
— Analyst				✓	
Finance:					S
— General accounting		✓			
— Cost accounting		✓			
Systems:					
— Analyst					✓
— Process controllers					✓
— Applications	✓				
Suggestion systems					✓
Factory production:					S
— Supervision	✓				
— Managers		✓			
— Shipping/receiving					✓
— Training			✓	✓	
— Traffic					✓
— Hourly specialists	✓				
Advanced engineering				✓	
Quality engineering					✓
Maintenance		✓			
NC programmers (process)		✓			
Tooling/fixturing		✓			
Facilities			✓		
Production control:					S
— Material	✓				
— Purchasing			✓		
— Scheduling/planning			✓		
— Order entry					✓
— Inventory		✓			
Personnel					S
Office administration					✓
Cell development	✓			✓	

✓ = Team Representative
S = Steering Committee Representative

Figure 4-2. *Roles included in the project team. An FMC project requires the involvement of many disciplines.*

54

TEAM DYNAMICS

By necessity, the project team is a cross-functional body. In the most successful FMC project teams, the representatives from different functional structures formally sever those reporting ties for the duration of the project. This may not always be possible. In fact, in a majority of project teams, members maintain two sets of reporting relationships. Consequently, the team may be faced with a matrix-style organization with dual reporting conflicts and multiple priorities. But by carefully establishing communication paths at the onset of the project, these problems can be effectively managed.

It is valuable to identify the three types of people within the organization that the team must be prepared to deal with during the project. These types of people can be categorized according to their attitude toward the FMC concept. They are:

- Believers,
- Doubters, and
- Blockers.

The *believers* are the eager participants; they are identified as team players whether on the core team or just by their overall positive assistance to the program. The *doubters* are usually not so eager initially, but they generally believe there is some value to the program. They usually have expertise and competence that would be beneficial to the team, and have a high impact on the organization. These people need to be converted into "believers." The champion should continually nurture these people until the transformation is complete. In most cases, these doubters, when converted, become strong team players and carry substantial clout in making things happen.

Blockers pose the biggest threat to the success of the FMC program. This group is composed of people who feel threatened by the new, changing environment. In all cases, these blockers must be handled in a manner that eliminates their negative impact on the manufacturing cell program. Their harmful attitude and undermining activities must be quickly identified and preparations made to remove their influence on the total factory population. Once these blockers have been neutralized, the champion can focus on handling the more important issues that may become barriers to program progress.

Turf Protection

Another obstacle to the FMC project may come from people trying to protect their turf. Within an organization, some individuals are concerned with the size, strength, or span of their areas; this is sometimes referred to as "empire building." Factory improvement programs, including FMCs, may be perceived as a potential threat to these organizations, thus reducing the power base of a manager. As a consequence, the team will have to deal with resistance to or manipulation of a program through a variety of techniques aimed at altering the course of the program or even discrediting its viability. Indications of managers trying to protect their "turf" include:

- Non-participation in the project by delegating the responsibility down to a subordinate and effectively elevating themselves "above" the program;
- Discrediting the benefits of a program by independently researching data and presenting a different view;
- Working around the project to develop or install other programs that would increase his/her power base.

The best approach toward the most difficult non-believer or turf protector is straightforward communication and involvement. Successfully surviving these conflicts usually comes from not walking away from the conflict and by working for win-win solutions. The most effective ways to develop win-win solutions are to do the following:

- Establish clear goals;
- Know your information and keep it up to date;
- Listen;
- Avoid absolute statements;
- Look for areas of agreement;
- Encourage consensus decision-making rather than the majority or the individual rule.

Dealing with turf or power issues is difficult even for the most experienced project manager; objectivity and professionalism must be maintained to defuse the emotional impact of the situation and portray the issues clearly for everyone involved. The negative risk of ineffectively dealing with turf or power issues can be enormous. "Power brokers" can diplomatically and systematically manipulate projects to protect domains, ultimately hiding or impairing the potential benefits of a program, making it unattractive to management.

Effective Teaming

Getting teams to work cross-functionally, even on a temporary basis, can be challenging. Greg Kesler[1], an organizational dynamics specialist, indicates that some of the common problems encountered in team-based projects include:

- Shared responsibility becomes an excuse for lack of accountability.
- Meetings and extensive negotiating get in the way of decision-making, virtually halting all progress.
- One function (frequently engineering) or geographic location dominates the other functions or locations to the detriment of the overall business or the customer.
- Organizational components may be unwilling to share information, expertise, and resources, creating disjointedness in the process:
  - Customer needs do not find their way into product designs.
  - Financial implications are not fully understood.
  - Impossible deadlines are set and missed.
  - "Best" practices are not communicated for the benefit of all.

- Overlapping responsibilities create duplicate staffing and other forms of waste, often hidden in the internal supply chain.
- Pooled utility functions often become large bureaucratic units unable to focus on or respond to new segmented demands.

To establish a sense of *shared* accomplishment in the FMC effort, the team should be aware of these pitfalls and make deliberate efforts to deal with them as they occur. Additionally, it is important to be aware of the characteristics of productive and unproductive teams; some of the most typical are outlined in *Figure 4-3*.

In Productive Teams:	In Unproductive Teams:
• All members are working together toward common, well-understood goals.	• Members are unclear of their own goals, others' goals, and overall priorities and goals of the team.
• The values and goals of the group are a satisfactory integration and expression of the relevant values and needs of its members.	• There is unnecessary duplication of effort.
• Clear work assignments are made and accepted.	• Some things just don't get done; they seem to "fall between the cracks."
• The team leader encourages supportive, cooperative, noncompetitive relationships among members. Mutual help is the norm.	• Team members seem to be pulling in different directions.
• The team frequently examines and discusses how it is functioning.	• Productivity is low.
• There is a high level of participation in discussion; discussions remain focused on the task at hand.	• Decisions are not completely followed up.
• Communication flows easily.	• Meetings are ineffective, dull, and stray onto tangents.
• Members listen to each other without jumping from topic to topic and are open in expressing feelings.	• Communication is blocked and messages are not received. Some people don't know what is happening.
• People are involved and interested; the atmosphere is informal and relaxed.	• Communication is primarily vertical; little lateral feedback.
• The team is comfortable with conflict, but carefully examines and resolves it.	• People are very careful about what they say.
• Criticism is constructive and directed toward resolving the remaining obstacles to getting the job done.	• Members of the team are frequently in conflict, and conflict is suppressed.
	• Team morale is low; people grumble.
	• All goals are not being met, or the process of setting and accomplishing goals is not completely effective.

Figure 4-3. *The characteristics of productive and unproductive teams.*

Communicating Progress

Communication is one of the most important elements in any project. To be effective, it must take place at all levels using a variety of methods. In addition to the administrative activities undertaken by the program team, frequent progress updates are necessary. This usually occurs in the internal meetings and the top management status reviews. This constant communication process is such a key part of the manufacturing cell program that it cannot be overemphasized. The primary communication in an FMC project takes place at three levels:

- Steering or executive management committee,
- Project team, and
- The rest of the organization.

Steering committee and management communication usually takes place in the form of a weekly update meeting to discuss progress, direction, and next steps. The project team leaders set the agenda and prepare overhead slides or handouts to illustrate and document important points. The nature of a major improvement project dictates the need for frequent management communication. Major decisions concerning financing, staffing, equipment, and schedules may be influenced and altered based on the weekly activities within the project.

Team communication—informal and formal—is required on a *daily* basis. Informal communication takes place during impromptu meetings, working sessions, coffee breaks, and casual conversations. These communications can be the most important because they generally cover a multitude of issues that may not be included in the formal structure. Casual conversations between team members working on different parts of a project may reveal common problems or issues that can be jointly addressed or resolved. For this reason, it is most important that the team members have a central place to meet and develop this rapport. This "war room" concept is elaborated on later in this chapter.

Formal inter-team communication includes team meetings, correspondence, wall charts or diagrams, meeting minutes, and project documentation. It is up to the project leader to determine the types of formal communications required, the frequency, and methods used. Methods of formal communication will vary from project to project. In a highly formal environment or culture, correspondence may need to be typed and flow charts drawn and documented on a CAD system. In less formal cultures, hand-drawn charts and handwritten notes may be acceptable. Regardless of the environment, communication needs to be structured and designed to keep everyone well-informed.

The entire organization needs to be part of the communication process so that the people are aware of the program *before* it directly affects them. It is essential to communicate project purpose, direction, and status. This can be accomplished through the company newsletter, monthly status reports by management, wall posters, or small group sessions. In any case, it is *essential* that management be honest, open, and sincere in what and how information is stated. Sensitive issues should be anticipated and addressed directly whenever possible. If something is asked and the presenter does not know the answer, then the presenter should simply say, "I don't know." Rule number one in communicating information about the project: Never attempt to bluff through an issue or speculate without qualification.

ROLES OF THE CELL TEAM

After the cell team has been selected and the program schedule has been solidified, the team begins the critical FMC activities. Initially, the team must determine several key program milestones, including the actions required of the core team and of the organization to successfully complete the program. Once these have been determined, the team must begin fulfilling the defined needs while becoming effective in:

- Establishing involvement and program ownership;
- Maintaining program accountability and responsibility;
- Efficiently dividing work;
- Obtaining/attaining the required skills and expertise;
- Providing accurate program checks and balances;
- Positively exchanging views and ideas.

These are the critical attributes of a successful cell team that are required for quality program implementation.

Involvement and Ownership

In terms of involvement and ownership, all highly successful FMC programs have two things in common:

- Top management fully supports the entire program and is 100% committed to it.
- The actual implementation process is accomplished by the factory people—from the bottom up.

Once top management has committed to the program, the project team can then assume the development and implementation duties. Each team member should have an impact on the project content and be an active participant in key program decisions. Additionally, each team member should play a part in setting the project's direction at the onset and in maintaining the focus for the entire program duration.

To reiterate, this total responsibility requires that each team member believes in the program's content and goals. Outside the project's core team, it is essential that the business disciplines provide assistance and support when needed.

Accountability and Responsibility

As the FMC program begins to take shape, operational changes required to complete elements of the project may become immediately apparent. In an effort to smooth the implementation process, the project team should clearly outline the changes in each functional area; this representation should be presented in technical terms.

Individuals with experience and/or responsibilities in the areas affected can usually provide the best indication of the requirements and impacts. If not already represented on the FMC team, individuals from areas highly affected may be assigned to the team, while those from disciplines impacted less overtly should participate on an as-needed basis. Eliminating or bypassing the organizational barriers, loosely depicted in *Figure 4-4*, is a necessity if the program is to be successful.

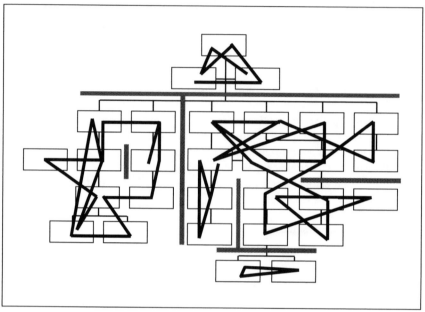

Figure 4-4. *Typical informal organization structure. Many organizations have developed "barriers" that inhibit communication and interaction.*

Divide the Work

FMC projects require substantial effort, and the size of the core project team will be determined by that level of effort. As soon as the program work plan (described in detail later in the chapter) is defined, the tasks should be assigned to the project team members. By balancing the work load between team members, each individual becomes more effective in a smaller arena, and team members are encouraged to interact to complete assignments. The proper combination of different skills in these small teams equalizes dominant strengths and minimizes individual weaknesses. Due to the specific skills resident in each small team, completion of the assigned task usually requires the involvement of non-core team members. This universal teaming will become the precursor for the large-scale teaming required for operation of the FMC.

In addition to equalizing the program work load, the project responsibility is allocated between the team members. All key task completion dates and program milestones become the individual team member's responsibility. This teaming process allows everyone on the team to gain invaluable experience in cell management and with other individuals/disciplines in the company. The increased ability of the individual team members to manage and digest change will greatly improve the company's internal strength and prepare it for many new business challenges.

Obtain/Attain Required Skills

Since this core cell team is the driving force for implementation activities, strong technical and interpersonal skills are needed. Earlier in this chapter, the technical expertise required of each team member was outlined. In some cases, the team may discover that a specific skill is not resident in the company's staff. Outside services may be brought in on a temporary basis to fulfill these needs.

Program management and interpersonal skills are just as important as technical skills. In most cases, these skills are not currently available to the degree required. Training and education, as well as outside assistance, can quickly supplement these requirements. Some of the more critical technical and personal skills that often need to be supplemented include:
- Program management abilities,
- Team building—from project kickoff through goal achievement,
- Motivating abilities,
- Problem-solving skills,
- Facility layout techniques,
- Presentation skills,
- Communication abilities, and
- Effective computer use.

Most of these skills can be learned but individuals only become effective with them after much practice. Basics can be taught by someone inside the company or training may be available through local technical schools or college programs. Specialists in the technical areas—like plant layout and project management—can be brought in to teach project team groups. These applications can even be valuable team-building exercises during the initial project kickoff period.

Checks and Balances

Each effective project team develops its own system of checks and balances, but most are based in some way on consensus-decision making. The understanding of project activities as integrated team efforts checks any individual tendency to dominate the team. Since each team member also brings his or her varied skills, strengths, and interests to the team, everyone will have different perceptions of the project objectives. This, too, balances individual contributions. All team members benefit from the process as individuals; it instills new ideas and broadens the experience base of each team member.

The team concept, by definition, encourages members to question and challenge each other, but this ultimately allows the members to develop a clear understanding of the team. As specific program issues are challenged, modified, and agreed to by consensus, the team members become a coordinated decision-making unit with a positive image. It is this coordinated body that communicates critical project information clearly and concisely.

TEAM OPERATION—GETTING THE PROJECT ROLLING

The project team's ability to plan the work is crucial to the overall success of the cell program. The primary operational components for this are:
- Getting the project kicked off properly;
- Communicating general information about the cell program to appropriate management and throughout the organization;
- Using the "war room" approach;
- Identifying project team tasks, sequences, priorities, resources, and responsibilities, and reflecting these in a comprehensive project work plan;
- Initiating and monitoring all team activities, and scheduling and tracking progress for the FMC program.

Kicking It Off

How a project gets "kicked off" establishes the level of organization, commitment, and perceived comfort with the process. There is a definite link between a significant, well-organized project "kickoff" and team progress, enthusiasm, and success. Key features of project startup include:
- Select a strong and experienced project manager as the champion—one who has totally "bought in" to the process.
- Establish the multifunctional team well in advance of actual project work.
- Take the entire project team, with the steering committee, off site for a two-day orientation seminar that includes:
  — Team building,
  — Project definition,
  — Cellular methodologies and examples,
  — Planning.
  This type of workshop usually uses experiential exercises to demonstrate setup reduction, cell concepts, pull production, scheduling methods, etc.
- Brainstorm ideas for the project planning—work through the logistics for priority setting, resource allocation, and assigning responsibility.

The facilitation methodology is vital for leading the project team along an independent path. It is also necessary to ensure that the team is fully aware of and understands the objectives and methodologies required for successful completion of the project.

External Communications

As stressed earlier, the most important activity the team will perform is communicating information about the manufacturing cell program. This may sound elementary, but the variety of audiences complicates the process. The same technique and style is probably not appropriate for both middle managers and machine operators. For instance, a quick internal update meeting between the team members may be detailed and highly specific, but the status update to the board of directors must be broad and only present critical events. The communication procedures must be identified and quickly learned to effectively transfer information to the usual audiences. These audiences may include:

- Top company management,
- Middle managers and staff,
- The entire work force,
- Full-time and part-time team members,
- Present and future customers,
- The industry in general, and
- The news media and trade press.

In addition to the verbal communication skills and procedures, written reporting mechanisms also need to be finalized. In most FMC implementation programs, the company requests a historical log clearly describing the before and after operating conditions, as well as the sequential steps taken during the implementation process. These reports are normally distributed and given high visibility; therefore, they are a key to establishing a strong positive image to the audience. In some cases, just a brief summary of the major points of the program, an "executive" summary, may be required; this reporting mechanism is short, clear, and easily understood. Other effective methods of communication that can be used include: video presentations, computer-generated cell simulations, and tabletop 3-D models.

The team also should contribute to the schedule and timing of the status update meetings. In most major manufacturing cell programs, the need for substantial communication occurs early in the program. As the program evolves from the conceptual to the implementation phase, the constant need for new information decreases. Subsequently, the interval between update meetings increases. The meetings may be required as key milestones are reached, so the timing of update meetings should be determined by the project team.

Creating and Maintaining a "War Room"

Early in the manufacturing cell program, the team meeting place must be established. This room is more accurately called the "war room" because of the highly intense activity that occurs within the team as differences of viewpoints are openly expressed. This room serves as the primary location for resolving troublesome issues by the team so consensus is reached before the issues begin impacting the actual program implementation quality. In addition, this room should be the primary location for project-related documentation. Up-to-date program schedules, selected FMC facility layouts, and each cell's material flow, as well as other key program information, should be displayed on the walls.

It is important for all views to be expressed by the team in this environment. It is here that the key communication and negotiating skills are honed to a razor's edge. Each team member has the opportunity to campaign, negotiate, "sell," and even battle for the acceptance of a particular viewpoint or belief. This "warring" process is extremely healthy, as team members benefit from the experiences and strong beliefs of the others. This brainstorming will help the team to resolve issues in a constructive way, and more importantly, ideas may be strengthened dramatically as a result of the intense examination.

The mechanism for monitoring team members' activities and responsibilities is also part of the war room environment and must be established early in the project. Normally an implementation chart is posted in the war room so participants can view the entire program status.

Early in the project, an issue requiring action may seem to be outside the responsibilities of the program team. Initially, the process for resolving such an issue may be difficult, but as the team members become more proficient in their interaction and creative problem-solving abilities, the subject will quickly be elevated to the appropriate level for resolution.

Possibly the most important activity performed within the confines of the war room is the development of the tracking and measurement system for the program's progress against the goals and objectives. In all cases, the data reported must tie to the company's strategic drivers and its accepted reporting methodology. In most cases, the FMC progress is a direct reflection of the present financial reporting and measurement system. In all program reporting documentation, the actual program costs and benefits must be reported; basically the program must stand on its own and record actual costs.

Work Plan Development

Sometimes the project work plan is created by the project manager with the team's input on tasks and internal dependencies. After timing for task completion is defined, the team members' roles will be assigned and negotiated according to each member's skills and availability. More often, however, the project work plan is a product of the team process.

The off-site project kickoff meetings frequently generate the initial work plan, including tasks, resources, priorities, obstacles, decision points, expected costs/benefits, and milestone dates. There are several tools that can facilitate the development of the work plan. One of the most successful is a brainstorming process developed by Walt Disney. "Storyboarding" encourages contributions from a variety of sources while the planning is still just that, planning. There are four basic kinds of storyboards:

- Idea storyboard—develops a concept or idea. The end result can be a drawing, a model, or simply a description of the concept.
- Planning storyboard—outlines the steps that are required to reach a desired end result.
- Organization storyboard—answers the questions "Who will be responsible for doing this task?" or "How are we going to organize our resources to accomplish the plan?"

- Communication storyboard—answers questions like "Who needs to know?," "What do they need to know?," and "How do we communicate it to them?"

An example of a work plan for the developmental phase is shown in *Figure 4-5*. The primary element of this schedule is the key milestone event requirements. Since top management usually concentrates on resource requirements and benefit achievement dates, it is crucial that the team maintains these dates. Total program credibility may be based on the actual performance to these plan dates. Identification and creation of the second-phase work plan is discussed in Chapter 17; it will include the detailed FMC development elements:

- Final detail layout,
- Equipment specification,
- Staffing requirements,
- Organization and procedures,
- Vendor specifications,
- Equipment purchase,
- Scheduling and sequencing,
- Facility and construction details,
- Resource planning,
- Material handling requirements,
- Systems, and
- Final financial impacts.

The key point to remember at this time is that the FMC project team is responsible for managing and achieving this program schedule, so its members should be aware of its longer term implications and impacts.

SUMMARY

Getting an FMC project off the ground requires well-organized project management programs. This chapter reviewed:

- The steering committee concept and establishing control mechanisms;
- Appropriate project team composition:
  - The leader or champion,
  - Functional representation,
  - Individual skills and attributes.
- Team dynamics as they relate to identifying the "right" players, dealing with blockers, *good* teaming, and the significant importance of internal project communication;
- Various cell team roles:
  - Involving technical experts;
  - Dividing the work;
  - Generating and evaluating ideas.
- Mechanisms for the project management, detailing the tasks, and reporting the task status.

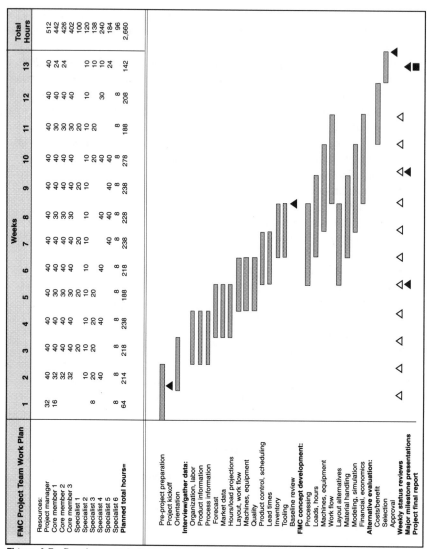

Figure 4-5. *Developmental phase elements of FMC project work plan.*

After the FMC project planning is complete, the team should be ready to actually develop the cell concept. Digging out the necessary information, processing it into meaningful conclusions, and proposing alternatives/recommendations is the topic of the next part of this book.

There is a distinction between the organization of the project team and implementation planning that results from the team's activities. At the onset, the project team requires development direction based on "how to" analyze and crunch data to propose specific details about an FMC. The actual FMC implementation planning (schedule, sequence, resources) is a separate activity involving many of the same processes. It is discussed in Chapter 17.

ENDNOTES

1. Kesler, Gregory C., "Building Team Organizations: Making the Matrix Work," Draft Paper, September 1990, p. 5.

IIIII 5 IIIII

MACRO FACILITY PLANNING

This chapter describes macro facility planning, the first step in an iterative process of designing the factory and the cells to work together. Chapters Six and 14 cover the other two steps, conceptual cell design and detail design of an FMC, as shown in *Figure 5-1*. The actual task of macro planning can be characterized as developing the answer to the question, "How should a factory look and operate if it is to produce our products at the expected mix and volume in a way that is consistent with our business objectives?" When the task is stated this way, it becomes clear how important it is that the planning team share a common vision of how the new facility will run, and that this vision be consistent with the business drivers developed as described in Chapters One and Two.

The quality of this macro planning will have a profound impact on the benefits the company will derive from cellular manufacturing. Most companies that are disappointed by cells have failed to move beyond islands of improvement on their factory floor. Cells are operating at the mercy of support services and product flows that were not designed to support cellular manufacturing. Development of an integrated macro plan will link the islands into a single, coordinated entity.

A team or individual beginning the process of redesigning a factory faces a sort of "chicken and egg" dilemma. Cells must be designed to be as self-contained as possible, to operate flexibly in response to customer demand, and to fit into available factory space. On the other hand, the factory must be laid out in such a way as to facilitate the production flow among cells and to give them good access to the services, utilities, and other support they need. Thus, the factory should be designed around the cells, and the cells should be fitted to the factory.

Details vary from application to application, but the following set of steps is helpful in organizing the team effort to create a macro facility design:

1. Establish operating parameters for the new factory.
2. Create a product database.

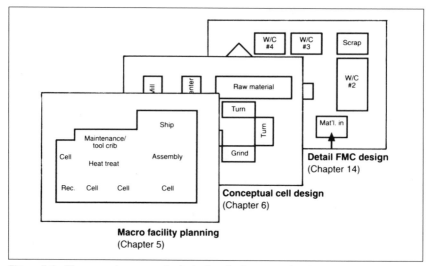

Figure 5-1. *There are three steps in designing the factory and cells to work together.*

3. Group products for cell design.
4. Make a process/machine load analysis for the cell groups.
5. Identify improvement opportunities.
6. Build the macro layout.
7. Select a candidate pilot cell.

The logical relationships among the steps are depicted in *Figure 5-2.*

This chapter includes a detailed discussion of these seven steps, as well as a section about the appropriate level of detail for macro planning, a discussion of cell types, and presentation of some analytical techniques that are appropriate to this stage of factory design.

STEP 1: ESTABLISH OPERATING PARAMETERS FOR THE NEW FACTORY

It is always a good idea for the team to review the business objectives, as discussed in Chapters One and Two, and then describe in general terms the operating parameters of a facility that will meet these objectives. This description should include the impact of business objectives on the size, staffing, and performance criteria of the redesigned factory. These can usually be reduced to half a dozen succinct points that will help guide the team. Depending on the product and environment, typical examples might include:

- Factory throughput time of three days;
- Lot size of one;
- Twenty inventory turns per year;
- Daily Kanban scheduling wherever possible.

These parameters should be posted prominently in the team work area, and should be reviewed frequently during the course of the design effort. Each of these parameters places some constraints on facility design, both at the macro

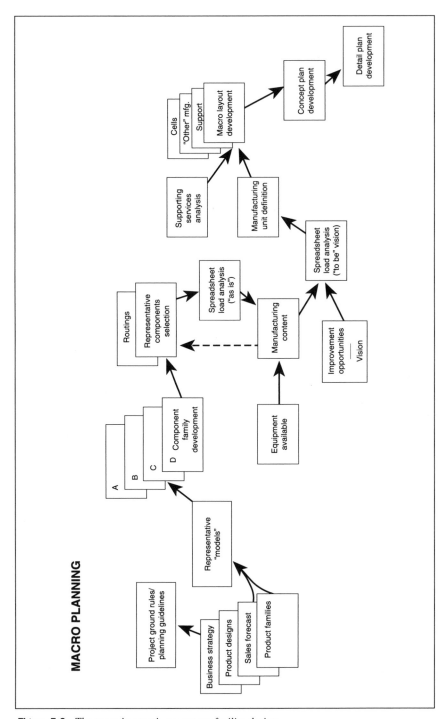

Figure 5-2. *The steps in creating a macro facility design.*

level and within cells. Some will also present cost reduction or resource conservation opportunities of their own. For example, "factory throughput time of three days" might provide the opportunity to design a macro layout with little or no finished parts storage, since no more than one or two days' worth of product would be stored. On the other hand, the same parameter might require the design to provide parts washing facilities within the cells, since trips to a central wash line usually consume about a day each.

Similarly, a requirement for Kanban or other physical scheduling methods makes it desirable to have component departments discharge production within sight of the assembly operation. This simplifies all types of communication and supports physical or visual scheduling of a product mix particularly well.

In addition to being design drivers, these criteria will become the benchmarks that the organization uses to determine the success of the cell design and implementation. If there is any doubt that the team and top management are in agreement on the factory performance criteria, a review should be scheduled to resolve the differences.

STEP 2: CREATE PRODUCT DATABASE

Having identified how the factory should perform, the team will turn its attention to describing what is to be produced. This effort begins with compilation of a product database that will follow parts and, later, part families through the entire planning and implementation process.

Cost-effective cellular manufacturing requires that the team build the production operation (facilities and procedures) around the characteristics of the products being produced. The better the database the team constructs, the better will be the chance of a smooth start-up and successful operation. This is another situation where teamwork will pay huge dividends. The information for the product family database should come from every functional group in the organization.

The database should include at least the following for each part to be produced:

- Product structure and design:
  - Current;
  - Future.
- Machine load data for each part or part group:
  - Key equipment utilized;
  - "As is" process times for key operations;
  - Identification of centralized manufacturing facilities (washers, platers, heat treatment);
  - Current changeover procedures (setup, tooling, fixtures).
- Sales volumes over time:
  - Current output;
  - Future year forecasts;
  - Seasonality.

72

- Part family groupings, if any already exist;
- Physically linked work steps or processes;
- Overall product and part flow:
  - Internal to the process;
  - External to the process.
- Area and overhead space requirements;
- Process and equipment capabilities;
- Concepts for improvement in processes and equipment;
- Make-versus-buy analysis of opportunities;
- Operational parameters:
  - Organization and training requirements;
  - Methods and procedures;
  - Systems.

Most applications will require additional information, due to the characteristics of the products and markets. For example, many product lines, particularly those with short product life cycles, will require cell structures that permit prototyping of new products in the production setting. The most cost-effective time to build simultaneous engineering into a product family is during the inception of cells. Other products may require clean rooms, special ventilation requirements, or special access for any number of reasons. Now is the time to identify these needs and build them into the facility design. Again, this sequence is iterative, as products drive processes, which, in turn, drive the facility design.

Particular care should be given to the machine load analysis, as it will determine all capacity data for this part of the exercise. If the project is a reorganization of an existing facility, good data should be available, either from production routings or actual manufacturing history. It is essential to use the most realistic information possible, even if it must be specially developed. For instance, standard hour routing information should be factored by actual performance experience in settings where large production variances are normal. This is important not only for predicting the capacity of the new operation, but also for establishing a performance baseline from which to predict and measure the improvements derived from cellular manufacturing.

The database will not be complete at this stage, but it should be packaged in such a way that the team can refer to it and add information as detail planning progresses. The best media for this purpose are PC-based spreadsheet and database programs.

Database Level Of Detail

Later in the planning process, it is necessary to consider the detail characteristics of most parts, but macro planning begins at a more generalized level. A useful tool for generalizing data is Pareto analysis, also called "the 80/20 rule." This principle states that 80% of the results are caused by 20% of the events occurring. At various steps in the planning process, it may be appropriate to perform a Pareto analysis of the product mix, ranking products according to one or more of the characteristics that govern how they are handled or produced.

Pareto analysis identifies how the bulk of the product mix behaves, leaving the exceptions to be dealt with in detail planning.

Good judgement is required in setting the appropriate level of detail for this phase. Major projects usually have a finite window of opportunity, and the team must deliver its plan while the window remains open. To do so, team members must develop enough detail to be sure of their approach and quantify potential benefits for management, but they must not give in to the well-known "analysis paralysis." On the other hand, the facility must eventually be designed to handle the exceptions as well as the high-volume items. This means that Pareto analysis and other generalization techniques must be used with care, and contingency funds must be added to the project budget to handle any uncertainty introduced by the use of generalized data.

The end result of the macro planning phase will be a facility plan (block layout) of manufacturing areas (cellular and traditional) and supporting services—everything in the factory that occupies space. Included will be a plan for installing the new factory design, covering:

- The main tasks to be performed;
- Their estimated implementation times, and
- A phased investment and benefit schedule for the project.

Keeping the project deliverables in mind should help the team determine the proper amount of detail for each step of the analysis.

Types Of Cells

Ideally, a cell should be designed as a small autonomous business, operated by a group of individuals, with the business objectives discussed above as the primary performance criteria. A well-conceived facility arrangement provides a physical setting where these objectives can be achieved. In most applications, different cell types will be knit together, forming the right combination or approach for the specific products to be made and markets to be served. Arranging a factory around logical product groupings is the foundation for cellular manufacturing. Understanding products and their processes forms the first step in developing a macro plan for an FMC environment. Nearly all cells are one of three basic types, as shown in *Figure 5-3*.

Product cell. Product cells produce finished products and ship them to a customer. Whenever a product can logically be manufactured complete by a limited team in one compact area, one has the basics for a product cell. This should not be confused with a "focused factory," which is normally a group of several cells with more global business objectives. The highest degree of control is achieved when all operations are thus linked together in close proximity. Of the three cell types, product cells typically return the greatest economic and performance benefits, since they permit elimination of nearly all nonvalue-adding activity. Examples might include automobile dash modules, personal computer chassis, or insert molded extension cords.

Process cell. A process cell manufactures components for one product or family of products requiring common processes that can be grouped or shared

74

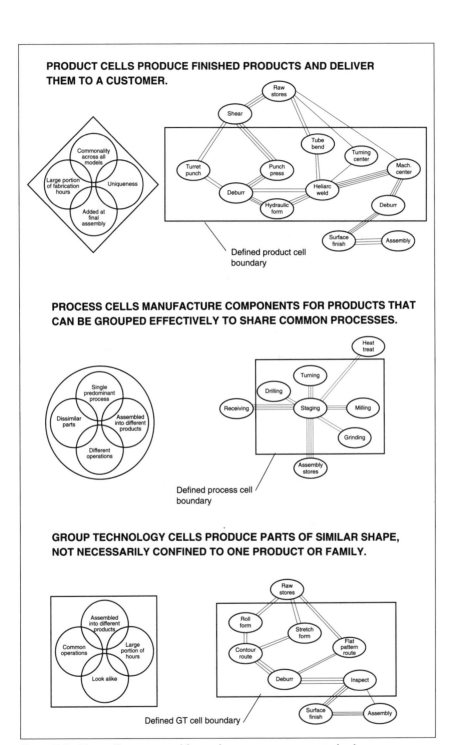

PRODUCT CELLS PRODUCE FINISHED PRODUCTS AND DELIVER THEM TO A CUSTOMER.

PROCESS CELLS MANUFACTURE COMPONENTS FOR PRODUCTS THAT CAN BE GROUPED EFFECTIVELY TO SHARE COMMON PROCESSES.

GROUP TECHNOLOGY CELLS PRODUCE PARTS OF SIMILAR SHAPE, NOT NECESSARILY CONFINED TO ONE PRODUCT OR FAMILY.

Figure 5-3. *Most cells are grouped by product, process or group technology.*

75

effectively but do not necessarily follow the same sequence of operations. The process involved is often one that, for reasons of equipment investment or environmental considerations, should not be installed in multiple product cells. Process cells can score impressive improvements in throughput time and cost for specific operations, but they function at the mercy of the operations that precede and follow them. For this reason, their speed and economic impact are generally less dramatic than that of product cells. Examples of process cells might include an NC lathe/screw machining cell for shafts, or a sheet metal processing cell, with a shear, NC punch, and forming/welding area.

Group technology cell. Probably the most advertised and promoted of cell approaches, the group technology cell occupies middle ground between dedicated equipment and a job shop environment. It produces parts of similar shape, not necessarily confined to one product or product family. Group technology has existed for over 30 years and is considered by many to be the forerunner of current cell thinking. When combined product line volumes will support a cell, but individual families cannot, a case can be made for group technology cells. Economic benefits from group technology cells are typically less than those produced by either product or process cells, since they require the same kind of scheduling and indirect labor support as a job shop. Examples of group technology cells include small miscellaneous machined brackets and levers for a variety of products.

Any other grouping of machines or processes, including those not producing a finished product or component, are not cells in the context of this book. Centralized heat treatment, plating lines, painting systems, and large main line assembly facilities may be considered as suppliers, customers, or "contract services" to cells, but they are not cells in their own right.

STEP 3: GROUP PRODUCTS FOR CELL DESIGN

The first step in macro plan development is establishing logical product groupings (families) which best meet business objectives. Technical considerations for family selection include:

- *Configuration*—the most common basis for product family definition. The team usually begins by intuitively grouping parts according to common-sense parameters:
  —Commonality or similarity of function;
  —Similarity of appearance;
  —Similarity in size and weight;
  —Commonality of raw materials;
  —Commonality of components, and
  —Similarity of quality or customer service requirements.
  A family selected by configuration data will usually be homogeneous in engineered characteristics, but this should not be taken for granted.
- *Common processes*—characteristic of very simple or single component products. The families may not be obvious.

- *Engineering content*—standard versus "engineered" or special products. A family may share configuration criteria, but be totally wrong in mix for an effective cell. The modification of a standard product can greatly alter the flow, sequence, time, and materials used. It is often desirable to create a separate area for specials even though they share common processes and material.
- *Volume*—a primary consideration for the cell planner. Each of the above considerations has a volume implication. While machine utilization typically is not used as a measurement of cell performance, there will be a capital value/usage ratio as a guide in the planning process. For example, a $1.5 million CNC machining center would not be placed in a cell if it had only 15% utilization. It may be necessary to share a high-value asset between cells. Separation between high- and low-volume business is a typical approach to facilities arrangement—even for the same product or a family sharing common components.

The technical criteria are well understood in the factory. Establishing product groupings (families) which best meet operating objectives, however, is not simply a manufacturing department exercise. Everyone affected by the decision must be part of the selection process. Marketing, sales, order entry, shop scheduling, purchasing, product engineering, manufacturing management, accounting, quality control/assurance, process planning/manufacturing engineering, and the executive staff all should be involved. The result will be a business decision supported by as much information as can be assembled.

Nontechnical considerations for family selection include:
- Markets served:
  —Specific products or variants may have unique characteristics which serve different markets and require different factory performance criteria. Example: Highly engineered nuclear power station pumps (high cost, long lead time, exotic materials) vs. agricultural irrigation pumps (low cost, short lead time).
  —Geography may be a consideration for products that are much more costly to ship than are their raw materials. Examples: Electrical panel boxes, molded plastic pipe.
- Customers served:
  —This includes customer-specific products or major variants. Examples: ABS valves for Ford versus those for General Motors. Or say that Global receives 70% of Acme's product C, and that the balance, with a lot of variations, go to 14 other customers. It may be wise to dedicate one cell or one line within the cell to meet Global's requirements.

Product Matrix And Anatomy

A product matrix will sometimes simplify product family identification, as shown in *Figure 5-4*. The team can prepare the spreadsheet by first listing all the candidate products down the left side, with key characteristics or other considerations listed across the top. The planning volume is placed in each box

Model no.	Market		Shipments		No. of orders per ship quantities			Present on-time percent
	Industrial	Agricultural	Direct	Distributor	1 to 4	5 to 19	20+	
VI-44-A	2,000		2,000		400	60		87
VI-40-AD	3,000		3,000		600	45	12	55
VI-36-A	4,000		2,500	1,500	1,000	25	4	88
VI-32-D	4,000		2,500	1,500	1,000	25	5	75
VI-30-AC	2,000		2,000		1,000			45
VI-26-B	6,000		4,000	2,000	1,500	100	10	60
VI-22-A	5,000		3,000	2,000	1,200	25	10	95
VI-18-C	3,000		3,000		800	100		94
VI-14-AB	2,000		2,000		1,000			97
VI-12-B	2,000		1,000	1,000	400	30	5	64
VI-10-A	3,000		1,500	1,500	600	60	6	85
VI-8-A	4,000		2,500	1,500	1,200	100	10	97
VI-6-A	2,000		500	1,500	500	30	50	98
VA-12-L		6,000		6,000		15	200	74
VA-10-LA		40,000		40,000		300	350	85
VA-8-L		22,000		22,000		200	200	80
VA-6-LB		10,000		10,000		120	400	88
TOTAL	42,000	78,000	29,500	90,500				

MATRIX ANALYSIS NOTES

1- Agricultural market is 65% of total volume and 25% of the variety

2- Agricultural market is 100% distributer shipments

3- Agricultural shipments are larger order quantities

4- Families not clearly defined by order patterns

5- Smaller models have fairly good delivery performance
(completion of large orders is a problem)

Figure 5-4. *Product family identification can be simplified with a product matrix.*

that represents a valid intersection of product and characteristic. The resulting matrix shows what family divisions could be considered, and what correlation may exist between groupings made according to different characteristics. In some cases, the appropriate product groupings become clear from this level of review. More often, the final decision will come from the output of the next steps in the macro study process—a look at the anatomy of the products to be produced.

The anatomy of a product is determined by how the product is processed or assembled. The "explosion" in an accurate bill of material shows product anatomy, but bills of material are often inaccurate or incomplete and they cannot always be trusted. Some team discussion and shop floor research should enable the team to establish the reliability of existing bills of material. If they are not accurate, it will be necessary for the team to create valid routing information for the representative parts.

An example of product anatomy (sometimes referred to as product structure) is shown in *Figure 5-5*. In this instance the team has taken a common product

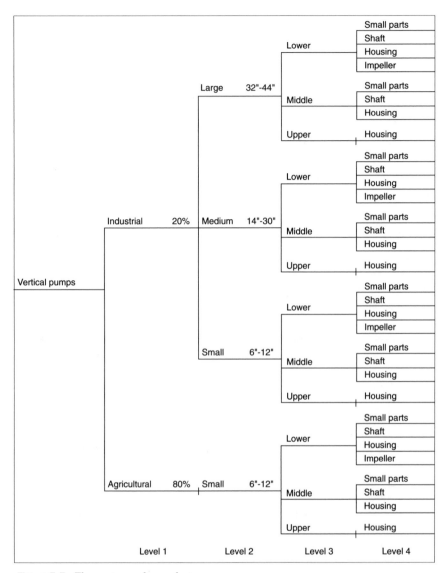

Figure 5-5. *The anatomy of a product.*

situation—a broad family of vertical pumps which divide into two distinct markets (Level 1). The industrial market pumps are divided by size into three groups (Level 2) according to process equipment, and all of the agricultural units are in the small category. Level 3 shows the subassemblies (modules) required for each size group. Level 4 identifies the primary component types for each module.

There is a size range within each group. At this level of study it is only necessary to know what the variety (the number of different sizes) of each component is. A representative part approach (discussed later in this chapter) will be used to quantify the manufacturing requirements.

Graphics visualization is a great help in assessing product volume and variety relationships. The data can be sorted and displayed in a number of ways to assist in the development of product families and the different ways they can be organized for manufacturing. The product anatomy can also provide a visual means of identifying logical groupings in various ways, as shown in *Figure 5-6*.

The Use Of Representative Parts In Cell Modeling

In most cases, it is not necessary at the macro level to use the entire product database for cell development. Instead, representative models can be used as the basis for generalized planning. If representative parts are chosen well, results will be accurate to 90%, with the uncertainty small enough to be covered by reasonable contingency planning.

It is also possible in most cases for the team to develop a set of simplifying assumptions about factory operations for macro planning and use them as design criteria. These assumptions may include things like the following, taken from the vertical pump example:

1. Annual volumes for all product families can be met within the existing factory.
2. Level production can be assumed, with agricultural pump sales operating countercyclically to industrial sales.
3. Assembly volumes, processes, and size groupings indicate three assembly lines:
   —Large industrial;
   —Medium industrial, and
   —Small agricultural and industrial.
4. Once subassemblies are produced, minimum inventory, fastest throughput, and tightest process control will be achieved by integrating the final assembly of all three lines. Most variation occurs at the component level.

One model is usually selected from each product family. The best candidate will represent the family's typical manufacturing content. The components in each representative model in turn become the representative parts used for cell development. The selection process is often iterative—balancing overall product knowledge and intuition with a disciplined approach. At this point some minor manufacturing requirements may be missed; adjustments will be made during detailed cell design. Each representative model must satisfy the three criteria— manufacturing process, size or geometry, and components.

Manufacturing process. The selection procedure begins with the team taking the highest volume model within the family and comparing it to the rest of the family for manufacturing process content. If the highest volume model represents 85% to 90% of the content, it passes the first test. If not, a lower volume candidate should be selected, and the first test repeated. It is often true that a lower volume model is a better representative of manufacturing content than the highest volume unit. If a family is characterized by special adaptations

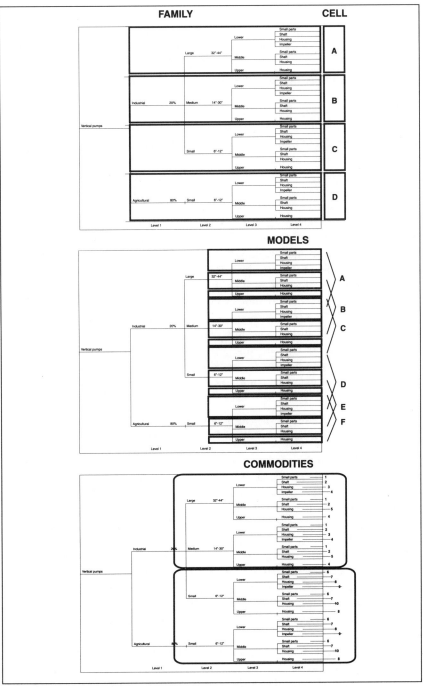

Figure 5-6. *Product anatomy can also be represented visually to aid in identifying logical groupings. This grouping shows product cells based on product anatomy, models, and commodities.*

81

or options, it may be necessary to "construct" a composite representative model or create a "special" component to the base representative unit.Typically, some variation from standard is expected and will be accommodated by the inherent flexibility of the cell. A major mismatch may indicate the need to perform a make-versus-buy analysis of nonstandard models, or to regroup the product families.

Size or geometry. This is the second criterion for making representative model decisions. Given that the volume is spread among several models, selection of a model of mid-range size is often appropriate for the equipment and FMC load analysis, although sometimes worst-case candidates are better. Briefly, the load analysis uses the manufacturing time multiplied by the total product family volume (with the representative model as the manufacturing database) to generate a total manufacturing load on key equipment. The methodology is explained in more detail in Chapter Six. Knowledge of how components are presently being manufactured and knowledge of improvements and/or alternatives in equipment or technology will help visualize the future and aid in sorting products into families. The team can test its decisions by verifying the match of size with equipment capabilities. It may be necessary to shift the largest or smallest models to another family, or to create a new family.

Components. This third decision point deals with whether the selected model is representative of all the standard components and additional options (when additional options are typical within the family). The description in *Figure 5-7* using the vertical pump example should help clarify this topic of representative models in terms of manufacturing process, size or geometry, and components.

STEP 4: MAKE A PROCESS/MACHINE
LOADING ANALYSIS FOR THE CELL GROUPS

Product data and the resultant family and (representative) component groupings define the starting point for cell development. The processes (machines and equipment) provide the cell content. A database containing process information can now be constructed to provide a parametric description of each cell. Typically, processing information is contained in some type of routing documentation. The process data for cell development should contain:

- Part numbers of selected representative parts for the macro phase;
- Workstation descriptions for key machines, equipment, or operations;
- Workstation codes for specific equipment when available;
- Operation sequence for each representative part;
- Setup time for each machine, or for parts if there is extensive variation, and
- Run time (machine, not operator) for each key process of each representative part.

The macro plan should be developed "bottom up," starting with the individual cells. A worksheet should be set up for each cell. This keeps the data in manageable increments. A Lotus-based spreadsheet for use in cell develop-

In the case of our example vertical pump manufacturer, most of the components are cylindrical, which require turning operations. Diameters range from six inches to 42 inches. Volume is such that several machines would be required to accommodate the load. Some key points to machine selection are:

- As equipment size increases, so does investment.

- Typically, accuracy decreases as equipment size increases.

- Horizontal turning equipment is more flexible, has better accuracy, and costs less than vertical turning equipment.

- Larger diameters (over 12 inches) begin to tax the capability of standard horizontal equipment.

- Large, heavy components are best suited to vertical applications, which take advantage of assistance from gravity.

- Standard medium-sized vertical turning equipment can accommodate parts in the 30- to 32-inch diameter range.

- Standard large vertical turning equipment can accommodate part diameters up to about 5 feet.

These points, used as guidelines, were an aid in selecting the size range groupings for each of the product families. Similar product and equipment knowledge would be applied to other parts and geometry requirements.

Figure 5-7. Determining representative models.

ment is discussed in detail in Chapter Six. It would be desirable to use this format at the macro level, then expand it for detail planning; the planning steps between the two are common.

A set of operating parameters is required to provide a time base for calculations and to define any allowances or constraints. The bottom line for driving the calculations is "planned available hours," usually expressed in hours per month or year. Typical considerations for allowances and constraints will include:

- Operating days per time period—month or year;
- Number of operating shifts per day;
- Working hours per shift;
- Machine operating efficiency—a percent of available or standard time usually based on machine capability or some adverse operating condition, and
- Machine downtime allowance—maintenance or tooling.

The final type of data required for calculating cell workload is the volume figures associated with each representative part number. *Figure 5-8* shows a print-out of a cell load calculation using the spreadsheet format discussed in Chapter Six. Combining the cell calculations with other manufacturing areas and the requirements for supporting services defines the manufacturing content requirements for the total facility.

Large & Medium Housings	Present Routing Data (Hours)									
	Annual Qty.	Run Qty.	Mill 534		Lathe 244		VTL 310		Drill 465	
Part Data			Setup	Run	Setup	Run	Setup	Run	Setup	Run
Housing U36	5,400	450	0.30	0.60			0.45	1.20	0.20	0.35
Housing L36	5,300	440	0.30	0.50			0.45	0.90	0.20	0.35
Housing U18	9,800	408	0.24	0.36	0.17	0.38			0.15	0.25
Housing L18	9,700	400	0.24	0.28	0.17	0.35			0.15	0.25
Hours available per resource = 4,000			19	12,134	8	7,119	11	11,250	12	8,620
				3.03		1.78		2.82		2.16
Resources needed and percentage used			4	(76%)	2	(89%)	3	(94%)	3	(72%)

Figure 5-8. *A cell load calculation.*

When product volume will not support product cells, when part family identity is not clear, or when it is necessary to subdivide a product family into several product flows, multiproduct process charts can be helpful. These charts provide a method for grouping components according to their common processes. The workstations are listed on one axis of the spreadsheet, with the selected part numbers on the other axis. The operation number or production volume is placed in the appropriate intersection. The spreadsheet columns or rows can be rearranged to better visualize logical groupings. The resulting groupings can be used for load calculations.

Once cell load calculations are completed, the team will know the generic type and quantity of each work center required to produce the planned mix and volume. Where planning volumes differ radically from current output, it is usually wise to produce a machine loading calculation for the current situation. This will provide a baseline figure against which to measure improvement. Planning volume load calculations will identify where focus must be placed for improvement and/or new equipment considerations. At this point, it is often true that more equipment appears to be required than is available. When this happens, adjustments are necessary.

The load calculation spreadsheet should include a utilization percentage for each work center. This is used in two ways:

- *Identification of bottlenecks or overloaded work centers.* Remedies for an overload condition include:
  - —Purchase of additional equipment;
  - —Changing the operating parameters (adding a shift, for instance);
  - —Shifting some of the load to similar equipment in the same cell;
  - —Moving some parts to another cell, if they are compatible with products produced there;
  - —Eliminating or reducing setup times;
  - —Improving fixtures, and
  - —Changing technology.
- *Identification of grossly underutilized work centers.* Opportunities posed by an underloaded condition are:
  - —Consolidating loads to similar work centers;
  - —Merging cells;
  - —Changing or simplifying technology;
  - —Combining operations.

Underloading of existing equipment is not necessarily bad. It often provides the opportunity to dedicate lines to high-volume products, or to equip a short-run area that will reduce the complication produced by short production runs in a high-volume production environment. The purchase of new underloaded equipment, of course, is to be avoided.

STEP 5: IDENTIFY IMPROVEMENT OPPORTUNITIES

A key step in the macro planning process is the identification of major improvement opportunities. Machine loading and other output from the cell grouping effort will show where improvement has the potential to help the team reduce cost and/or avoid investment. These changes usually have an impact on the physical arrangement and space required for production, so they should be discussed at this point in the project. This is also the time to set some high-potential targets for detail planning. If an improvement does not reduce the plant payroll, increase shipments, or reduce inventory, it is probably not worth the time and money that will be spent to get it. Improvements on noncritical operations are seldom beneficial.

Targets for improvement in the cellular factory might include things like:
- Setup on machine no. 3426 reduced from 2.30 hours to 0.25 hours to remedy overload.
- All fixtures to be stored at workstation, with primary setup to be performed external to machine operation.
- In-cell labor equivalent reduced from eight to six operators per shift.
- Cross training and task sharing in-cell.
- Overall inventory reduced by 85%.
- Throughput time (factory lead time) reduced by 90%.

Operating parameters should be reviewed at this time, and any changes incorporated and submitted for the necessary management approval.

The team should identify individual improvement elements that impact each cell, adjust the routing content to reflect work center changes, revise routings to reflect new process times, and re-run the load calculations. This will provide the "to be" process scenario, which, in turn, can provide a first approximation of the project's financial potential.

STEP 6: BUILD THE MACRO LAYOUT

The development of the macro layout should follow an organized, systematic approach. One excellent, widely used methodology is the 9-task Muther systematic layout planning process, shown (adapted) in *Figure 5-9*.

The input used to drive layout planning (P, Q, R, S, and T in *Figure 5-9*) is the information developed in the macro planning steps that were covered earlier in this chapter:

- *Product*—families, subassemblies, and components.
- *Quantity*—planning volumes for each product category.
- *Routing*—the operations, work centers, sequence, and times.
- *Support*—identification of all required supporting services.
- *Time*—operating parameters and/or schedules.

Every activity area occupying space must be included in the macro plan. While the main focus has been on the development of cell content, there are usually other production areas that may need to be included, such as:

- Central heat treatment;
- Central plating lines;
- Special processing or fabrication lines;
- Central painting facilities, and
- Main line assembly facilities.

To be identified next are all of the supporting services for the plant. These may be divided into the following categories:

1. *General*: Those functions serving the entire plant, consisting mostly of the "general" office areas and related activities.
2. *Production*: Those activities primarily serving the production organization(s), like shop offices, tool cribs, inspection, etc.
3. *Personnel*: Those functions operating primarily for serving or handling the needs of people, such as first aid, cafeterias, or rest areas.
4. *Physical plant*: Those services primarily concerned with the needs of the physical facilities, including building, equipment, utilities, maintenance, land, etc.

A key feature of cellular manufacturing is an *organization* designed around the "work." This means extensive integration of the supporting services into, or close to, the area served. A lot of traditional departmentalized functions can and should be incorporated as an integral part of the cell. Obviously, the extent to which integration can be achieved is dependent upon many factors, not the least of which is cell size and content. When extensive integration is achieved, functions such as these are in-cell: production control, purchasing, quality

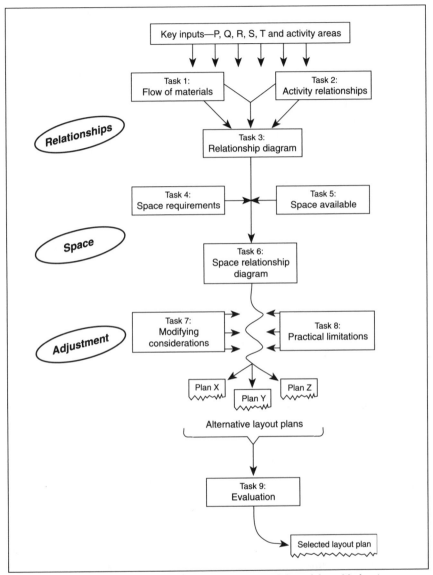

Figure 5-9. *Macro layout process for facilities planning. (Adapted from Muther.)*

control, manufacturing/industrial engineering, maintenance, tooling and fixture crib, gage control, cost accounting, and order entry and dispatching. Chapter Three elaborates on organizational effectiveness and how FMCs are impacted.

The macro plan is a "vision" of the future, with a planning horizon of at least three years. Major improvement means major change in the way a company does business, and such programs will take a number of years for most manufacturing plants. The macro facilities plan needs to reflect the changes seen for the organization across the planning horizon as well. Again, change shouldn't be forced for the sake of change—the planned integration must be supported by suitable data, logic, and rationale.

Task 1: Flow Of Materials

Task 1 in the macro layout process is to develop the flow of material patterns between the production areas. In a product-based cellular facility, flow of material analysis is relatively simple—direct flows from a "supplier" to a cell, and from a cell to its "customer." A measure of magnitude—quantity of moves during a selected time period—is used to determine the proximity desired between each pair of activities. The method used for analysis is guided by the complexity of the move patterns and the number of activities involved. *Figure 5-10* shows some flow diagramming applications for macro planning. The primary analysis methods and their applications are:

- Process flow diagram—operation process chart:
  - Single product or simple product family;
  - Common operation sequence;
  - High-volume/low-variety line concepts.
- Multiproduct process chart:
  - A few (five to seven) parts with common processes;
  - Medium to low volume;
  - Aids in grouping similar parts for a process cell;
  - Develops flow within a process or group technology cell.
- From-to chart:
  - Complex multipath flows;
  - Large number of activity areas;
  - When no other method is suitable.

Operation process and multiproduct charts provide the routes (and in some cases the amount of movement) for overall macro flow analysis. A from-to chart will contain both the route and the amount of movement during a selected time period. Whatever the method used, these data need to be put in a format for identification of proximity requirements.

The first step is to calculate the movement across each move path (between each pair of activities where there is measurable material flow). The total of all movement between activity pairs is charted, then sorted in descending order for the proximity diagraming in Task 3. A typical method for documentation of the flow data is in a from-to format. *Figure 5-11* shows a sorted set of from-to data.

Even though it is necessary to make all calculations using a common unit of measure, some businesses handle a wide variety of loads. A 25 pound, one-cubic-foot container or component moved from cell A to final assembly is not the same as a 500 pound, 15-cubic-foot pallet or component (unless moved one at a time by the same means). Material handling equipment must be planned accordingly.

88

OPERATION PROCESS CHART

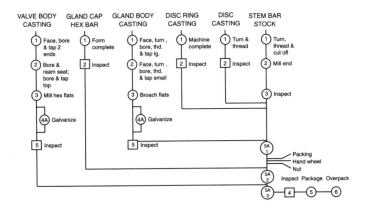

FROM-TO CHART

Item(s) Charted _____ Pump components _____ Basis of Values _____ Moves/day

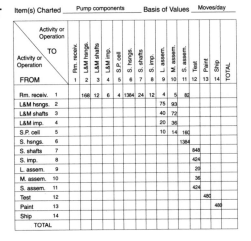

Activity or Operation FROM	Rm. receiv. 1	L&M hsngs. 2	L&M shafts 3	L&M imp. 4	S.P. cell 5	S. hsngs. 6	S. shafts 7	S. imp. 8	L. assem. 9	M. assem. 10	S. assem. 11	Test 12	Paint 13	Ship 14	TOTAL
Rm. receiv. 1		168	12	6	4	1384	24	12	4	5	82				
L&M hsngs. 2									75	93					
L&M shafts 3									40	72					
L&M imp. 4									20	36					
S.P. cell 5									10	14	160				
S. hsngs. 6											1384				
S. shafts 7											848				
S. imp. 8											424				
L. assem. 9											20				
M. assem. 10											36				
S. assem. 11											424				
Test 12												480			
Paint 13													480		
Ship 14															
TOTAL															

MULTI-PRODUCT PROCESS CHART

	Part Number														
	101	201	102	202	103	104	105	106	107	108	109	110	111	112	113
1 Receiving	1	1	1	1	1	1	1	1	1	1	1	1	1	1	1
2 Storage	2	2	2	2	2	2	2	2	2	2	2	2	2	2	2
3 Turret lathe	3, 4	3, 4	3, 4	3, 4	3										
4 Mill and broach	5	5	5	5	4		5							5	
5 Automatics						3			3	3		3			
6 Galvanize	6		6												
7 Assembly	7	6	7	6	5	4	3	3	4	4	3				1
8 Packing												3	3	3	2
9 Warehouse															3
10 Shipping															4
Handling Units Per Year	545	385	251	218	469	236	3,718	135	156	156	890	451	113	226	4,000
Inspection is part of last operation															

Figure 5-10. *Flow diagramming applications for macro planning.*

FROM-TO ANALYSIS						
Work-center Path	Intensity*					
	10	20	30	40	50	60
447-420	**********	**********	**********	**********	**********	*********
455-I57	**********	**********	**********	*****		
I57-246	**********	**********	*********			
I57-229	**********	**********	*******			
I57-245	**********	**********	******			
420-I57	**********	**********	****			
260-447	**********	**********	**			
229-447	**********	**********	**			
420-455	**********	**********	**			
246-260	**********	**********				
245-260	**********	*****				
260-455	**********	*****				
223-229	**********					
245-420	**********					
M10-I57	*********					
246-223	********					
447-I57	********					
453-I57	*******					
447-453	*******					
Z10-M10	******					
223-420	****					

* "Unit" moves per day between activities.

Figure 5-11. *A sorted set of from-to data.*

It is necessary to establish a unit of measure that represents the difference in the move effort or move method, and pick a configuration that represents a large percentage of the movement and call that one a ''move unit.'' Smaller configurations are probably still one move unit (or fractions if necessary). Larger configurations become multiples of the selected move unit. Since this is a macro plan, it isn't necessary to split hairs. Five or six move unit categories should be adequate to differentiate the typical facility move requirements.

Task 2: Activity Relationships

Task 2 is the development of relationships that are not material flow-related and merging them with the flow relationships. Even with a high level of supporting service integration (within the cells), there are activities that occupy space and relate to other activities in the facility, as well as some "general," mini- or focused-factory, or centralized services that may modify the proximity desired between various activities.

Figure 5-12 shows an example of a relationship chart combining all activity relationships. Proximity requirements are based on the degree of closeness desired between each pair of activities. A method for ranking relationships has been devised to identify different levels of relationships and to simplify the diagraming in Tasks 3 and 4.

Task 3: Relationship Diagram

Task 3 diagrams the relationships to establish the desired proximity between all activities in the facility. The closeness desired is drawn with line values

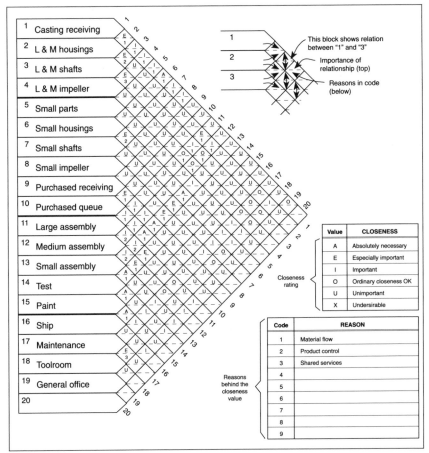

Figure 5-12. *A relationship chart that combines all activity relationships. (Adapted from Muther.)*

between each ranked activity pair. The line values follow the relationship ranking codes and are drawn with different line lengths to "visualize" the desired proximity, the key for which is shown in *Figure 5-13*.

Relationship/proximity diagramming is an iterative process, starting with the "A" relationships, then the "E," etc. Redrawing the diagram (and rearranging alternative positions) with each added group of relationships will help to maintain the proper visualization. The resulting diagram, *Figure 5-14*, provides a picture of the proximity desired within the overall facility. Since there is probably more than one solution, alternative trials should be made until one or more satisfactory results is achieved. As a guide for appropriate application of a process flow diagram or a proximity diagram, see *Figure 5-15*.

Closeness	No. Of Lines	Suggested Length (Inches)
A = Absolutely necessary	////	1/2 to 3/4
E = Especially important	///	3/4 to 1
I = Important	//	1 to 2
O = Ordinary closeness O.K.	/	2 to 3
U = Unimportant	No lines	
X = Undesirable	M\M	2-plus

Figure 5-13. *A key for relationship/proximity diagramming.*

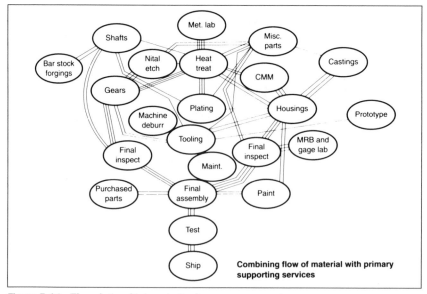

Figure 5-14. *The relationship/proximity diagram.*

92

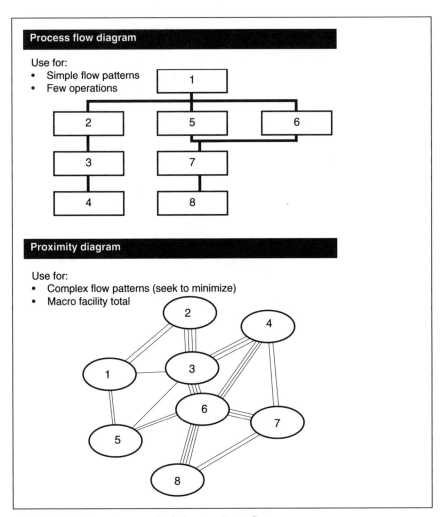

Figure 5-15. *Two primary methods for visualizing flow.*

Task 4: Space Requirements/Task 5: Space Available

Tasks 4 and 5, determining space requirements and the space available, provide the space data for the macro layout. High-level layout density is a major benefit of cells, so extra care must be taken in the method used to calculate space requirements. It is best to start with a complete list of the machines and equipment to be used in the facility, and to group the machines and equipment by cell or other activity area. The machine and equipment footprints are then established by careful measurement and calculation. Allowances need to be made in the footprint calculations for maintenance and operator access.

Added to the footprint information will be any requirements for material handling. Component or load size is often a consideration, and set-down space will probably be required in a process or group technology cell. When data is not available, every means possible should be used to obtain a good estimate.

An aisle allowance can be added to each area by calculation (determine a logical area space shape and add one-half of an aisle width to one side) or to the sum total of all areas by a percentage. The first method is preferred for initial macro layout preparation. There is no definitive calculation for aisle allowances: a traditional factory layout will have aisles that occupy anywhere from 15% to 35% of total space. A lot of factors influence the allocation: building configuration, machine and equipment size, fire control access, and material handling to name a few. A key fact to remember is that a traditional aisle layout is designed for a multitude of fork trucks running about, which should no longer exist in the cell environment, so the aisle space should be reduced accordingly.

Next, a spreadsheet format is organized to include the following types of activity area information:

- Activity area identification (name);
- Activity area designation number (used for diagraming later);
- Space allocation for each activity area;
- Layout/location considerations;
- Special features such as:
  —Bridge cranes in a high bay;
  —Access to external transport;
  —Fixed facility (something that can't be moved);
  —Special utilities, and
  —Environmental requirements.
- Space shape constraints.

Extra attention to the balancing of the space required with the space available is necessary in three situations:

- Multibuilding site;
- Multifloor building, and
- Building with major cells or wings.

Arriving at an acceptable balance is a trial process; it involves successively matching groups of space requirements with the units of available space until a fit is obtained, without losing sight of the proximity requirements.

Task 6: Space Relationship Diagram

Task 6 is simply the addition of space to the proximity diagram. Space will impose some modifications to the original diagram; the result, however, should begin to shape the macro layout. *Figure 5-16* shows an example of a space proximity diagram.

Task 7: Modifying Considerations/Task 8: Practical Limitations

Every layout is a compromise. There are a number of factors affecting the final arrangement. Task 7 identifies the modifying considerations and Task 8 considers the practical limitations. Practical limitations are things that impose specific constraints on the arrangement of facilities. Modifying considerations are those things that create alternatives in the way activities are arranged.

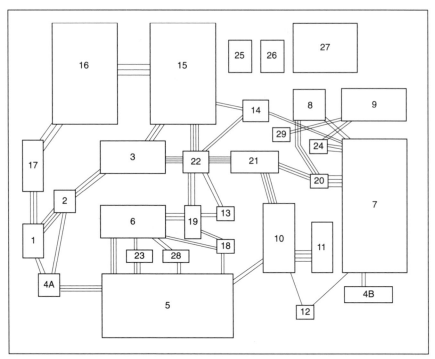

Figure 5-16. *Space relationship (proximity) diagram.*

Alternative macro layouts are developed by applying modifying considerations and practical limitations to the space proximity diagram. Common factors in these considerations and limitations are shown in *Figure 5-17*.

Visualization can be done manually or with CAD software such as ComputerVision or AutoCad. Again, this is an iterative process that should generate three or four viable alternative arrangements.

It is recommended that initial macro trials be prepared with aisle allowances incorporated within the individual cells and other activity areas. Area space

Modifying Considerations	Practical Limitations
Material handling system	Building configuration
External access	Fixed facilities
Services and utilities	Site arrangement
Scrap/chip handling	Floor loading
Safety/security	Environmental needs
Special management needs	
Aisle patterns	

Figure 5-17. *Common factors that require modified macro layouts.*

shapes can be adjusted to achieve alignment to provide logical aisle locations, and the physical aisle pattern can be added to the selected layout.

Cells typically create more individual manufacturing areas than traditional process-oriented departmental arrangements. This can mean more individual relationships between the manufacturing cells and activities such as:

- Receiving and shipping;
- Raw material storage;
- Pre-assembly purchased parts staging;
- Shared assets or processes;
- Maintenance and tool services, and
- Quality and gage labs.

Care must be exercised to ensure their logical assignment and placement in the macro layout. Initial concepts may require modification to obtain realistic solutions. Several macro layout examples are shown in *Figure 5-18*.

A logical question at this point in the process is: "Are there computer programs for making macro layouts?" The answer is "yes," there are a number of computer algorithms which can be used to aid layout development. The

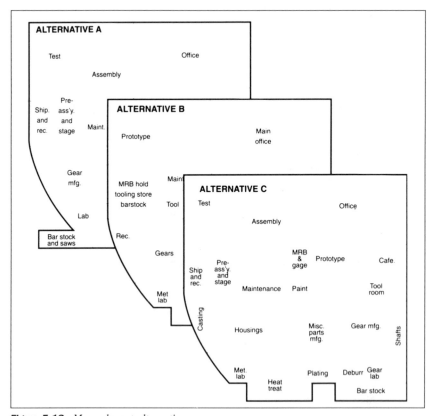

Figure 5-18. *Macro layout alternatives.*

96

emphasis is on *aid*. Algorithms do not have thinking capability or knowledge and understanding of the necessary modification or adjustment factors—an algorithm will only handle one set of input data at a time. Two types of algorithm routines are available: improvement and construction routines.

Manual adjustment is required with all of the algorithms to eliminate unrealistic locations, shapes, and alignments. None will cope with negative relationships or architectural influences. Some of the other limitations of the routines are shown in *Figure 5-19*. With modification, CRAFT can be used to check the most likely alternative as insurance against having missed considering all logical alternatives.

Task 9: Evaluation

The last step in the macro layout process, Task 9, the evaluation of alternatives, is discussed in Chapter Seven.

LIMITATIONS	
Improvement Routines	**Construction Routines**
Requires a building shape	Will not consider building shape
Hard to consider relationships other than flow	May ignore critical relationships
Can't cope with multi-floor or multi-building conditions	Only a limited number of programs will consider multi-floor or multi-building conditions (and not very well)
Iterative process; results depend on quality of initial layout	

Figure 5-19. *Computer algorithms can aid layout planning, but they have these limitations.*

STEP 7: SELECT A CANDIDATE PILOT CELL

In the process of developing the macro facility plan, the actual cellular opportunities should become clear. Based on that knowledge and the ability to match those opportunities to the best chance of success, the project team should be prepared to make a decision on an appropriate pilot cell to begin the detail planning exercise.

The selection criteria most commonly used for this decision are:
- The space to implement the FMC is immediately available (as identified in the block layout, the area can be easily cleared).

- The benefits can be immediately achieved, documented, and utilized to great advantage.
- The level of complexity is low enough to encourage a first-time project team to "experiment" or discover the process.
- Adequate existing machines and equipment are readily available for installation, avoiding investment decisions.
- Maybe because of several of the above, the enthusiastic support of the shop floor people is ensured.
- In some cases, a logical candidate for the pilot FMC is just plain obvious, or the company already has some specific new equipment ordered or in the budget.

The project team should agree on the meaningful criteria, review the candidate pilots, and make a decision based on documented data.

SUMMARY

Ideally, the cell environment is customer-driven and product-oriented. An organized approach to macro planning is required to ensure optimum results, and the steps outlined in this chapter are aimed at guiding the cell planner toward achieving them. Well-planned macro layouts have these objectives:

1. Overall *integration* of all factors influencing the layout;
2. Material moving a *minimum distance*;
3. Work *flowing* through the plant;
4. All space effectively utilized with high *density*;
5. *Satisfaction* and *safety* for workers, and
6. A *flexible* arrangement that can be easily readjusted.

The ability to achieve the last objective is very important in today's ever-changing business environment. The facility layout cannot be static—it must be dynamic. Response to change is a characteristic of the continuing improvement process demanded in the world marketplace.

The next step is detail cell development and detail layout planning. A well-conceived macro plan is the basis for effective detail planning. The same set of procedures outlined in this chapter is applicable to detail layout planning at the individual cell and process level.

CONCEPTUAL DEVELOPMENT
OF AN FMC

In Chapter Five, the overall macro facility concept was developed by data manipulation, and a parametric description of each cell was created. A block layout and manufacturing plan were developed to describe the most appropriate manufacturing environment for the product mix to be produced. As a result of that exercise, the planning team should have the following available to begin actual cell design:

- Part family groupings for the entire product mix;
- At least one block layout of the entire facility;
- Overall part/product material flow characteristics for the redesigned facility;
- A determination of which cell types to use for each product group;
- General processing sequences and/or opportunities for process changes for each product group;
- Selection of an appropriate specific cell to develop first, perhaps as a pilot for several larger projects.

This chapter provides the general process steps for designing individual manufacturing cells at a conceptual level. Since most FMC applications (about 75% of the total) are product-oriented, and since this kind of cell usually delivers the most value for the team's effort, this chapter will assume the conceptual development of a product-driven FMC. Most steps for FMC development are the same for group technology and process cells.

In conceptual cell development, the team is looking for ideas, trends, and benefits to be associated with the cell rather than the finite details of measurements, machine brands, and foundation designs. *The objective at this planning stage is to position the team to prepare and present the cell alternatives that best fit the overall strategic objectives for the factory change.* Once the cell concepts are developed, the team will need to prepare a financial analysis of the opportunities in cellular manufacturing and a presentation of the cell concepts to management. The need for more engineering detail will arise after the concept and design specifications are presented and approved for implementation by the steering committee.

The distinction of the effort and objectives between the macro facility planning described in Chapter Five and the cellular conceptualization elaborated on in this chapter will often become blurred. The majority of applications will result in both steps being combined and intertwined to a degree that the objectives are inseparable, unless there is a specific desire to develop only one pilot cell.

The following eight-step process is a good way to approach the task of cell concept development. Some shortcuts are available in most situations, but the eight-step discipline will make results intelligible later, and will help the team to avoid overlooking important points in the early analysis:

- Step 1: Gather and record data.
- Step 2: Develop process flow within the cell.
- Step 3: Identify equipment required and compare it to what is available.
- Step 4: Balance machine/manning workloads.
- Step 5: Select/assign equipment.
- Step 6: Build a relationship diagram and an initial cell layout.
- Step 7: Recapture data in the "cell" structure.
- Step 8: Conceptualize support relationships.

STEP 1: GATHER AND RECORD DATA

Data gathering should be simplified by the work performed in creation of the macro facility plan. Output from this effort includes a parametric description of the work each cell must perform. The parametric description includes:

- The product mix and volume that will flow through the cell;
- The physical process or other characteristics that determine the part family or group that will run in the cell—this may include:
  —Size and shape;
  —Material;
  —Lot size or volume, and
  —Process steps or unique processes.
- Routings or process sheets for representative parts from each family or group, including:
  —Equipment, tooling, and gages for materials used;
  —Potential technological or process changes or upgrades.
- List of existing equipment available for the cell, with indications of capacity considerations and identification of common equipment needed in two or more cells;
- Planning guidelines for the family:
  —Product life cycle information;
  —Future volume assumptions;

100

— Management-imposed constraints;

— Key design changes or new product information.

- Operation guidelines:

  — Annual working days;

  — Contractual or other constraints;

  — Square footage currently assigned to the cell;

  — Special utilities or other facility considerations, and

  — Material flow and service relationships with other areas.

- Cost and staffing baseline data for the parts assigned to the cell.

This is obviously a great deal of information for each cell, but it is needed to support optimum cell design. Some of this data will not be needed until detail cell design which is discussed in Chapter 14, but all of it should be captured following the product family development step, and an information package should be developed for each cell group that will follow it through the cell design process.

The most efficient projects use worksheets or some other kind of data standardization to enhance information packaging. Data manipulation required for cell design is best accomplished on a PC spreadsheet or similar tool. The alternative is a great deal of tedious writing and rewriting of the same data. The time required to learn such a tool is usually paid for by the second week of computer-assisted cell design effort.

STEP 2: DEVELOP PROCESS FLOW WITHIN THE CELL

Process flow diagrams or tables must be developed for each part flow pattern within the cell, as shown in *Figure 6-1*. In a product or process cell this may be one diagram, and in a parts family cell there are often several. In situations where a wide variety of flows exist, as is often the case in process and group technology cells, it is often appropriate to perform a Pareto analysis.

Strategic planning takes seriously the fact that a few critical activities achieve most of the results produced by an organization. Or, as Vilfredo Pareto, a

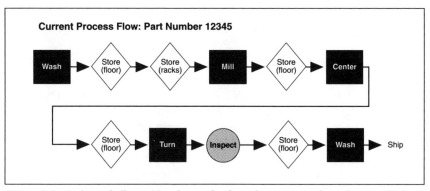

Figure 6-1. *A chart of all activities shows why throughput time is often a large multiple of process time.*

101

nineteenth-century Italian economist stated, "Eighty % of the value lies in 20% of the elements, while the remaining 20% of the value lies in the remaining 80% of the elements." This is also referred to as the 80/20 rule. The art of strategic planning is to focus on the 20% that will accomplish most of the results. A Pareto analysis will help to determine whether a few flows comprise the majority of product movement in the cell. When this is the case, high-volume flows should receive most of the team's analytical attention.

Flow analysis will sometimes require breaking the products down into subfamilies or modules; if this is necessary, volumes must be determined for each of the various flow alternatives. If the product or process mix in a cell is truly complex, the same analytical techniques described for the factory proximity analysis in Chapter Five can be used to sort out equipment placement and product movement within the cell. Usually, though, this is not necessary. In a conceptual exercise, the time and cost required for exquisite planning of each part number may cause the team to exceed its schedule and budget, and worse, may cause top management to lose interest. The detail planning phase will be the time for this kind of precision.

It is often true that annual averages are not adequate to describe production flows. Variation occurs in three areas, which must be thoroughly reviewed by the cell team:

- Variations in capacity due to vacations, holidays, and number of workdays in each month.
- Product life cycles and introduction of new products and product styles (another color, for instance).
- Market seasonality (as with lawn furniture or snow blowers).

Any of these can invalidate cell capacity planning if it is not clearly addressed in early planning. Again, the multifunctional team will help by providing access to market, new product, and capacity information.

When the effort turns to developing the physical relationships among machines and people, it will be necessary to understand the various flow patterns and their relative importance. For this reason, the tables or diagrams should include some indication of the amount of production (not just the quantity of part numbers) that flow in each pattern. This way the cell can be optimized for high-volume production, with compromises being limited to low-volume production items.

The current flow diagrams should include all of the steps that the products take, including trips to holding areas (buffers) or storage racks, and operations movement to washing or plating operations, as well as any other nonvalue-adding operations. By including all of these handlings and operations, the team can be sure not to omit any of the seemingly minor tasks that must be included in the cells. It is very easy to forget a cleaning operation (or two) at this point and neglect to include equipment in the cell budget to replace it.

This discipline, in addition to being a cell design step, will help demonstrate the changes that the cell will create in the plant. By recording the reduction in the number of handlings that parts undergo, the team can demonstrate the validity of the huge reductions that cells will deliver in throughput time and support headcount.

This is no small matter; the team may be called upon to convince management that half of all inventory can be eliminated, along with a third of the workforce. This kind of claim calls for impressive backup data. Such data are derived in this step. Cellularization is the best tool for reducing the nonvalue-adding content of production jobs, and once cells are installed throughout a factory, the need for several nonvalue-adding groups goes away. The most frequent nonvalue-adding targets for reduction are finished parts stores, material handling, expediting, and stand-alone repair areas.

Once the flow diagrams or tables are complete, they should be challenged for unnecessary or redundant costs, such as:

- Nonvalue-adding operations of any kind;
- Operations that can be eliminated;
- Operations that can be combined;
- Operations or materials that can be standardized, and
- Operations that are performed on a small percentage of the parts in the cell.

Another benefit of cellular manufacturing emerges at this point. Since all of the products that will flow across a particular group of machines are captured in this step, some kinds of changes that would not have been practical when done one part number at a time become fair game for the cell team. Part family changes can be made, and instructions can be left with product engineering that the changes should apply to all future designs. Here is another opportunity to apply the power of a multifunctional team to part families. Typical improvements include:

- Cleaning and/or deburring operations can often be eliminated by changing machining techniques, saving time, cost, and, in the new layout, capital outlays.
- Simplification of raw material can often be achieved by adding unique characteristics at the beginning of an assembly line. In products built from castings, forgings, or injection moldings, part numbers often differ by only one hole, or an orifice size. This kind of feature can often be installed at very little cost during an operation in the cell. It is often possible to reduce raw material inventory by 50% using this technique alone.
- In cold headed or stamped parts, it is often possible to standardize material sizes coming into the process by simple changes in die design, or by the addition of in-line drawing, annealing, or rolling processes. Again, 50% reductions in raw material inventory from this technique are not unusual.
- Careful selection of coolants, combined with performing operations in rapid succession, sometimes makes cleaning between operations unnecessary. In other cases, it permits parts to be cleaned by flushing with machine coolant. This often works between metal removal operations, or between metal forming and finishing.

Any reductions or improvements that are made in the work content of products at this stage will constitute savings in time, money, and equipment wear every time a product is made in the new cell. Once the improved processes or process alternatives are developed, they should be tabulated as the current process was, as shown in *Figure 6-2*. This affords the first look at the simplification that is possible from cellular manufacturing; the potential is often startling. Introduction of automated process planning to document the improved process routing will simplify the task of ongoing standardization. The next part can then be designed and processed in a consistent manner time after time.

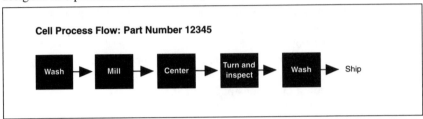

Figure 6-2. *A good cell design can achieve throughput velocity* $\left(\frac{Time\ in\ Department}{Process\ Time} \right)$ *of two or less.*

STEP 3: IDENTIFY EQUIPMENT REQUIRED AND COMPARE IT TO WHAT IS AVAILABLE

Each alternative cell description will require an equipment list. This list can be easily developed from *Figure 6-2*. It is only necessary to insert equipment numbers under the operations. Once the actual machines performing the operations are identified, it will be possible to add production volume by machine, as shown in *Figure 6-3*, and to generate capacity data. *Figure 6-3* shows the average time per piece produced and per setup. It will be necessary to establish the production lot sizes that are expected to run through the cell. This determines the number of setups required, and the amount of production capacity lost to setup.

It is important to remember that the number of setups currently performed on a machine may increase dramatically in a cellular environment. If the cellularized factory is to be more flexible than the current layout, or if major reductions in WIP inventory are planned, the cells will have to run smaller lot sizes than the conventional layout. Workload tables should include both current and future setup loads, so that the need for SMED or other setup reduction programs can be highlighted and this effort can begin.

Setup reduction is easily pursued in a cellular operation, but it can also pay benefits in a traditional layout. Moreover, this is an outstanding way to make the reality of change felt on the factory floor. It is good to start on this kind of improvement as soon as the team identifies the equipment that will constitute a bottleneck as a result of increased setup activity. This is also a good opportunity to identify change agents; individuals who stand out in accomplishing setup reduction goals will also be leaders in continuous improvement after the cells are in place. Since these change agents constitute only about 10% of the average factory population, the team must find and nurture them.

When the chart in *Figure 6-3* is complete, it will identify the assets that are over- or underutilized in the new environment. Underutilization of paid-for equipment has no financial significance, so, depending upon the availability of machinery, it may be acceptable to underload some assets. If, however, the cell design calls for the purchase of major equipment which will be underutilized, opportunities for eliminating operations, moving a few parts to some other cell, or selecting low-capacity, low-cost equipment should be explored. In a few cases, casters that allow movement have even been installed on machines that are used infrequently in a number of locations. This is an opportunity for teamwork and innovative engineering. Of course, when equipment overloads will constrict cell capacity, they must be engineered out or duplicate equipment assigned.

It is also important to avoid any operations that take place outside the cell. "Washer worship," for instance, in which all parts leave production areas and travel to a central cleaning area, can easily double the WIP requirements and throughput time in a factory, and they reduce part quality control. Simple, low-cost equipment should be substituted wherever possible for out-of-cell processes that can't be eliminated. In questionable cases, like using in-line induction heat treatment instead of a standard combustion furnace, both alternatives should be developed and a cost/benefit analysis performed for the

PROCESS VOLUME CHART							
PART NUMBER	VOLUME	WASH	MILL	CENTER	TURN	BROACH	GRIND
12345	550,000	1,100,000	550,000	550,000	550,000		
23456	235,000	235,000	235,000	235,000	235,000	235,000	
34567	178,000	356,000	178,000		178,000		178,000
45678	43,000	43,000			43,000	43,000	
56789	3,000	3,000	3,000	3,000	3,000		3,000
TOTAL	1,009,000	1,737,000	966,000	788,000	1,009,000	278,000	181,000

MACHINE NUMBER	BT-1234	BT-5678	BT-9123	BT-4567	BT-8912	BT-3456
OPERATION TIME:						
STD. MIN. PER PIECE	0.34	0.62	0.28	2.73	0.68	2.03
STD. HRS. PER DAY	39.72	39.93	14.71	183.64	12.60	24.50
SETUP TIME:						
SETUPS PER WEEK	0	8	7	8	2	2
MINUTES PER SETUP	0	20.5	10.0	37.7	60.0	48.5
SETUP HOURS PER DAY	0	0.55	0.23	1.01	0.40	0.32
TOTAL HOURS PER DAY	39.72	40.47	14.94	184.64	13.00	24.82
OPERATORS PER MACHINE	2	1	1	1	1	1
MACHINES REQUIRED	0.8	1.7	0.6	7.7	0.5	1.0
OPERATORS REQUIRED	1.7	1.7	0.6	7.7	0.5	1.0

Figure 6-3. *The process volume chart forms the basis for capacity, equipment, and staffing requirements. The plan for this cell is to run through the product mix twice per week.*

change to in-line processing. Making the investigation at this time will also provide time for process development work, if any is required.

Where equipment is shared by numerous cells, as in the case of central washers, plant-wide alternatives should be analyzed. Through the cooperation of multiple cell design teams, the central resource can often be eliminated, realizing a major cost reduction. This kind of cost reduction escapes traditional machine-by-machine investment analysis. Again, the first step is the generation of multiple scenarios for the cells involved. These situations all point toward the value of the fully developed macro plan illustrated in Chapter Five to visualize the detail impacts each cell might have on investment decision-making.

STEP 4: BALANCE MACHINE/MANNING WORKLOADS

When the sequence of operations has been laid out and the workload added, manning should be applied. Again, a spreadsheet is a good tool for the application (see *Figure 6-3*). Where one person will operate each machine, this step is trivial, but opportunities for staff reduction should not be overlooked in this step.

The cellular environment often makes it possible to perform deburring and other minor operations simultaneously, particularly those involving little or no equipment, by utilizing what would otherwise be machine cycle wait time. In cases where container loading and unloading steps are eliminated by the new layout, free time is often generated. Using this time does not constitute an increased operator workload, and it should cause no problem even in a union environment, as long as the new job is properly engineered and introduced. Balancing of operations and people takes planning, but the cell environment encourages the people to accept and, in many cases, propose their own balance.

The manning table will identify any areas where job consolidation should be investigated to minimize requirements for partial people. In some cases manning may dictate that it is appropriate to share a resource between two cells. Paint lines or shot blast cabinets often fall into this category. Wherever the decision is made to share, the same team should design both cells if possible, or the two teams should work very closely together.

STEP 5: SELECT/ASSIGN EQUIPMENT

Cells built with existing equipment are always designed for specific machines. Once the cell concept begins to mature, however, it is necessary to "check out" the needed assets from a master list for the entire facility. Otherwise, key assets are likely to be assigned to more than one cell.

Teams are often surprised at this point by how much the company underutilized its equipment. Theoretically, cells tend to reduce machine utilization. In reality, however, traditional layouts keep machines and work apart, subtly reducing the effective capacity of the equipment. A cellular layout eliminates much of this effect. Nevertheless, some resource conflicts are usually generated

in this step of cell design. They must be resolved through teamwork and creative planning. Techniques that apply include:
- Eliminating the contested operation through material or process changes.
- Positioning two or more cells to share equipment.
- Trading large equipment for multiple smaller units, new or used.
- Recombining product mix to allow all parts using a particular machine to flow through the same cell.
- Redesigning the process to make the shared operation the first step in both cells, so they are joined at the beginning instead of in the middle. This configuration is much easier to schedule and manage.
- Redesigning the parts to take advantage of common characteristics or process steps already resident within the cell design.

Unless both teams are convinced that sharing will cause no inconvenience, two alternatives should be developed for both cells, one with and one without the investment needed to duplicate or eliminate the contested resource.

STEP 6: BUILD A RELATIONSHIP DIAGRAM AND AN INITIAL CELL LAYOUT

Once specific machines and staffing levels are identified, they should be loaded into a conceptual flow diagram. This creates the first pass at a physical cell layout.

There are several methods of analyzing process flow. The method used most frequently by many professional consultants involved with cell planning is straightforward and objective, since it uses all known product and process parameters. It is built on a series of three tables, shown in *Figures 6-4, 6-5,* and *6-6. Figure 6-4* presents the flow sequence for each part as defined by routings. When this table is completed, it shows all possible routes through the sequence of operations in the cell. The column headings are the operation sequence numbers for various product flows. Processes or machine numbers representing the resources needed to produce the parts are listed within the body of the table. The numbers at the left of each row represent either the specific part numbers to be manufactured or groups of parts that follow the same process flow. In *Figure 6-5,* each resource is analyzed for the specific number of "transactions" it has with other resources. Note that the raw material transactions are included as resources.

PROCESS FLOW TABLE						
PART NO.	VOLUME	Op. 10	Op. 20	Op. 30	Op. 40	Op. 50
12345	550,000	Wash	Mill	Center	Turn	Wash
23456	235,000	Mill	Center	Broach	Turn	Wash
34567	178,000	Wash	Mill	Turn	Wash	Grind
45678	43,000	Turn	Broach	Wash		
56789	3,000	Mill	Center	Turn	Grind	Wash

Figure 6-4. *The first step in proximity analysis is establishment of a process flow table for the cell.*

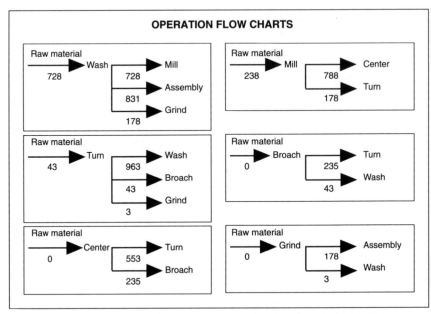

Figure 6-5. *The second step in proximity analysis is building flow charts for all operations. Figures under the arrows are production volumes in thousands of pieces.*

The preceding steps have established the relative proximity requirements for all the resources in the FMC. They have also identified those resources which need to be accessed for delivery of raw material, and those which should be as close as possible to assembly. It now remains to construct a diagram showing how these requirements can be met by a physical layout. Often several alternatives for the FMC will be developed, but all will stem from *Figures 6-4 and 6-5*. It is worth noting again that the diagramming technique described in Chapter Five is applicable for multipath conditions seen in complex process and group technology cells. Product cells can usually be designed more intuitively.

It must be stressed that *Figure 6-6 does not* provide the FMC layout. Some assessment can be made of the area requirements, but the layout can only be completed when a suitable site is selected and the FMC is supplemented with the addition of tool storage, building characteristics, and with other "reality" things that reflect real-world planning when it reaches the factory. In most cases the output of the macro analysis will have ensured site suitability. When new information is revealed during cell design, like space requirements, space shape, or utility requirements, the process must loop back to revisit the impact on the overall factory design. Again, the multifunctional team will pay its way by injecting realism where engineering might be encouraging wishful thinking.

Where process alternatives exist, as in the case of major production technology changes, this step should be repeated for each alternative. This will provide the basis for an "apples-to-apples" comparison of all of the various approaches.

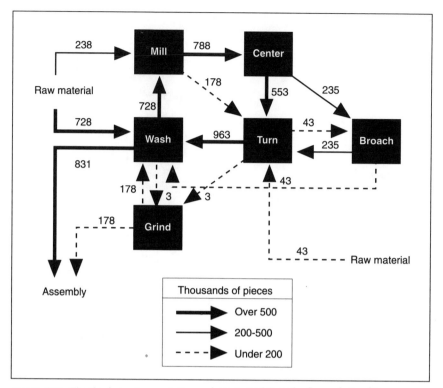

Figure 6-6. *The third step in proximity analysis is modeling the flow relationships among operations. Volume numbers permit the team to optimize high-volume flows.*

STEP 7: RECAPTURE DATA IN THE "CELL" STRUCTURE

A balance must be engineered at this point between flexibility and throughput speed. A maximum-speed cell design has machines located in fixed sequence, very close to one another. Maximum flexibility calls for an open layout and wider variety of potential flow paths. The tightest layout will nearly always be the fastest way to run one part configuration, but it may be the worst for running the others.

Figure 6-7 shows the same cell product mix used in the proximity analysis. Here the mix is divided into a homogeneous flow of one high-volume part number and a more flexible structure to accommodate the other four products. Flows are depicted with volumes indicated as an aid to designers. *Figure 6-8* illustrates the flow intensity of the block layout that is the direct precursor to the final conceptual block layout. *Figure 6-9* shows the same cell as a complete concept, with material staging, support, and tool storage areas.

The next step is detail, followed by more detail. Depending upon the level of detail the team wishes to develop prior to a management presentation, a large number of items must be designed or allowed for in contingency figures within the project budget. It is usually a mistake at this point to develop a precise project

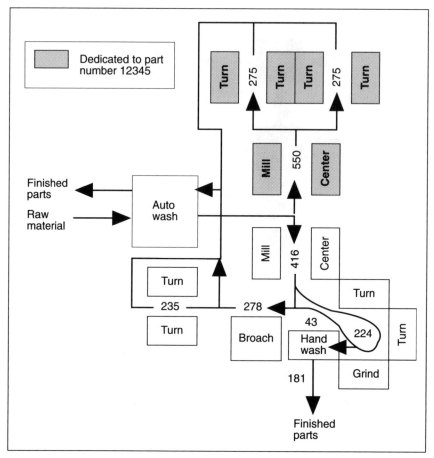

Figure 6-7. *Cell concept is developed using proximity relationships and equipment data. Note that the top section is dedicated to one high-volume part. Also, a second centering machine has been added and operator count is reduced.*

budget. Some contingency funds make it clear that further design is required. At this point the team will not have covered all of the detail required for completion of the cell design, nor should it have. It is enough that a sound basis is established for a business plan; more engineering on a project that may not be approved would be wasteful.

Detail items should be listed, however, and it should be made clear how thoroughly they have been evaluated in the cell concept stage. These items will include:

- Material handling between operations;
- Tool, fixture, gage, and document storage areas;
- Tool setting, die rework, or other tool preparation facilities, as dictated by products and processes;
- Paperwork areas and PCs, if the scheduling and documentation functions demand them;

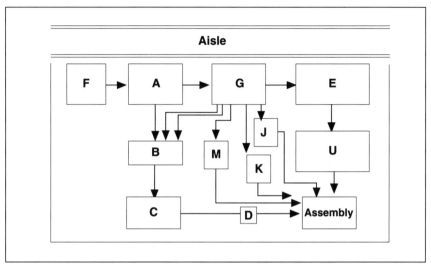

Figure 6-8. *Machine block layout as shown from part flow relationships.*

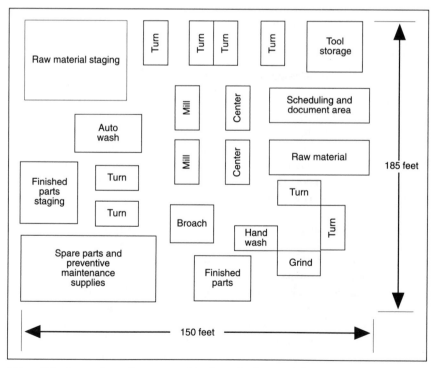

Figure 6-9. *A complete cell concept, with all required space and support.*

111

- Inspection benches, deburr and repair areas as needed, and all the other minor operations that impact budgets if they are ignored, and
- Meeting, training, and other people facilities needed in the new factory.

Once these issues have been reviewed, the team is ready to add its contribution to the data package for the new cell. This package must include:

- One or more layouts of major equipment showing square footage requirements for the cell, as well as all key access points for product shipment in or out of the cell.
- A budget for creation of the cell, including capital, tooling, and expenses required to bring the cell on-line. This should be prepared in accordance with local practices and formats.
- A capacity plan, identifying production bottlenecks and their productive capacities, and comparing these with immediate and future production requirements. For cells with near-term capacity problems, the cell design should include floor space and other provisions for increasing capacity to meet anticipated need.
- A manufacturing plan showing the staffing and working hours needed to meet the production goals for the cell. This information should be very conservatively derived, as it will be the basis for setting the scorekeeping system for the project and the new factory operations. Excessive optimism here can be dangerous.

By this time, the team will have a clear idea of how the new cell will run. If other teams have been working on their respective areas, it is time for the teams to get together and discuss the assumptions they have been making about how the new factory will run.

STEP 8: CONCEPTUALIZE SUPPORT RELATIONSHIPS

A large number of equally valid ways exist to organize a factory. Scheduling can be performed within each cell or on a factory-wide basis. Similarly, maintenance and tooling support can be centralized or distributed. These decisions should be made in light of the specific business, products, and processes that are involved. The important thing is that all cells be designed with the same kind of support structure in mind. That is, one cell cannot plan on doing its own cutter grinding if another is planning to send dull tools outside to be sharpened.

Organization of support organizations is one of the key leverage points for reducing production costs through cellularization. It should be the aim of every team to eliminate staff and staffing structures that support nonvalue-adding activity. On the other hand, a defense plant that wants to eliminate the quality assurance department is in for a rude awakening. Again, as in all estimating, a healthy dose of pessimism is appropriate. It is all very well to assign preventive maintenance activities to the cell team, but someone will have to teach team members their new tasks. If the maintenance budget is cut to the starvation level, teachers may not be available. The same will apply for tooling, quality assurance, repair, and all the other services provided by support groups.

Cellular manufacturing cuts costs and improves quality and customer satisfaction by focusing the workforce on products that are needed by the market *now*, but it does not call for elimination of the support that is required for quality operations. Each of the support groups that remains after the factory redesign must have a plan with the same elements as a cell design. These investments and manpower requirements will be recorded for inclusion in the factory manufacturing plan and the overall project funding proposal described in Chapter Eight.

SUMMARY

This chapter has outlined the steps in the path to the creation of an FMC design. The particulars of each product line and organization will dictate some changes to the procedures, but the general structure remains constant.

By building the cell designs and the overall factory structure around the flow of products needed to satisfy the customer, the team will have described a lean operating factory and organization that should be a formidable competitive entity in the future. By carefully including every product now being produced, every change or new product now under consideration, and all support groups required, the team should provide a workable factory design and a project description that, properly handled, will provide management with a clear picture of what is proposed for the future and the returns that structure should provide.

EVALUATING ALTERNATIVE FMCs

There is seldom a single, obvious solution to optimum cell design. Numerous equipment options generate a series of alternative layouts differing in content and physical arrangement. Because the selected FMC must meet a number of financial and operating objectives, an effective evaluation and selection methodology is needed.

The evaluation process is usually best handled in two steps. First, identify the cell contents that best meet the project's technical requirements. Evaluate and select the content with approved financial justification techniques. Second, evaluate the arrangements using a factor analysis technique. Use financial information to finalize decisions and refine planning. Rigorous financial evaluation of alternatives that don't meet functional expectations wastes a lot of team effort and resources.

This chapter describes a proven approach for the evaluation and selection of FMC physical arrangements that defines and quantifies each evaluation factor, rates each alternative against the factors, and scores each alternative for comparison. Included also are some comments on the financial justification process as an aid to putting cost justification in an appropriate perspective for evaluating alternatives.

TECHNICAL EVALUATION METHODOLOGY

Making decisions can be arduous and frustrating, especially when a number of unquantifiable factors are involved. The process is further complicated by different points of view. Determining cell content or cell arrangement requires the involvement of people from all levels of the organization, and the people involved need to work as a team. No one point of view can be allowed to dominate if optimum results are to be achieved. Weighted factor analysis, a common approach with easily understood conventions, provides the structure for alternative evaluation by a diverse group of people and produces decisions everyone can agree with.

Getting Started

Preparing to evaluate alternatives requires careful identification of information, data, and resources. In the context of this discussion, evaluation of alternatives implies:

- A clearly stated objective for the FMC:
  –Strategic drivers,
  –Technical objectives:
  - *Flow* of material through the cell characterized by having no alternative routings or backtracking, and straight, unidirectional paths;
  - *Density* of equipment in the cell characterized by having minimum space between machines, opportunity for automation, and minimum inventory;
  - *Velocity* of communication and production in the cell characterized by quick problem resolution, value-adding activity throughout the process, and satisfied customers.
- A series of viable alternatives,
- A clear definition of each alternative,
- A list of criteria or factors to be met,
- An accepted methodology to evaluate alternatives.

Before starting the evaluation process, revisit the purpose of the planned FMC. Prepare a clear and concise statement defining specific goals such as:

- A dedicated FMC for bevel gears will eliminate a critical bottleneck in meeting our customer-driven assembly schedule.
- A widget FMC will:
  –Reduce lead time 80%;
  –Reduce work-in-process inventory 50%;
  –Consolidate raw material for reduced handling and control;
  –Simplify scheduling;
  –Provide quick feedback on quality performance, and
  –Fit our business unit organization.
- The manifold assembly FMC will provide quick response to final assembly and reduce work-in-process stock by an estimated $65,000.

Restating the FMC's purpose will provide an initial list of factors for the evaluation of alternatives. It also will test to ensure that the original strategic drivers and technical objectives were not lost or misinterpreted as alternative layouts were developed.

Each alternative will have specific identifying characteristics. Using these to create a simple definition of each alternative (key features) will aid communication and add clarity during the evaluation process (see *Figure 7-1*).

Specific location within the factory and internal configuration of the FMC are separate issues. Location selection is a function of macro or block layout

116

evaluation. Individual cell arrangement selection results from the evaluation of concept or detail layouts. The evaluation methodology is the same for all levels of plant layout detail. The arrangement alternatives listed in *Figure 7-1* could be for any one of a series of content options (see *Figure 7-2*). It is important to recognize that there are content options that are as varied as the arrangement options—and that carry a much greater financial impact on a business. The evaluation process should be kept as simple as possible. It is awkward to perform an evaluation of both cell content and cell arrangement in one step unless the machines and equipment are common to all alternatives.

Arrangement Alternatives	
1	"U"-shaped flow, manual handling
2	Straight-line flow, manual handling
3	"Z"-shaped flow, manual handling
4	"U"-shaped flow, mechanical handling
5	Straight-line flow, mechanical handling

Figure 7-1. *Typical descriptions of FMC alternatives by type of arrangements.*

Factor evaluation or decision analysis techniques are used for location and arrangement evaluation. Organizing the evaluation process in steps will ensure "apples to apples" comparisons. Minimizing the number of alternatives to be evaluated at one time (select the best five, or less) will also simplify the process; more than five alternatives makes the decision-making process much more difficult.

Cell Content and Arrangement

The first step in defining alternatives is to recognize which planning phase is being addressed—cell content or cell arrangement.

Cell content is essentially the equipment selection alternatives under consideration. The evaluation of these cell "content" alternatives usually emphasizes the financial aspects, which may include:

- Using only existing equipment, which requires little capital investment and can achieve quick results.
- Purchasing sophisticated, new equipment that requires more training, debugging, start-up time, and money to become productive.

117

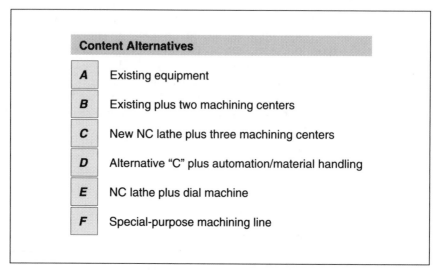

Content Alternatives

A	Existing equipment
B	Existing plus two machining centers
C	New NC lathe plus three machining centers
D	Alternative "C" plus automation/material handling
E	NC lathe plus dial machine
F	Special-purpose machining line

Figure 7-2. *Typical content alternatives for FMCs.*

- Placing low-cost, stand-alone (existing or new) machines into *dedicated* operation, which might yield a quick return on investment.

A number of other important issues may be affected by the application of various content selections. These include labor utilization, material control, changeover flexibility, operation balance, level of technical support, etc. In addition, effective evaluation requires thorough documentation of each alternative:

- Organize cost and benefit data for financial calculations and criteria for factor evaluation.
- Prepare clear, scaled drawings.
- Begin listing and separating evaluation criteria or factors— those things the FMC must achieve and things it would be "nice" to achieve.
- Organize evaluation formats, identify who will participate in the evaluation process, and establish the time frame.
- Understand the methodology and obtain management approval for its use—don't hit people cold with an evaluation technique that differs from traditional procedures; communicate the agreed procedure to individuals involved in the evaluation process.

EVALUATING PHYSICAL ARRANGEMENTS—THE LAYOUT

There are *always* alternative arrangements to be considered in planning FMCs and overall facilities. Effective evaluation toward subsequent selection will include a number of factors that influence the layout. A list of typical factors is provided in *Figure 7-3*, but each project will have its own, both in terms of content and "value."

Evaluation Factors	Considerations
Effectiveness of material flow	— Shortest path for greatest velocity — Access, internal and external
Layout flexibility	— Ease of rearrangement — Ease of adding or replacing equipment
Material handling effectiveness	— Simple versus complex — Relationship with external sources
Space utilization	— Density — Configuration — "Fit" with overall layout
Support service integration	— Tools and fixtures — Information — Maintenance — Quality — Utilities — Scheduling — Human resources — Supplies
Supervision and control	— Line-of-sight management — Cell-to-cell communication — Environmental factors
Working conditions	— Workstation access — Access to amenities — Environmental concerns
Ease of implementation	— Production interruption — Sequence and timing — Building alternations — Utilities — Access for riggers
Cost of implementation	— Building alterations — Temporary moves — Required overtime — Equipment costs (new, used, rebuilds) — Utilization of existing equipment

Figure 7-3. *Effective evaluation factors.*

A key factor for evaluating FMC alternatives is the *effectiveness of material flow*. Flow of material between work centers is developed to guide layout planning. However, each alternative will have differences which influence layout selection; a means of measuring the differences is required for the evaluation process.

A unit of measure must be established—typically distance times quantity per move during a selected time period—and the quantity per move must be defined as a unit of measure with a "common denominator" (such as the most-used container size or representative part size). Greater or lesser move quantities can be expressed as fractions of the standard unit for calculation. Material and move characteristics need to be considered—a pound of feathers is a lot different from a pound of steel. This approach provides a specific, weighted value which can be "scaled" for inclusion in a factor analysis exercise. The flow analysis identifies how well each cell layout alternative meets the primary objectives of flow, density, and velocity (or more precisely, the ability to achieve them).

It is often necessary for some batch process and/or shared operations to be external to the cell. These could include activities such as heat treat, plating, or painting. Since a cell by definition does not function in isolation, external operations need to be evaluated for optimum location as a part of the macro layout evaluation, which looks at proximity between all plant areas. Cells have operating relationships with many external activities and areas; supporting services also must be included in the external and internal planning and evaluation activities.

Since material flow is usually the primary consideration in the development of proximity requirements within the cell, supporting services become more important at the factory level. In a plant environment where service relationships are especially important, those relationships should be surveyed and rated separately, and then merged with the flow relationships to establish an overall proximity. A block layout should be created for each alternative. The differences between each block layout should be quantified and used to rank the evaluation factors presented in the weighted factor analysis.

Weighted Factor Analysis

Prepare the evaluation matrix by listing the appropriate evaluation factors in the first column and alternative identifications (usually letters) across the top, as shown in *Figure 7-4*. Each participant will have a different idea of the weight of each factor, as well as the ranking of layout alternatives, but a group average establishes the numeric value to be used for the evaluation. (In this process, any fraction should be rounded to the nearest whole number.) There are four easy steps to completing the matrix:

1. Weight each factor on a scale of 1 to 10 (10 being most important). The same weight can be assigned to more than one factor.
2. Rank each alternative for one factor at a time on a scale from 1 to 5 (5 being the most important).

Factor	Weighted Value of Factor	Layout Identification			
		A	B	C	D
Internal flow	10	4/40	5/50	4/40	3/30
Future flexibility	8	3/24	3/24	4/32	4/32
Space utilization	9	4/36	5/45	5/45	4/36
Product autonomy	10	5/50	5/50	5/50	4/40
Cost of quality	5	4/20	4/20	2/20	4/20
Customer appeal	6	3/18	4/24	4/24	5/30
Employee access	5	4/20	5/25	4/20	3/15
Implementation ease	7	5/35	4/28	3/21	3/21
Implementation cost	5	3/15	4/20	3/15	5/25
TOTAL SCORE	N/A	258	286	267	249

Figure 7-4. *Weighted factor analysis. The first number in the pairs of numbers under each alternative layout represents the rank for the factor. The second number is a score made up by weight multiplied by rank.*

3. Calculate the value for each factor in each alternative by multiplying weight and rank.
4. Total the values obtained for each alternative, arriving at an overall numerical value comparison between alternatives.

If one alternative is a clear winner, the decision is easy. If the values are very close, the evaluation process must continue. After conducting the first evaluation without determining the most appropriate alternative, the next steps include:

- Adding more factors and re-evaluating;
- Eliminating the obvious losers and re-evaluating;
- Creating a new alternative (combining the best features of several low scoring alternatives) and evaluating against the previous top candidates.

The evaluation process is often quite revealing; features with a high score may be incorporated in a new or revised alternative to produce a winner. Since time spent planning and evaluating *will* pay off, the alternative process should be seen positively. One key is to evaluate the plans as they are presented and not as what they *might* become with modification. Make the modification and then evaluate again.

In general, implementation cost factors should not weigh more heavily than other factors in the factor evaluation process. The ranking score derived from the weighted factor analysis should rule out any prohibitively expensive alternatives. When cost differentials are significant, it is best to do a financial evaluation as a separate exercise. Just as an alternative that cannot perform to the functional requirements of the FMC will undermine the success of the project, so too will selecting an alternative which does not meet the financial justification criteria. To meet this requirement, it is important that a key financial representative participate in *all* phases of the project.

IDENTIFYING COST FACTORS

Cost justification provides the basis for most business decisions, especially when capital expenditures are involved. The company has guidelines for the procedure and rules of attainment (such as hurdle rates or ROI). One major problem exists in FMC planning: Traditional justification methods do not recognize many of the benefits of cells. Still other benefits do not peak until most or all of a business unit is cellularized. The best way to deal with this is to build the project around the company's financial drivers. If the project has not been built around financial objectives, it may be rejected if the numbers come out wrong. The planning team needs to understand the financial requirements, work to quantify every cost and benefit element, and then include them in the justification process.

Elements of Justification

FMC cost documentation. A thorough, detailed listing of all cell investment and cost items is the first step in preparing data for cost evaluation and justification. A spreadsheet should aid in the identification and collection of data for each alternative. The list should include:
- Capital requirements for:
  - Machines and equipment,
  - Buildings and utilities,
  - Tooling,
  - Additional inventory (new product introduction).
- Expense requirements for:
  - Rearrangement of facilities,
  - Production adjustments,
  - Training and education.

FMC benefit documentation. Benefit documentation requires some creativity, since traditional accounting systems mask some financial benefits *and* completely ignore nonfinancial items like faster throughput, improved quality, and better customer service. These improvements often equip the company to build market share and *really* boost profitability. Chapter Eight deals with the issues of capturing and presenting this information; this chapter will only list the benefits:

- Inventory adjustment,
- Manufacturing personnel reduction:
  –Wages,
  –Fringe benefits.
- Support function reduction:
  –Wages,
  –Fringe benefits,
  –Material.
- Quality improvements:
  –Internal cost,
  –External cost,
  –Scrap level.
- Customer service improvements:
  –Throughput time,
  –Ease of customer interface,
  –Market share.

Quantifying Cost and Benefit Elements

In this process, it is essential to use the tools available in an analytical and thorough manner. All of the hidden benefits as well as costs should be quantified. Justifying an FMC investment should be considered in the same vein—flexible. As shown in *Figure 7-5*, the "real" break-even point occurs when the cost of doing nothing equals the benefits of installing the FMC. Typical return-on-investment calculations assume that the market share, volume, and life cycle of that product are unchanging and predictable over time. Today's marketplace, however, is less predictable and is evolving toward rapid change— contrary to these ROI assumptions.

Again, investments in cell development and implementation should be compared against the net decline in market share, profits, and quality that would (and will) result in a negative baseline, not against the status quo of a break-even line. The use of traditional "cost accounting" to justify new manufacturing systems often is very difficult during the first FMC presentation.

As indicated earlier in this chapter, it is critical to the entire evaluation process to clearly define the alternatives. Financial justification, specifically, focuses on FMC content evaluation—the equipment and how well it meets the business objectives in financial terms. Maintaining the current situation may be offered as an alternative, but at the very least an alternative based on as-is equipment must include optimization to make it viable. It may, however,

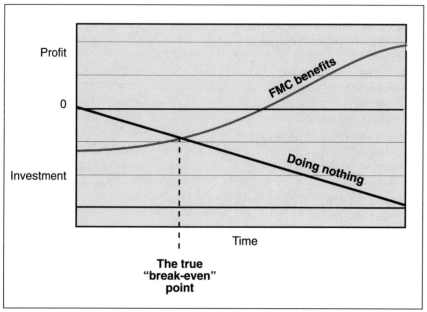

Figure 7-5. *FMC justification.*

provide a powerful baseline to demonstrate the need for the new concepts. Finally, the basic litmus test for each alternative should be that it meets the planning parameters for volume and variety.

At this stage, dollar values should be developed for the alternatives. To accurately measure investment performance, all benefits must be quantified. The team's financial representative should begin establishing values for intangible benefits. Some known benefits—increased flexibility, reduced lead time, reduced floor space, the real cost of carrying inventory, reduced cost of quality, and the potential for gain in market share with the new system—will be difficult to quantify. These should be estimated and presented to management for validation. Intangibles like these often significantly impact cash flow.

These dollar values should be established within a specified period of time, the investment window. The same time frame, five to 10 years, must be used to establish the dollar values and, in turn, to evaluate all alternatives. This investment window should not be confused with the working life of the equipment. If equipment has a three-year life span, the investment required to replace it every three years should be included in the dollar value attributed to that alternative.

To compare the alternatives' investment costs and benefits, the discount rate should be considered. Most companies have a specific discount rate that project planners must comply with; if it is higher than 10%, this imposed rate may be too restrictive for FMC justification. The actual comparison should be conducted using an established discounted cash flow method. Many companies prefer the present worth or internal rate of return methods, but annual worth, future worth, or cost-benefit ratio methods also should be considered valuable resources. The simple payback method may be attractive because of its simplicity. However, it does not adequately reflect the complexity of high-tech alternatives.

In addition, where a range of values exist for various analysis parameters, the alternatives should be tested with a sensitivity analysis. Risk analysis can be applied when there is sufficient knowledge of the values of potential parameters and their likelihood of occurrence.

Finally, this detailed analysis of the investment economics is critical to the selection of the FMC alternative. Engineers tend to avoid in-depth involvement in the justification process, but they should be encouraged to be active participants. Since FMCs are cross-functional, it is essential that team members understand why the best engineering solution was not necessarily best for the business.

SUMMARY

Evaluation of alternatives is a process that requires that all involved accept the idea that there is more than one solution to a problem. Economic evaluation is focused on cell content and the overall business cost benefits of candidate alternatives. Factor evaluation techniques are applied to the arrangement of cell elements, both internal and external. Together, the tools available for evaluation provide the entire planning team with the means to select the best option for a sound business decision.

The process described for comparing alternative FMCs should result in selecting the "best" alternative. But even if the second or third "best" alternative is selected, the process itself will have accomplished an important objective—consensus! If all the people involved have participated in the process, they are more likely to support the outcome. Certainly, even doubters will better understand the benefits because of their participation, and they will work more constructively in the implementation of the selected alternative.

||||| 8 |||||

SELLING THE CONCEPT
TO TOP MANAGEMENT

The project team may have been commissioned to look into possible manufacturing improvement programs and report to management. Top managers may not even be aware of the FMC concept. These project teams that put together cellularization programs as the way to improvement and *then* try to "sell" them to management will have an additional task—they must begin by educating their audience about the values of cellularization. Usually, this is not very difficult, but it must be addressed specifically. It is likely that there will be two points at which management's approval will be sought:

- First, for approval to plan how the FMC concept should be applied to the business. The primary resource involved in this process is people's time.
- Second, to implement the FMCs. Resources involved include: money for equipment rearrangement and purchase, training, possible inventory build-up to avoid delivery interruptions, etc., as well as more time from more people and management's commitment to the change plan.

The concepts presented in this chapter are concerned primarily with the second approval—the one that necessitates opening the corporate checkbook and changing the corporate culture. If the communication, planning, and project selling process is managed all along the way, the second approval should be granted as easily as the first. Countless companies have lost years in the race to remain competitive because, although they have made many plans, none ever made it to the implementation stage. The most important communicating the project team may be doing is keeping management apprised of all aspects of the project; the motto "no surprises" is a good one. With proper communication, the actual selling of the project will be a normal milestone along the way.

127

AUDIENCE AWARENESS

It is most important to define and identify who will make up the various audiences that will hear about the FMC project throughout its duration, understanding what their sensitivities are, and adjusting to them as much as possible. This extends beyond top management. At different times, the audience may be cell operators, middle managers, labor union stewards, or specific technical support people within the company. External audiences, like the board of directors, stockholders, bankers, and the press, also exist. Several communication mediums will probably be employed to appropriately address these different audiences.

The viewpoints of each of these audience members should be taken into account at the beginning of the project, before the language or structure of the project is determined. For example, if the company's former president is now on the board of directors (which must approve the project's capital expenditure request), highlighting the "wasteful" inventory practices of the past could be detrimental. Also, referring to the potential for "head cutting" (especially if it is not up to the FMC team to decide what to do with excess labor) will not be as effective in winning employee support as identifying "freed-up labor that can be applied to additional production" or "the number of positions that will disappear through attrition." Messages about FMCs must be carefully, unanimously, and truthfully crafted to sell the entire organization on the project and solicit ownership. This type of communication is beyond "playing politics." It's vital to the success of making major changes in the way the business is run.

During the initial phases of the project selling job, the primary audience will undoubtedly be the top division and company managers. The rest of this chapter will concentrate on this audience.

The Historical Approach

One way to demonstrate the improvements possible through cells is to use historical data. *Figure 8-1* demonstrates how historical data can be used. This approach works well with conservative and financially oriented managers. When using the historical approach, the team may find that it is best to use well-known cases, such as other divisions of its own company. These cases should emphasize the actual numbers if possible—before and after amounts, and the percentage of improvement. The most effective presentation of these results would probably be made by the people who were involved in accomplishing the task. Making a one-day tour of one of these FMC facilities also can be illuminating. A note of warning, however—the team should only use cases that it has researched well enough to answer basic questions about, such as "How big is the plant?" or "What do they make, and with what processes?" If the team cannot answer these questions, the audience will conclude that the project has no bearing on the company's situation and may not even be valid in itself.

Properly handled, this approach reduces feelings of anxiety besetting managers when they consider employing new techniques. Phrases like

128

CELLULAR MANUFACTURING PAYS OFF

Type of company	Inventory	Quality cost	Direct labor	Cost of sales	Mfg. lead time	Floor space
Aircraft			-40%		-80%	-40%
Aircraft engines			-85%	-30%	-85%	
Turbines	-50%		-25%	-25%	-75%	-30%
Hand tools	-50%			-25%	-50%	
Hand tools				-30%	-65%	
Motors	-15%	-20%		-10%		
Oil pumps	-70%	-30%	-37%		-96%	-25%
Exhaust systems	-75%	-30%	-55%	-25%	-99%	

Figure 8-1. *Cells allow manufacturers to achieve results.*

''proven solutions'' and ''established techniques'' will help the new concepts avoid being labelled as radical; they should appeal to all managers instinctively.

The Technical Approach

Figure 8-2 shows an approach that works well with technically oriented managers. Combining production operations and eliminating storage operations can cut working capital, shorten throughput time, and eliminate nonvalue-adding work. This can be amplified by tracing the actual flow of material through the factory, as shown in *Figure 8-3*, and then tabulating the distance two or three

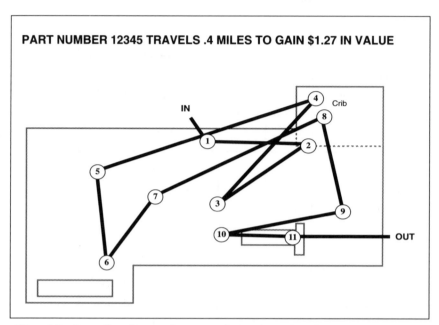

Figure 8-2. *A part flow diagram shows a product path requiring layout improvement.*

specific parts will travel and the number of times they are rehandled in the present manufacturing arrangement. This data can then be compared to the same indicators in the proposed cellular manufacturing system.

This is a good technique for preparing managers to accept the deep cuts in indirect labor that cellularization usually produces. It demonstrates the necessity of cellularizing on a factory-wide scale to capture the benefits. It also has the advantage of being built on data generated in the course of cell design, making additional analysis for presenting a strong case unnecessary.

The Marketing Approach

Marketing- and sales-oriented managers will appreciate hearing about the impact FMCs could have on the customers they serve. For example, managers know that the market is becoming more attuned to requiring parts or products on a Just In Time basis; they will see that the FMC's potential to easily provide delivery on shorter notice and in exactly the right amount will be appealing to customers. The team should play up the FMC benefits of quick changeover, high flexibility, fast throughput, and small lot size capability.

Consistently high quality is also becoming a market standard. The fact that FMCs demand high quality to function and make quality more visible as a whole should be a selling point. Another part of selling an FMC under the marketing

80 PERCENT OF WHAT WE CURRENTLY DO TO A WIDGET ADDS NO VALUE

Operation	Total handlings	Value-added handlings	Distance traveled (feet)
1 Saw	2	1	20
2 Store	4	0	600
3 Mill	2	1	12
4 Store	4	0	500
5 Drill and tap	2	1	8
6 Wash	4	0	350
7 Store	4	0	500
8 Stock line	4	0	150
9 Assemble	2	1	10
10 Pack	2	1	40
	30	5	2,190

Figure 8-3. *Only five value-adding operations were performed. Reduction of nonvalue-adding operations is needed.*

approach might be the potential for reducing stocking inventory throughout the distribution chain (often a responsibility of the marketing manager) because of quicker response times.

With an external publicity campaign to highlight the company's improved customer service (quality, delivery, flexibility) through FMCs, it should be able to capture a greater share of the market. Although it's often difficult to get a commitment from marketing staff members on how much this gain could be, increased market share is a viable justification measure for FMCs given an aggressive marketing team. The audience must be convinced that cells are not purely an engineering exercise, but the combined application of engineering, financial understanding, and common sense to everyday manufacturing needs.

BUILDING YOUR CASE

The job of selling the project has four parts.

Explain the project to establish the connection between it and the strategic drivers. This may include educating the audience about the general benefits of cellularization. Effective top managers convey their goals to their organizations in terms of strategic drivers, guidelines that permeate decision-making and communication throughout the company. To easily sell an FMC project to management, the project's potential for improving the business should be stated in terms of well-defined corporate strategic drivers. In effect, these drivers are what management has said are the allowable reasons for spending time and money to change production operations. Ford Motor Company's "Quality Is Job #1" slogan is a widely publicized example of this kind of driver. Of course, most companies don't do such a thorough job of publicizing their drivers, so the FMC team may have some clarification to do (see Chapter Two).

Congruency of corporate goals and FMC project goals helps ensure the project will move the company in the direction top management wants it to go. Strategic drivers are often oriented around markets, finance, and/or quality, but they may have unique directions as well. For example, one Midwestern pump manufacturer uses a "no layoffs" policy as a strategic driver; anyone promoting significant change in this manufacturer's plants is aware that ideas involving only reducing the number of employees will not be well received. A successful cellularization program is underway there, however, because it was justified based on its impact on customer service. This understanding of corporate drivers, particularly when applied at the beginning of project design, can help simplify the task of obtaining management approval, not to mention improving the quality of the project and its outcomes.

Quantify the costs and benefits of the project. The team must reassure itself that the cost/benefit projections are realistic, including a substantial contingency fund in the investment plan (something like 20% of the investment for first-time project teams, and 10% thereafter). This financial planning is covered elsewhere in the book.

Set a project schedule and show how the team will later demonstrate the planned level of progress against the schedule. The team should set a realistic implementation schedule showing how the project will progress through the steps toward fulfilling the original objectives, with clearly defined milestones. Time is always in short supply during an implementation. Frequent tactics to avoid running out of it include making liberal time estimates, and having adequate manpower available for engineering, equipment rearrangement, machine cleaning, and other tasks that make up a cellularization effort.

An auditing approach should be identified that will quantify progress to management's satisfaction. Numbers are the best, even if they're nonfinancial: pounds shipped per dock hand, percentage of on-time deliveries, order throughput time, or reduction in scrap rate all qualify. If the team sets up this kind of scorecard, it will be able to show that it is delivering on its promises. If the project is geared toward only the final result, it will be a disappointment until its conclusion.

All too often, a team develops and executes a project that provides exciting benefits to the company but it is subsequently branded a failure because the effort took longer or cost more than was originally estimated. The best estimates of time and resource requirements are developed by the people who will have to do the work; there is risk in estimates that come from any other source. (Refer to Chapter 17 for more on implementation scheduling.)

Identify the human elements crucial for project success and explain the implications of each, demonstrating how each element is built into the project solution. These elements are discussed further in other chapters in this book. A review here will help serve as a reminder of the "soft side" issues (and their associated costs and resources) that must be resolved for best implementation:

- Developing skills (planning for the education and training necessary to succeed in the new environment);
- Promoting high involvement and teaming (to give everyone a chance to participate and learn to share responsibility);
- Developing a flexible work force (having fewer organizational barriers to the flow of work);
- Determining how to measure performance in the new environment (these methods will probably be nontraditional);
- Changing the reward system to elicit the new desired performance (this encourages group success), and
- Promoting continuous improvement (an ongoing process of change to keep the business agile).

DEVELOPING A MEANINGFUL PRESENTATION

Once the team is sure of its ground, it should create a presentation of the kinds of changes that the project will create and their impact on the strategic drivers. A presentation can take many forms, including written and oral. The best is probably a combination of the two presented to management by the team in a meeting. This allows the decision-makers to accumulate adequate written detail for further review, and also benefits the team members by allowing them to supplement their data as necessary and respond to special concerns or points of confusion.

Here are some guidelines for organizing the presentation itself:

- Identify the decision the team wants management to make, and who the key decision-makers are; design the presentation along the clearest line to elicit that decision. The importance of "storyboarding" cannot be overemphasized.
- Make sure the presentation has a beginning, a middle, and an end, and give the audience a preview of its structure. This will help lead them through the presentation and help them know when to ask questions.

- Say it with pictures. Wherever possible, use graphics to illustrate points. There are many good texts available on preparing and making presentations (outlining information, selecting charts and graphs, preparing visuals, making oral presentations); inexperienced presenters should take advantage of these.
- Keep it simple. Use one topic to a chart, and no more than three or four main themes to a presentation. This doesn't mean that the detail doesn't have to be available for those interested; if some points need to be supported by complicated data, prepare this information as handouts and backup material and use them when necessary.
- Put the bottom line up-front. If the project will save $4 million a year, the presentation should make this point early. The presenter will then have the audience's full attention as he covers the details. A rule of thumb is that any member of your audience should be able to leave your presentation after five minutes (and many times in business may have to!), carrying away the main points and conclusions you wanted to convey.
- Be conservative, but specific. Remember, it is more acceptable to promise $4 million and deliver $6 million than the reverse.
- Realize that large dollar amounts may be involved. The team shouldn't let fear of top management's reaction to the investment numbers force it to reduce the project scope to ineffective levels or, worse, promise deliverables without an investment that will make them achievable. As long as the pluses outnumber the minuses, the project has value.
- Don't ruin a well-prepared presentation by trying to answer questions that the team doesn't have answers to; a response of "That's a good question, but I don't have the answer right now. Is it all right if we get back to you tomorrow on that?" will earn more respect than an obvious bluff, which could make people question the rest of the information.
- Don't get bogged down in specific technical discussions of no interest to the audience as a whole. Allow the concerned person to express his thoughts and address them in general. If a detailed technical debate seems to be developing, ask the person if he can discuss the topic further with someone from the team after the meeting.
- End the presentation with a summary of the major conclusions, a list of specific recommendations, and an identification of the next steps in the project. Ask that the decision the team is waiting for be made; if it isn't made at this time, find out when it will be made. Make sure no further information is required from the team.

PRESENTATION CONTENTS

Each project will have its unique story to tell. An outline of an effective top management presentation for selling an FMC project might be:

- Up-front summary (benefits/costs);
- Project objectives;

134

- Planning guidelines/assumptions;
- Planning/project methodology (the actual principles to be applied) and approach (how the principles will be applied to the specific company);
- Current cost/situation breakdowns;
- Before and after part flows and/or groupings (families);
- Cell layout/flow and equipment (don't forget how it fits into the macro plan);
- Improvements (savings, inventory, speed, space, etc.);
- People impacts (manning, training, etc.), and
- Implementation steps.

Once the presentation is written, the team should review it with everyone who will be involved in making the project happen. If someone is going to say, "Hey! I didn't mean that," it should be in a practice session, not an open meeting. It is especially important to include the financial staff members in this review. Their support is required for any major project, and the cellularization effort will have a lot better chance if they aren't surprised in an open meeting. (Besides, they're supposed to be on the team, remember?) Some teams find it helpful to have team members participate in a role-playing exercise while the presentation is being practiced by the presenter. They assume the motivations and concerns of specific people who will be in the final audience (cautious president, supportive manufacturing and human resources managers, concerned union representative, contrary controller, etc.) and help to prepare for questions and special points of view.

The team should plan on editing the presentation after this meeting, and presenting it to the group again if necessary. It's often better to use penciled overheads for the review, so that people are comfortable suggesting changes. This is the team's last chance to develop project ownership among the broadest possible spectrum of the organization; nobody will argue with a presentation they helped to design.

GOING FOR APPROVAL

Once the team is in agreement, it is time to solicit management's approval. The presentation may have to be exposed to one management level at a time, or it may be presented to top management immediately. There is no "right" way. The company's culture will be the determining factor.

If the presentation is structured properly and the meeting well planned, the team should leave the meeting with the following from management:
- Agreement with project objectives, authorization to proceed, and the commitment of resources;
- Acceptance of the measurement criteria set forth in the presentation;
- Acceptance of the project schedule and update approach outlined in the presentation, and
- Satisfaction that the crucial human elements of project success are built into the project solution.

After the presentation, the presenter(s) should call a meeting and report back to team members who did not attend. It also may be a good idea to review everyone's impressions of the meeting to see if any corrective or further actions are necessary. Management may come back to the team with some concerns needing further information; then it's back to the drawing board, presumably with a better understanding of management's objectives. There are some nontechnical issues the team may need to consider in establishing its case:

- *Do managers feel that the potential improvement is worth the effort, expense, and risk?* Does management believe that this team can provide the benefits it is promising at the stated costs, within this organization?
- *Is there something conflicting with the project that would prevent top management from supporting it?* For example, managers may not be interested in investing in this part of the business if it is late in the products' life cycle, if the business is a candidate for divestiture, etc. Or, it may be that someone has already signed up for an MRP upgrade or some other approach that is supposed to achieve the same objectives the team has set. How does this affect the FMC project?
- *Does the presentation address management's "hot buttons?"* What if the hot buttons do not match the strategic drivers?

Ineffectual or preoccupied managers sometimes fail to make the right decision, even when faced with an excellent presentation. Dictatorial managers may dismiss any idea they did not come up with themselves, no matter how good. When team members face this situation and conclude that it is permanent, they probably have some serious career planning to do! More often, the conclusion will be that a better presentation of the cellularization project will achieve the necessary support. If management subsequently approves the project, it is time to establish near-term objectives; set reporting relationships, responsibilities, and timetables; and return to work.

APPLYING TECHNOLOGY

Selecting the type and level of technology that is appropriate for a specific FMC application and determining where to apply it are directly related to the mission (goal) of the business as discussed in Part One. To perform this evaluation, an understanding of existing technology must be established; from this point, recommendations can be made for the specific application—how much and what kinds of technology should be implemented into the cell(s).

This chapter outlines approaches to help those implementing cells identify the appropriate technologies for their FMC(s). This technological assessment and the future recommendations will need to be developed further into financial and/or corporate justification to ensure that the recommendations will help the FMC achieve the company's objectives such as increased market share or technological superiority.

This chapter does not deal with specific technologies, although it does refer to some of the most recent applications. Instead, the information reviews the needs to be considered before investing in technology.

WHAT IS TECHNOLOGY?

Technology usually is defined as the application of science, especially in industry or commerce, or a method or process of achieving a practical purpose. Some companies, however, look at technology as "anything newer and different than we have now." In this case, an NC lathe is a technological application if the only tools in the shop are manual lathes.

For this discussion, technology has been categorized in three levels:
- *Low technology* is that technology which is in general use throughout industry. It has been tried and proven in numerous situations, carries a low risk factor, and usually requires a low level of investment. While the individual operator skills required can range from low to high, low

technology is usually found in a one-worker, one-machine operation. For example, a single-spindle drill press that uses simple jigs to drill holes in individual piece parts is low technology.

- *Now technology* is widely used, but it may not be proven and carries a medium risk factor. The investment for this level of technology and the operator and support skill requirements can range from moderate to high. An example of now technology would be CNC machining or turning centers applied in an environment that has never seen them.
- *High technology* is state-of-the-art equipment, and it has a medium to high risk. The investment required here is often high. The skill level requirement of operators and support personnel is more advanced than at other levels of technology. For example, maintenance troubleshooting may require programming logic and circuit board expertise instead of a background in basic electronics. Applying this type of technology also may require significant modifications to the facilities to provide the required environment. An example of high technology is a flexible manufacturing system with automated material handling and higher level computer control. This technology is tied to other operating computers within the company. The physical environment may require a "clean" area, a cooling system, and vibration-free foundations.

Figure 9-1 shows how common manufacturing and fabrication applications fit into these categories. With any application of technology there are two inescapable facts—technology costs money to acquire and it costs more money to implement effectively. The higher the level of technology, the higher the cost, and of course the higher the risk.

If technology is applied as part of an overall business strategy, however, the rewards, expressed as improved market share, improved quality, and reduced costs, can be outstanding. The selection of technology direction may match an overall business strategy, such as maintaining current market position or diversifying into new areas. Some companies, for example, place heavy emphasis on developing new technology. Motorola stresses evolutionary improvement and manufacturing efficiency. Through aggressive acquisition and application of external technology, Japanese manufacturers extended their businesses. General Electric traditionally develops new technology internally; so does Emerson, but the company has also broadened its technological base through acquisition.[1] Realistically looking at the business drivers and what the cell requires to achieve its objectives will indicate *what* technology should be applied and to what level.

IS TECHNOLOGY REQUIRED?

When planning for cells and their technological requirements, the first question should be: "Is technology *really* required?" And if the company can utilize technology to *improve* business and customer service, the answer is yes.

138

	Technology Level		
	Low	**Now**	**High**
Manufacturing			
Single-spindle drill press	✗		
Milling machine		✗	
Drilling machine		✗	
CNC lathe		✗	
CNC machining center		✗	
CNC grinder		✗	
Flexible manufacturing system			✗
Fabrication			
Saw	✗		
Shear		✗	
Brake press		✗	
Punch press		✗	
Other			
CNC punch center		✗	
CNC water jet			✗
CNC laser cutting			✗

Figure 9-1. *Technology level—how some typical processes for cellular manufacturing rate.*

The primary business factors that should be considered are:

- *Market share*. Before considering technological improvements, a thorough understanding of the business drivers and market must be achieved. Each company must define, in realistic terms, how much market it wants, how much it needs, and how much it can afford not to have.
- *Products*. The company's present and future products and the expected benefits from technology must be established. Will technology result in cost reductions, improved processes, improved material flow, and short-ened lead times? What will manufacturing requirements for future products be? Marketing, sales, and product development people should be involved in selecting manufacturing capabilities shaping future products.
- *Quality*. The level of quality required to satisfy customer's expectations also must be considered. What the customer perceives as quality can differ significantly from in-house quality standards. For example, all domestic automobiles meet in-house quality standards but foreign automobile manufacturers have still gained significant market share based on the customer's *perception* of higher quality.

Remember also that the customer may be internal as well as external. This orientation—where the customer is the next operation—is critical as it directly influences the type and level of technology required to satisfy quality expectations. Quality must be a planned part of the FMC's manufacturing process.

ARE YOU READY FOR TECHNOLOGY?

Several issues that should be addressed to determine readiness to invest in technology are the company's attitude, cost effectiveness, competition, and human resources (people).

Attitude

Do company policies inhibit the successful introduction of technology? Are the measurements, training programs, accounting methods, systems, functional groups, standard operating procedures, processes, and pay scales set up to take advantage of technological improvements? In short, are the people prepared to take advantage of technological improvements?

Cost Effectiveness

Can the company obtain, install, and staff the level of technology required to achieve the *necessary* quality and capacity levels within the capital restraints imposed by the accounting structure or corporate policies? Technological improvements need to be viewed for their long-term contributions to the business, not just for the short-term improvements.

Many of the U.S. cost accounting methodologies applied today actually undermine the perceived value of applied technology. Seldom do they pick up the added "actual" expense incurred to support higher technology devices in a cell environment, as well as the more expensive support. Subsequently, the cost structure will make the highly technological FMC look so attractive that it is loaded with parts that do not fit the structure and will bog down its effectiveness.

Since the era of Henry Ford, American industry has used technology to improve the performance of its organizations through increased efficiencies. Although this justification has diminished somewhat, the drive to maximize short-term profits should be replaced with the recognition that technology will sustain value for the company over the long term. Any technology considered should correspond with the business drivers. For example, applying high-tech solutions to an FMC where the FMC's projected life span is limited will simply waste resources. Conversely, using simple machines with a two-year life span in an FMC that is projected to operate for 10 years is also an inefficient use of the cell's resources.

Competition

Will the planned level of technology increase the company's competitiveness or just bring it even with the pack? One of the major mistakes in the application of technology is to simply copy the competition. At best, this will put a company even with the competition, duplicating the competition's errors. The chances are that the competition is already on its way to further improvements.

Survival is a prevalent strategy for U.S. industrial manufacturing organizations. Increased foreign as well as domestic competition has propagated apprehension and concerns about our once formidable manufacturing facilities. For that reason, U.S. manufacturers will often inappropriately assume that the reason for the foreign competitor's success with cells was due to the higher technological equipment installed. In reality, success is usually the result of the planning and organizing that preceded the FMC implementation. Remember that the competition will be working to:

- Concentrate on overhead reduction;
- Reduce diversification of product lines (part design standardization/ rationalization);
- Establish a niche (usually through product design);
- Revise pay and incentive systems that reward team or cell productivity;
- Invest in employee involvement programs;
- Enter new markets (internally or externally motivated);
- Develop new products.

Few, if any, of these improvement programs are directly addressed by the application of technology. In fact, higher technology equipment will normally increase overhead and suppress broad employee involvement efforts. Sometimes a high-tech machine tool will force some part standardization to take advantage of limited tool holding capabilities.

Human Resources

Involving the human resources function in FMC planning, especially for the introduction of technology, is essential. Before adding new technologies, the following questions should be asked:

- Are there enough people in the organization to assume all of the leadership roles necessary to ensure success?
- Are the technical skills available to effectively employ technology?
- Have the gaps been identified?
- Is there a recruitment program to fill the gaps?
- Is there a training program?

WHERE SHOULD YOU INVEST?

After deciding that technology is feasible for the business and that the company is indeed ready to handle it, there are a number of areas within the business that would probably benefit from an infusion of technology. The problem becomes one of identifying which area gets the attention. Technology *does* cost money, usually scarce capital, and which area gets the attention is normally reflected in the budgetary process. Of course, there is always the option to improve all of them.

Determining where to apply technology requires objectivity. Everything that is involved in customer service, for both internal and external customers, must be evaluated to determine where the most benefit can be attained. Here are some of the typical considerations:

- Products:
  - —Which are most profitable?
  - —Which are the products of the future?
  - —Which will benefit the most from technology?
- Facilities:
  - —Is the condition acceptable?
  - —Are they repairable?
  - —Is there enough floor space and crane capacity?
  - —Do they conform to state and federal requirements?
  - —Is there adequate power, water, and air?
  - —Is the location of power, air, and water lines suitable?
  - —Is there adequate floor loading capacity?
  - —Are the surroundings suitable for electronic control systems?
  - —Is the factory in a location that will attract employees with the required skill levels or where training is readily available?
  - —Is the lifestyle of the area around the factory desirable?
- Layout:
  - —Is the present flow of product and information (paper, etc.) designed to support cells?
  - —Are there monuments (major fixed barriers) in the way of improving material flow?
  - —Can cells be organized?
  - —Is there enough aisle space?
  - —Is there room for expansion?
- Equipment:
  - —What is the present condition?
  - —What is the repair history?
  - —Are support requirements, such as maintenance, tooling, and programming adequate?
  - —Is relocation (especially with regard to foundations) complicated?
  - —What is the equipment's capacity?
- Manpower:
  - —What training programs are locally available?
  - —How will training be funded? (Training may be as high as 20% of the cost of technology.)
  - —What are the current hiring and interviewing practices?

—What skill levels exist (in all disciplines)?
—What are the cultural attitudes and backgrounds of the employees and how do they impact on:
  • Company goals, measurements, rewards?
  • Customers?
  • Ability to change?

• Systems:
—Are all systems set up so they can measure, monitor, and control performance to the goal?
—Can they support the higher technology integration, especially within the confines of a stand-alone FMC?

HOW MUCH SHOULD BE INVESTED?

Invest in technology only if it will help attain the goal of the business drivers. Unless the business is heavily driven by research and development, it doesn't make economic sense to invest in technology just for technology's sake.

Once the business drivers have been evaluated and the company goal and master plan to achieve the goal have been established, the decision of whether to invest in technology can be made. The questions then become:

• What level of technology (low, now, or high) and how much of each is appropriate?
• Where should this technology be applied?

And since this book is about cells, the subject is applying technology to cells. However, cells can be anything from simple to complex, and technology can be applied to any segment of the cell or to the entire cell.

Now is where the company goal and master plan become important. During formulation of the master plan, it will become apparent that there are several alternative routes (strategic plans) that will achieve the goal. Each of these alternatives will require a set of tactical plans and a complete cost/benefit analysis to determine which alternative is foremost for business success.

Some typical reasons for technology investments include:

• *Material handling/flow:* Includes parts and raw material into and within the cell as well as parts out of the cell. Investments in material handling will always benefit the cell through reduced inventory, increased speed, and greater ease of handling. This might include straight-through conveyors, specially designed queue stations, transfer devices, robotic placement devices, automatic elevation adjusters, or just proper containerization.

• *New equipment:* There are many opportunities to invest in new but simple equipment that will result in more effective processing, less maintenance downtime, and generally quicker processing times. New equipment could be automatic cleaning or washing machines, high-output induction single-station annealers, CNC multispindle drill presses, or a versatile multi-tooled vertical mill/drill/bore machine that can replace multiple machines in an FMC.

143

- *Group technology/part families:* Sorting the characteristics of products being manufactured in the cell may point out the feasibility of certain technology solutions that was not apparent before. For example, defining all parts with characteristics of hollow, cylindrical shapes; diameter of less than two inches (50.8 mm); and end turning, may illustrate the applicability of a new automatic bar cutoff lathe instead of sawing and turning.
- *A green field facility:* A new facility design will normally not have barriers (or monuments) already established that prevent installation of appropriate material handling equipment, large turning centers, or the smooth floors required for automatic guided vehicles. It may make more economic sense to pre-install central scrap or coolant distribution systems during construction than to do it retroactively in an existing building.
- *Computer-integrated manufacturing (CIM):* The necessity of tieing the total business and manufacturing closer with real-time information to support cellular manufacturing is all the more reason to potentially invest in technology. Much more on this topic is presented in Chapter 19 on integrating technology.

WHAT LEVEL OF TECHNOLOGY SHOULD YOU PURSUE?

Applying technology to a cell is dependent on the cell itself. Cells can be formed around parts, products, departments, or the entire business. A cell can be anything from a one-worker saw/deburr operation to a computer-integrated, fully automated, flexible manufacturing business. Each of these requires a different kind and level of technology and will incur different costs and risks.

The first step in determining the level of technology to apply is to discover the technology's availability. Some large companies rely on an in-house group usually called Advanced Manufacturing Engineering or Technology. The function of this area is usually to investigate and maintain knowledge about new technological advances. This area is generally the source for recommending new processes, equipment, and technologies. Companies without an established technology research area generally assign the research function to a task team. This team should consist of multifunctional disciplines to gain maximum benefit in the minimum amount of time.

Some resources and methods the team might use to obtain information about technological advancements are:
- Literature searches (manufacturers, professional societies, and technical libraries);
- Colleges and universities (research programs, professional knowledge, libraries);
- Technical associations, conferences, tool shows, seminars;
- Factories that are using new technologies (visits, interviews);
- Leading suppliers of technology.

A thorough understanding of the features and applications of the currently available technology is required before developing a framework of what might be *practical.* Then one must identify the need and apply technology to *it,* rather than buy technology based on the manufacturer's promises. This will ensure that the most appropriate technology to meet the business goal is employed.

WHAT APPLIES TO FMCs NOW?

In the technological arena, there are exciting things happening that may not immediately apply to cellular manufacturing. A brief list of recent technologies that may be appropriate in cellular applications is shown in *Figure 9-2.*[2] These range from low tech to high tech; the chapters in Part Three of this book provide more specific details on some of these concepts—CIM, PLCs, machine controllers, robots, simulation, and material handling innovations.

Lasers

Industrial lasers are tooled machines capable of concentrating high-energy levels of light into very small areas, principally for cutting, machining, welding, and selective surface treatment. Because of the predictability of the laser light beam, it can be used effectively for accurate measurement as well.

At recent trade shows lasers were the hottest topic, but do they really apply to FMCs? They are getting very close to becoming *now* technology and can usually be integrated. Today's lasers are very articulated with five- and six-axis maneuverability, power conservation, and precise controllability. Their cost is still very high, but with increasing application and acceptance, lasers are expected to find increased industrial usage.

There are two basic types of laser mechanisms in industrial use: solid state and gas-based. Both provide light energy in the infrared spectrum. The YAG solid-state system is generally used in lower-power applications, using mirrors to reflect the beam in varying directions. Gas lasers use gas-filled tubes to excite photons by electrical discharge. Carbon dioxide is the usual gas medium for most high-power cutting and welding applications.

For FMCs, the most feasible laser use will be for marking, etching, or part identification. Vision-equipped inspection lasers will be valuable for maintaining accurate quality checks within the cellular environment. Traditionally, the most common application for lasers has been the cutting, welding, or fusion processes, mostly in sheet metal or fabrication cells. A relatively new use for laser technology in FMCs is in fine laser machining applications, where they frequently replace awkward electrical discharge machining (EDM) equipment.

Self-Guided Vehicles

The precursor to self-guided vehicles (SGVs), automated-guided vehicles (AGVs), have been a part of the manufacturing scene for several years with varying success. Typically used for part handling between cells, the results have been less than sensational. The most noted complexities with AGVs have been the facility terrain, the in-floor guidance installation, and safety hazards.

Technology Applications	Levels Of:		
	Technology	Risk	Cost
Handling			
Automatic retrieval systems	2	4	10
Automatic guided vehicle systems	1	4	8
Self-guided vehicles	5	4	7
Robotics	2	6	10
Machine Tools			
Automatic tool selection	1	1	5
Self-diagnostic programming	4	4	3
Probing, offset selection	1	1	4
Quick change/setup innovations	1	1	2
Self-programming/artificial intelligence	9	8	8
Custom, flexible special machine systems	6	10	10
Quality/Inspection			
Automated coordinate measure machine systems	1	1	9
Vision inspection	4	7	6
Voice recognition systems	8	9	2
Laser inspection (metrology)	3	4	7
Alignment (e.g., machine spindles)	1	3	3
Automatic identification systems	1	2	3
Lasers			
Machining systems	6	8	9
Cutting	2	3	4
Fusion/welding	1	3	4
Drilling	4	5	6
Marking/etching	1	3	1
EDM (electrical discharge machining) capability	7	7	10
Tooling			
100 to 150 tool changer systems (AS/RS)	1	1	5
Modular tooling systems	1	1	4
Multi-pallet changers	1	3	3
Auto tool clamping	1	1	2
Live tooling	1	3	2
Articulated tooling	5	6	8
Diamond	3	3	4
High-speed steel cutters	4	1	3
Control Software			
Expert (software) systems	10	10	7
Self-diagnostic software	3	3	5
Electronic data interchange	3	3	4
Simple, twisted-pair cell controller (networking)	5	5	3
Simulation technology	2	4	3
Engineering			
Integrated engineering (concurrent)	1	1	2
CAD/CAM (design/modeling/tooling/mfg.)	2	2	4
3D modeling (prototyping)	4	1	7
Stereolithography (solid object modeling system)	8	9	9
CASE (computer-aided software engineering)	6	7	4
CAPP (computer-aided process planning)	1	1	3

Figure 9-2. *Technology/risk/cost assessments.*

Scale: 1 = Low
        10 = High

146

Within the cell itself, part delivery between machines has normally been accomplished with fixed handling devices like conveyors or rail-guided carts which are inflexible and costly to change frequently. A new alternative for this application is the SGV. This cart, a powered handling truck, does not follow a guide wire or painted path; instead, it navigates using internal navigational controls, responding to signals from a computer. The vehicle verifies its position from laser-detected bar-coded targets located along its path. The SGV's advantages include its ability to utilize existing floors, lower installation costs, and flexibility to changes in path, parts, or locations.[3]

Artificial Intelligence

Artificial intelligence (AI) is a technique for emulating human thought through logic patterns which result in action-based decision-making. This type of programming, based on "real time" inputs, allows the computer to take action rapidly, but predictably. When AI was first introduced it was oversold and underdesigned, resulting in an image that shunned practical manufacturing. Today, the integration of normal language style inquiry makes it user-friendly and more acceptable. This evolution has led to a manufacturing "expert systems" application for focused use; machine diagnosis is probably the most widely known application, used for automobile troubleshooting.

Another common application for AI is with robotics and vision systems, to enable them to recognize patterns, paths, or model comparisons and to make instant adjustments and changes in the robotic actions. Expert systems are also effectively used for welding, assembly, and statistical process control applications. Pick-and-place robots will use the logic decision-making capabilities to analyze the part presented to them and place it reactively according to that input.[4]

Electronic Data Interchange

First introduced during the 1970s, electronic data interchange (EDI) has begun to grow throughout manufacturing. EDI is an electronic exchange of common business information, like purchase orders, invoices, shipping instructions, production status, or order status. It is usually established between regular trading partners, such as suppliers and customers. In everyday experience, EDI is used in banking—automatic payments, automatic deposits, etc. In manufacturing, EDI is playing a major role in facilitating Just In Time (JIT) planning between the FMC's material requirements and supplier planning. By linking the entire chain—from the customer to internal databases to supplier/vendor operations—the information flow has been accelerated and redundancies, errors, and paper have been eliminated.

The EDI evolution has enabled many more manufacturers, suppliers, and customers to accomplish development of standards, integrated information systems, and affordable software compatibility. More than 3,000 North American manufacturing companies are using EDI on a daily basis. Communicating

147

the daily/hourly material requirements instantly to suppliers and sharing real information through EDI will enable companies to remain competitive. One purchasing department, for example, claims that EDI has resulted in a 60% reduction in the cost of a purchase order, a more than 35% savings in buyers' time, and a 25% increase in efficiency of the administrative staff.[5]

These are just a few of the latest "innovations" in the manufacturing technology field today. Although they have all been around for many years, new technology takes many years to fully mature and become less risky.

SUMMARY

This chapter challenged the appropriate application of technology in the FMC design. This challenge is necessary since the early reputation of FMCs was usually synonymous with high-tech installation of expensive state-of-the-art technology. The chapter's intent is to encourage consideration of the real values and reasons for buying that technology before making the purchase decision. The number one reason for implementation start-up delays is the difficulty in getting new, untried technology up and running, usually after several delivery delays.

Do not be afraid of technology, but do not expect more from it than it is capable of giving. It is not wise to add more bells and whistles if the basic objective is elimination of waste. Singing and dancing robots will not net savings on the bottom line. There are "right" uses for some of the latest that technology has to offer, while being realistic about the pain that it sometimes takes to be pioneers in applying it.

Refer to Chapters 19 and 20 for information on *integrating* technology and recognizing the latest trends for FMC application.

ENDNOTES

1. Steele, Lowell W., *Managing Technology*, New York: McGraw-Hill, 1989, p. 255.
2. This information was based on an informal survey of Ingersoll Engineers' consulting staff, taking into consideration their most recent experiences in manufacturing.
3. Hahn, Robert, "AGVs Self-Navigate Around Cell," *American Machinist*, November 1990, pp. 74-76.
4. Rasmus, Dan, "AI in the '90s: Its Impact on Manufacturing—Part 1," *Manufacturing Systems*, December 1990, pp. 30-34, and Rasmus, Dan, "AI in the '90s: Its Impact on Manufacturing—Part 2," *Manufacturing Systems*, January 1991, pp. 32-39.
5. Kalashian, Michael A., "EDI: A Critical Link in Customer Responsiveness," *Manufacturing Systems*, December 1990, pp. 20-28.

MATERIAL HANDLING
AND STORAGE

Material handling and material storage can be viewed as dynamic and static activities in and around a manufacturing process, providing prompt placement of material when and where it is most needed. A primary objective of cellular manufacturing is fast throughput; the cliche "the right material at the right place at the right time" has never been more appropriate. The design of the material handling system is a critical link in cell design. The foundation of the cellular concept of integrating internal and external interfaces is also at the core of material handling and storage in the cell.

In this chapter, we define the role of material handling and storage in the cell environment and touch on the key material handling principles. The key relationships between material handling and facilities planning are discussed, and an organized approach to developing material handling and storage solutions is presented. Material handling and storage equipment selection guidelines are also provided. Since material storage for cellular manufacturing is considered a liability, storage should be kept to a minimum during the cell design. Details on storage, queues, and automatic storage and retrieval systems (AS/RS), and requirements for cellular applications, are discussed later.

DEFINITION

Many material handling textbooks offer comprehensive definitions of material handling. Whatever words are used, the cell designer must have an awareness of the overall implication of the events and activities surrounding the processes which control movement.

For our purposes, material handling is the management of the continuous, synchronized flow of products and information through the manufacturing process, beginning with the material source (internal or external) and finishing with the delivery of the completed product to the customer.

This definition goes beyond traditional thinking in three important ways.

1. It is not limited to the internal manufacturing sequence but extends from the original material source through to packaging, labeling, shipping, and delivery to the next customer.
2. It includes not only the physical flow of parts and materials but also feedback about the movement which permits intervention and corrective action; it is a closed-loop activity.
3. The movement of goods and information must be continuous and synchronized; it requires integration.

MATERIAL HANDLING DESIGN DRIVERS

Material handling and storage activities add cost—not value—to a product. Quickly and accurately positioning material where it's needed in a cellular environment is essential. Minimizing material handling reduces waste in the production process by:

- Moving material directly to production (value-adding) operations;
- Ensuring that the distances between operations (material moves) are as short as practical;
- Eliminating stops (storage, queues, etc.).

Innovations in technology and equipment have made this necessary activity respectable, even fashionable. AS/RS, automated guided vehicles (AGVs), laser scanning devices, and robotics can make material handling and storage exciting. Note, however, that automating material handling for the sake of automation or for labor reduction usually fails to address fundamental problems, such as geography, distance, or ineffective shop/plant layout. Streamlining the physical movement of materials simplifies the control and reporting systems needed to determine where material is, where it must go, and by when.

As outlined in Chapter Nine, *technology is not always the right answer*. In fact, large expenditures on material handling and storage equipment are usually warning signals that the flow and layout of the factory are inefficient. Sometimes products require automatic part identification, and the cell designers should be concerned with the appropriate technology to satisfy those requirements. In this case, technology for labeling, bar coding, or part identification in the cell should be purchased.

Another consideration at the design stage should be financial investment; material handling is expensive and represents a significant portion of the cost to do business. Some cost accounting methods have demonstrated that material handling can amount to 20% of total manufacturing costs when *all* costs (like direct and indirect labor, fork trucks, pallets, inventory, transaction costs, and other less direct contributions) are included. Cell manufacturing, as a concept, recognizes this. In practice, cellular applications must minimize material handling cost. The handling design and equipment selected will impact overall cell cost.

Material handling and storage equipment *can* provide quantified returns under specified conditions. First, we can see tangible improvement to product cost; in some applications, the appropriate equipment can improve the transfer speed of material between production operations, thereby improving throughput. Damage and breakage cost can be lowered for certain types of fragile goods, such as printed circuit boards, china, and glass products. Quality costs can be improved when material handling methods provide special product protection.

Investment in this equipment is usually justified when it will improve working or environmental conditions. Automation techniques are often used to eliminate safety hazards, or when hazardous toxic conditions are prevalent. Storage facilities may be justified when there must be a "balance" or buffer between a source and customer due to cyclic (usually seasonal) variation. In this instance, it might be less costly to build to inventory than attempt to build to direct demand. Buffering also can occur between operations where there is an imbalance between input and output (for example, when a cell's daily demand may be re-sequenced by the cell operators or when internal output storage exceeds customer demands). (See the discussion on storage and cell ownership for storage of inventory).

The most fundamental factor in the definition of material handling (and storage) is the acceptance of cost, since we are dealing with nonvalue-adding operations. Significant evaluation of minimum movement and the associated cost of material handling and storage equipment should govern cell design. For example, picking up small aluminum parts by hand and having cell operators move them on hand carts is more economical than employing AGVs.

MATERIAL HANDLING PRINCIPLES

The design of material handling systems for the cell environment can be guided by 20 basic principles. These were established by the College-Industry Council on Material Handling Education, under the auspices of the Material Handling Institute, Inc. While these principles are not absolutes or substitutes for thorough analysis or judgement in a specific application, they do have great value as a checklist to facilitate the design process. These principles, slightly modified by Tompkins and White[1], are:

1. *Orientation Principle:* Study the problem thoroughly prior to preliminary planning to identify existing methods and problems, and physical and economic constraints, and to establish future requirements and goals.
2. *Planning Principle:* Establish a plan to include basic requirements, desirable options, and the consideration of contingencies for all material handling and storage activities.
3. *Systems Principle:* Integrate those handling and storage activities that are economically viable into a coordinated system of operations, including receiving, inspection, storage, production, assembly, packaging, warehousing, shipping, and transportation.
4. *Unit Load Principle:* Handle the product in as large a unit load as practical.

5. *Space Utilization Principle:* Make effective utilization of all cubic space.
6. *Standardization Principle:* Standardize handling methods and equipment wherever possible.
7. *Ergonomic Principle:* Recognize human capabilities and limitations by designing material handling equipment and procedures for effective interaction with people using the system.
8. *Energy Principle:* Include energy consumption of the material handling systems and material handling procedures when making comparisons or preparing economic justifications.
9. *Ecology Principle:* Use material handling equipment and procedures that minimize adverse effects on the environment.
10. *Mechanization Principle:* Mechanize the handling process where feasible to increase efficiency and economy in the handling of materials.
11. *Flexibility Principle:* Use methods and equipment that can perform a variety of tasks under a variety of operating conditions.
12. *Simplification Principle:* Simplify handling by eliminating, reducing, or combining unnecessary movements and/or equipment.
13. *Gravity Principle:* Utilize gravity to move material wherever possible, while respecting limitations concerning safety, product damage, and loss.
14. *Safety Principle:* Provide safe material handling equipment and methods that follow existing safety codes and regulations in addition to accrued experience.
15. *Computerization Principle:* Consider computerization in material handling and storage systems (when circumstances warrant) for improved material and information control.
16. *System Flow Principle:* Integrate data flow with physical material flow in handling and storage.
17. *Layout Principle:* Prepare an operation sequence and equipment layout for all viable system solutions, then select the alternative system which best integrates efficiency and effectiveness.
18. *Cost Principle:* Compare the economic justification of alternative solutions in equipment and methods on the basis of economic effectiveness as measured by expense per unit handled.
19. *Maintenance Principle:* Prepare a plan for preventive maintenance and scheduled repairs on all material handling equipment.
20. *Obsolescence Principle:* Prepare a long-range and economically sound policy for replacement of obsolete equipment and methods, with special consideration to after-tax life cycle costs.

No single set of principles can be applied to all situations, but these are an appropriate set of guidelines for the cell designer. The challenge of defining the *best* levels of handling, storage, flow, serviceability, safety, and technology can be met within this framework. Any cell layout should exemplify simplicity, density (closeness), and minimization for material handling design.

GETTING STARTED

Designing material handling and storage systems *into* the cell environment encourages the building block approach to *total* integration. Today, an operator would load and unload a lathe, but in five years (with the necessary financial investment) a robotic arm may perform that task. Preparing for future growth and expansion in material handling and storage systems design is an important part of cellular design.

Material handling and storage are the inseparable partners of facilities planning (plant layout), and different material handling and storage options will usually produce layout alternatives. Chapter Seven provided the basis and methodology for properly evaluating those alternatives, as well as the criteria that are regarded as most important.

The base data used for cell or overall arrangement planning should also be used for material handling and storage design. Material handling and storage are not "hardware" solutions to a problem; the roots of material handling design, described as the "3Ms" are:

- Material;
- Movement;
- Method.

These basic factors have been the focus of most material handling theory developed by the recognized experts in the field over the last 40 years. *Figure 10-1* depicts the relationships and considerations in these key factors.

In practice, material handling and storage design takes place at three levels:

- Plant level:
  —From cell to cell to assembly to stores;
  —Delivery of raw materials;
  —Shipping/receiving.
- Cell level:
  —Carts;
  —Conveyors, queues, racks.
- Machine or workplace level:
  —Conveyors;
  —Robotic pick and place.

The 3Ms have their own characteristics for analysis at each level. It is desirable to organize all pertinent data at the appropriate level. This helps to define specific tasks and keeps worksheets as simple as possible. *Figure 10-2* shows an example of a manufacturing cell material handling data worksheet.

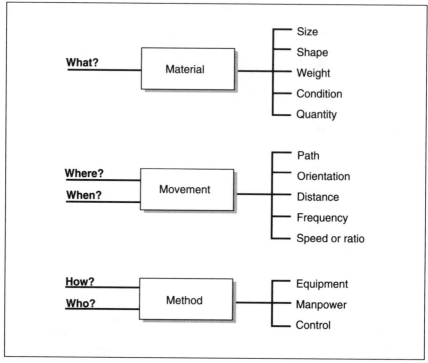

Figure 10-1. *Key factors of material handling.*

AN ORGANIZED APPROACH

The design of a material handling system for the cell environment is an engineering effort of sorts and should follow an organized, systematic engineering process. The following steps or phases can guide the design process:

1. Define the objectives and scope for the material handling system. Include the operating parameters and specific guidelines for the. level being addressed.
2. Analyze the requirements for handling, storing, and controlling material. Construct a database based on criteria for cell design.
3. Develop alternative designs which meet the requirements for handling and storage.
4. Evaluate alternative material handling and storage designs. (See Chapter Seven for common guidelines.)
5. Select the preferred design for handling, storing, and controlling material.
6. Implement the selected design as a coordinated element of the cell environment.
7. Audit on a regular, scheduled basis. Follow-up and feedback are crucial to effective material handling and storage systems design.

There are some general characteristics of cellular manufacturing that will influence the material handling and storage systems design. In addition to fast

Representative Parts			Characteristics								Handling Intensity			
Part Number	Description	Qty. Per Unit	Length	Width	Height	Weight	Handling Unit	Parts Move	Value	Total Variety	Part Qty. Per Month	Moves Per Month	Intensity Per Month	Remarks
DGD	Case and frame assem.	1	84	24	15	88	UL	1	4	1	82	82	328	Final assem.
D4X	Case and frame assem.	1	60	31	14	66	UL	1	4	1	161	161	644	Final assem.
D8L	Case and frame assem.	1	48	14	12	45	RX	1	5	1	15	15	75	Final assem.
966	Engine frame	1	28	20	6	115	SD	1	4	2	105	105	420	Final assem.
(A)	(B)	(C)	(D)				(E)	(F)	(G)	(H)	(I)	(J)	(K)	(L)

A Part number— Representative part/module or group
B Description— Commodity/module general name
C Quantity per unit— Pieces, pounds, assemblies
D Physical characteristics— Length, width, height, weight
E Handling unit— Type of container or unit load such as:

Code	Description
SD	Special handling
RX	Rack 26" x 42" x 24"
UL	Pallet (flat) 30" x 40"
CT	Standard tub 30" x 40" x 24"
BT	1/2 standard tub
AT	Small bin up to 1/4 standard tub

F Parts per move— Based on delivery frequency rules
G Value— Weighted score assigned to part move configuration
H Total variety— Number of parts represented by representative parts
I Part quantity per month— H multiplied by planning volume
J Moves per month— I divided by F
K Intensity per month— J multiplied by G
L Remarks— Destination of transaction

Figure 10-2. *A material handling data worksheet.*

throughput, identified as a primary objective in the introduction to this chapter, there are other characteristics which need to be considered, such as:

- No in-process inventory (inventory is a liability);
- No fork trucks;
- Batch size of one (or approaching one);
- High density (of machines and equipment);
- No containers;
- Close-coupled activities;
- Zero setup time;
- Integrated quality checks (no inspectors);
- Simple control systems;
- Team tasking (no such thing as "not my job"), and
- A philosophy of "don't put a part down unless someone is going to do something to it!"

The material handling principles coupled with manufacturing characteristics can be influential tools for material handling systems design. All available resources for material handling and storage systems design should be used. (A brief bibliography of textbooks and trade journals which can provide help throughout the design phase can be found at the end of this chapter.)

In-Cell Handling Concepts

A flexible manufacturing cell can be one of three basic types (as discussed in Chapter Five):

- *Product*—machines and equipment grouped together to produce a product, components for one product, or similar parts for a well-defined family of products;
- *Process*—machines and equipment grouped together to produce parts which have common processing requirements (within product family);
- *Group technology*—machines and equipment grouped together to produce parts of common shape (no family orientation).

Each has features that influence the material handling system design, as shown in *Figure 10-3*.

Ideally, operation sequence (flow of material) defines the proximity between workstations within the cell and produces the physical arrangement of facilities, but this is not always possible. Cell content features and building configuration, together with material handling design considerations, can influence layout in several ways:

- Quantity of workstations in the cell to be served;
- Material size, shape, and weight;
- Batch sizes;
- Distance between workstation load and unload points;
- Horizontal and vertical orientation across workstations;
- Labor assignments and utilization;
- Shared facilities (between cells);

Cell Type	Key Features
Product	— Small lot size (goal of one) — Dedicated/linked operations — Direct transfer between stations — Continuous flow through cell
Process	— Small batch sizes — Multi-path through process — In-cell balance queue
Group technology	— Large batch sizes — Unit load or container — Conveyors between stations — Palletized fixtures (machining center)

Figure 10-3. *Cell type influences the material handling system design.*

- Material source(s);
- Bay spans and height;
- Building's capacity to carry weight;
- Physical barriers;
- Cell customer location(s);
- Space required for material handling equipment.

As indicated in Chapter Six, plan the cell layout arrangement with maximum density first, then modify it as the material handling system is developed. One method is to visualize the handling elements as the cell layout takes shape. The use of a proximity diagram shown in Chapter Five is interrelated to material handling considerations in basic cell layout. Flow process charts or flow diagrams aid immensely in this process; the diagrams in *Figure 10-4* show typical examples of layout and handling relationships.

There is seldom one obvious equipment solution for each move. Several options need to be considered. *Figure 10-5* provides a simple reference guide to the selection of material handling equipment within a manufacturing cell. Simple, low-tech material handling solutions are the *norm* in the FMC setting.

157

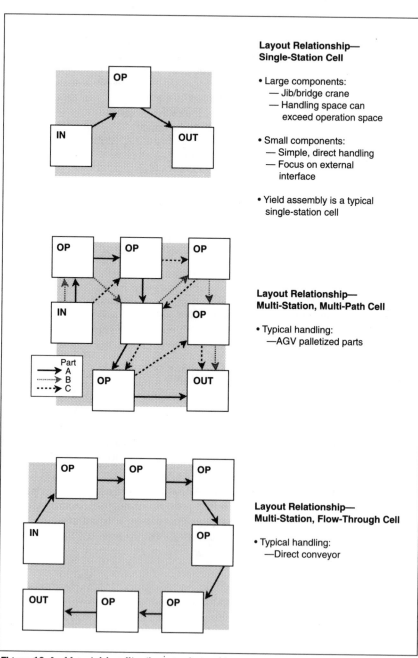

Figure 10-4. *Material handling/layout relationships.*

	Material			Move		Path		Control		
	Parts	Container	Pallet	Vertical	Horizontal	Fixed	Variable	Automatic	Manual	Programmable
Vehicles:										
— AGV	X	X			X		X	X		X
— Cart	X				X		X		X	
— Lift truck	X	X	X		X		X		X	X
Conveyors:										
— Belt	X	X		X	X	X		X	X	
— Slat	X	X	X		X	X		X	X	
— Gravity:										
• Roller/wheel		X	X		X	X			X	
• Chute/side	X	X	X		X	X			X	
— Lift and carry					X	X		X		
— Trolley				X	X	X		X	X	X
— Cart-on-rail					X	X		X		X
Crane/monorail:										
— Jib	X			X	X	X				
— Bridge	X			X	X	X	X			
— Power trolley monorail	X			X	X	X				

Figure 10-5. *Guidelines for cell material handling based on key factors.*

Storage

As stated in the introduction of this chapter, storage (inventory) is actually a liability in FMC design. Ideally, there is no storage in the cell environment, but reality means that something may have to be stored somewhere. Raw material usually comes in "batch lots" for economic reasons, or because of source or supply line distances, low usage, or some other constraint that recognizes reality. For example, bar stock is now frequently available from steel service centers in daily precut batches. Thus, storage requirements can be kept to a minimum, as can container size.

Purchased parts are received and stored for many of the same reasons. The general trend in industry today is to work closely with suppliers (at any quantity level) for the timely delivery of materials to match demand. In-bound material handling and storage is a prime place to apply the unit-load principle. In this case, fork trucks are a reasonable equipment solution. Wherever feasible, material supplies should be located at point of use. In-process storage (parts or assemblies produced in the factory) should, for all practical purposes, be engineered out of the manufacturing system. Three cases do exist for in-process storage:

159

1. *On the input side*: To satisfy the requirement of sequencing input between supplier (the cell) and customer (another cell, assembly line, or external). This typically occurs when small batches are produced across common equipment within the cell.
2. *In processing:* In-process buffer where the costs of synchronizing the output of one operation to another is prohibitive (usually the bottleneck process or operation).
3. *On the output side:* To provide a shift differential buffer between ''suppliers'' and ''customers.'' This exists when parts are produced on two or three shifts while assembly operates on one or two shifts.

In concept, the supplier cell owns its production output until the customer needs it. This helps to force the necessary discipline on the cell team conducting business. The cell pays the penalty for creating inventory beyond what is planned. If there is true ownership and involvement by the cell team, it will work to drive inventory down, yet still serve the customer. Storage must be provided for various supplies, tools, and fixtures that support the manufacturing operations. These storage items should be located close to the point of use and should be given the same attention to minimum inventory as is given to production materials.

The last type of storage is for finished products. Finished goods storage is a complex issue whose requirements are often dictated by company policy or a business decision. Ideally, they should be dictated by the customer. The world class manufacturer operating in a cell environment will attempt to minimize requirements for finished goods storage. Finished goods storage is the *most* expensive way to store material.

Storage requirements should be designed as part of the overall material handling system design, following the steps in the organized approach outlined earlier in this chapter. The data for each storage requirement should be documented on a worksheet similar to the example shown in *Figure 10-6*.

Storage equipment comes in a wide range of configurations, ranging from simple racks to sophisticated automatic storage/retrieval systems. The right solution requires careful engineering and analysis. Use ''expert'' resource help wherever possible, but be cautious of storage equipment supplier's off-the-shelf solutions. Be sure the requirements are clearly defined when seeking help. A general guide for storage equipment selection is shown on *Figure 10-7*.

SUMMARY

Material handling and storage are important elements providing the right material at the right time and right place within the cell environment. The 20 material handling principles previously referenced help the cell designer gain a perspective and set priorities for material handling and storage design. Following an organized approach will ensure that all the bases are touched throughout the planning process. ''Keep it simple'' should be the philosophy that guides the design and operation of the material handling system. Minimum moves, minimum distance, and minimum inventory are the goals.

| Part Number | Product Data | | | | Path | | | Move Data | | | |
	BOM Level	Description	Qty. Per Unit	Weight (Lbs.)	Supplier Mfg.	Supplier Purch.	Point Of Use	Cont. Code	Qty./ Cont.	Deliv. Freq. In	Deliv. Freq. Out
673452	1	Case assem.	1	45	W1		CD	P1	1		A/O
673441	2	Cover	1	5		610	W1	C	20	2/W	
542667	2	Screw	1	.02		610	W1	A	500	1/M	
673433	2	Case subassem.	1	31	W1		W1	—			
673421	3	Case, lower	1	14		610	W1	C	20	1/O	
673422	3	Shaft, A	1	6		610	W1	B	20	2/W	
672674	4	Gear	1	4		610	W1	B	26	2/W	
542236	4	Key	1	.02		610	W1	A	500	1/M	
67462	3	Shaft, B	1								

Figure 10-6. *Storage planning worksheet.*

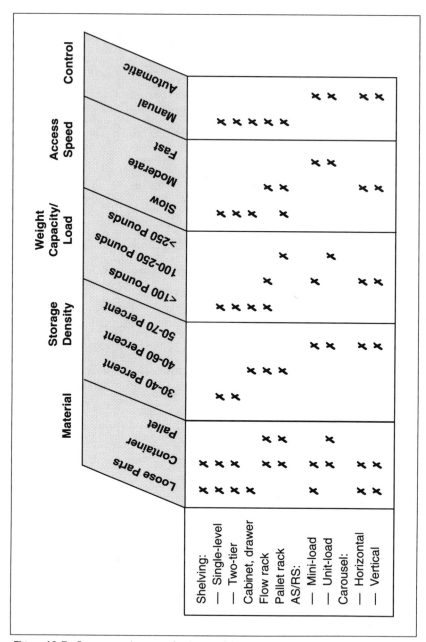

Figure 10-7. *Storage equipment selection guidelines.*

ENDNOTES

1. Tompkins, James A., and White, John A., *Facilities Planning*, New York: John Wiley & Sons, 1984.

11

INDUSTRIAL ROBOTS:
STEEL-COLLAR WORKERS IN
FLEXIBLE MANUFACTURING

A robot can be either a key resource in a manufacturing cell or an impediment to the performance of that cell. Thorough and thoughtful engineering analysis is crucial to the profitable application of a robot in any situation. This chapter guides a cell designer to answer the following questions:

1. How can I recognize high-potential opportunities for implementing robots?
2. What design factors must be considered for successful installation of robots?
3. How do I assess the performance (production and economic) of the selected robot before and after installation?

Today's vision of a robot as a mechanical human is totally inappropriate as a starting point to determine where and how to implement robots. A robot should never be considered just a tireless, uncomplaining replacement for an operator in the workplace. The only way to integrate a robot into a manufacturing cell is to consider how well—with its advantages and disadvantages—it satisfies the cell requirements. These pros and cons—part processing, material handling, inspection, etc.—leverage the unique features of the robot to maximize cell performance. Accurate production and economic performance analyses will then quantify the suitability of the robot as an integral component of the cell.

During the cell concept and design effort, the potential robotic applications should be examined against the following criteria:

- Is the operator adding value to this operation?
- Are boredom and fatigue likely to result in scrap, rework, or other quality problems? Would I want to do this task all day?
- Has this activity led to claims for worker's compensation for back, arm, or leg injuries? Are other medical problems (repetitive wrist or finger syndromes, skin rashes and reactions, heat exhaustion) contributing to absenteeism? Is this task simply undesirable or hazardous for humans?

- Is an operator too slow to keep pace with equipment functioning at optimum speed?
- Is it cost-effective to perform this operation on two or three shifts?

This chapter discusses the following issues of applying robotic technology to a manufacturing cell:
- Robot applications;
- Robot types;
- Ancillary equipment:
  —Sensing;
  —Fixturing;
  —End effectors (robot hands—also known as end-of-arm (EoA) tooling or grippers);
- Safety;
- Controls and programming (see Chapter 13 for information on cell-level controls and programming).

The information presented here assumes the reader's familiarity with guidelines for applying technology to the manufacturing cell presented in Chapter Nine.

ROBOTS IN MANUFACTURING CELLS

Typical Applications

The mechanical "humans" of literature and movies are idealistic, conceptual relatives of today's industrial robots. Industrial robots are no more than machines performing repetitive tasks and motions within a limited range of movement. There are very few tasks robots cannot physically be adapted to. However, the following situations typically provide the best opportunities for cost-effective application of robotics in manufacturing cells:
- Simple, repetitive motion activities, where the dexterity and intelligence of a human operator are not required. Loading and unloading parts from machinery, transferring parts continually between two or three machines, and repetitive assembly of simple components are examples.
- Repetitive motions involving heavy tools, tooling, or parts. Examples include spot welding, machine loading and unloading, part transfer between machines, palletization of parts, and exchange of tools and/or fixtures.
- Manufacturing operations and material handling in hazardous environments. Examples include spot welding, painting, plating, loading and unloading forging and extrusion presses, working in nonoxygen or explosive environments, and transferring or applying toxic or reactive chemicals.
- Operations where precision in process movement, speed, approach angle, etc., from part to part is important. These would include such processes as deburring, spray painting, simple gas and arc welding, water jet and laser cutting, and sealant applications.

- Movement of material, tools, and similar objects in operator-paced and/or continuous operations. Typical examples include tending equipment running 24 hours per day, loading and unloading machines with short cycle times, process and assembly operations where material handling or part manipulation is time-critical (such as epoxy application in assembly or unloading compression and transfer molding machines), and assembly operations in synchronous assembly machines.
- Complex operations that require equipment that is not cost-effective. These would include laser and water jet cutting of free-form or three-dimensional shapes, riveting large panels to structural skeletons, and drilling and tapping odd-angle holes, etc.

Basic Types of Robots

There are six basic robot types or "configurations" which are classified primarily by how the robot end effector (or end-of-arm tooling) is moved through the robot's workspace.

"Pick and place" robots. These robots *(Figure 11-1)* are simple, non-programmable versions of two-axis robots. They are frequently applied to simple

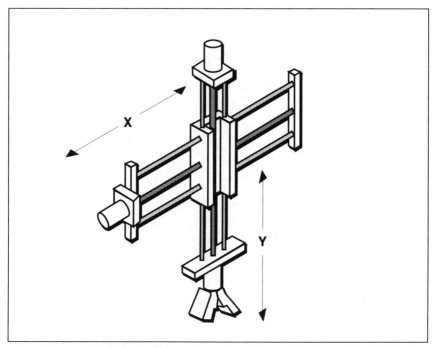

Figure 11-1. *Pick and place robot.*

material transfer tasks that do not require changes in path or start/finish locations. More complex, programmable, two-axis robots often load and unload machine tools. Most often they are utilized with cell-serving lathes, because they can only perform material positioning tasks in one plane. Positioning accuracy is typically ±0.005 in. (0.13 mm).

Gantry robots. These robots have three axes (or more) and are capable of servicing any workstation or location within a large cell area. The gantry, a frame erected on side supports to span an area, holds the robot *(Figure 11-2)*. If vertical machine and workstation access is possible, gantry robots are often the most practical and cost-effective robots for machine tool cell tending. Examples would include operations on large, typically flat parts. A significant advantage of gantry robots is that when they fail (and robots *do* break down), the vertical arm can be moved out of the workspace and operations can be resumed manually. This is the only robot configuration that will allow this level of access to a cell. Positioning accuracy is typically ±0.010 in. (0.25 mm) to ±0.025 in. (0.64 mm) or greater, depending on payload. More accurate gantry robots are available, but these are usually significantly more expensive and more limited in their range of access.

Articulated arm robots. These robots *(Figure 11-3)* are most appropriate when complex three-dimensional (four, five, or six-axis) tool/part motions are necessary in the cell or workstation. Other applications include vertical

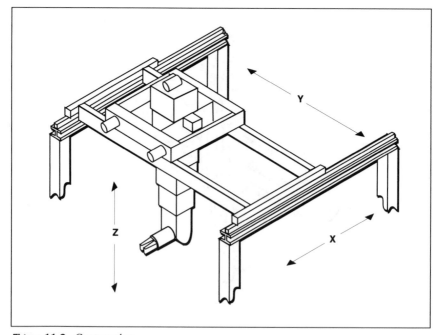

Figure 11-2. *Gantry robot.*

166

machines/workstations, where access is limited or unavailable. These robots require an oddly shaped workspace that typically limits the number of workstations/machines that can be accommodated by a single robot. Articulated arm robots are also the most difficult to program off-line. Positioning accuracy is typically ±0.005 in. (0.13 mm). Particular attention must be given

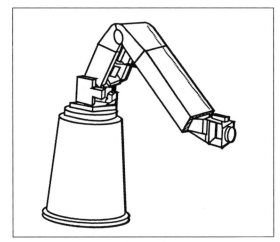

Figure 11-3. *Articulated arm robot.*

to the angular accuracy of the tool as well as the positioning accuracy with this type of robot.

Cylindrical coordinate robots. These robots combine some of the machine access capability of the articulated arm robot with a wider, and often more practical, work zone *(Figure 11-4)*. These robots are equal in payload capacity and speed to the articulated arm robots. Additionally, they are simpler to program and are often significantly less expensive if complex end effector control in the cell is not necessary. Positioning accuracy is ±0.005 in. (0.13 mm) for standard versions of this type of robot.

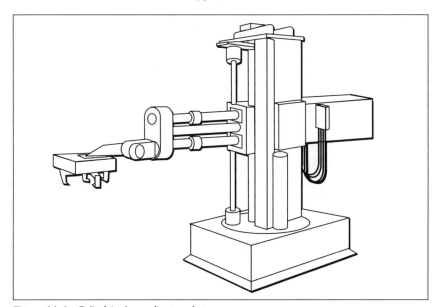

Figure 11-4. *Cylindrical coordinate robot.*

167

SCARA robots. Selective compliance robot arms for assembly (SCARA) robots are excellent for high-speed, close-tolerance operations *(Figure 11-5)*. Primary applications include vertical axis mechanical and electromechanical product assembly, as well as printed circuit board assembly. These robots typically require a small, oddly shaped workspace, and as a result, SCARA robots are usually not considered for material handling within a manufacturing cell. Positioning accuracy can be from ± 0.001 in. (0.025 mm) to ± 0.005 in. (0.13 mm), and this can be improved for certain assembly tasks when the robot has a remote center compliance (RCC) wrist *(Figure 11-6)*. Remote center compliance provides a low-cost, fairly rugged alternative to a force sensor and adaptive control system to provide small corrective motions in response to error forces.

Cartesian coordinate robots. These robots *(Figure 11-7)* are also excellent for assembly of close tolerance mechanical and electronic products. However, they can be somewhat slower than their SCARA counterparts. Cartesian robots are better suited to larger prismatic products than SCARA robots and can usually access more part presentation locations within their rectangular workspace than can SCARA robots. The positioning accuracy of this robot can also be ± 0.001 in. (0.025 mm) to ± 0.005 in. (0.13 mm).

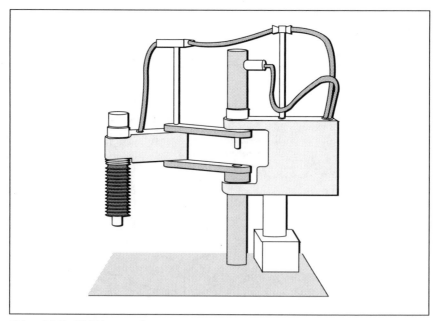

Figure 11-5. *SCARA-type robot.*

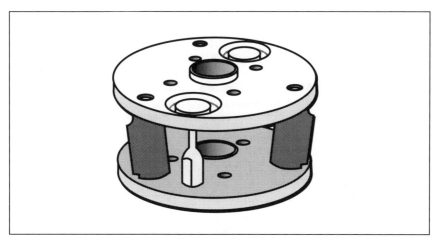

Figure 11-6. *Remote center compliance device.*

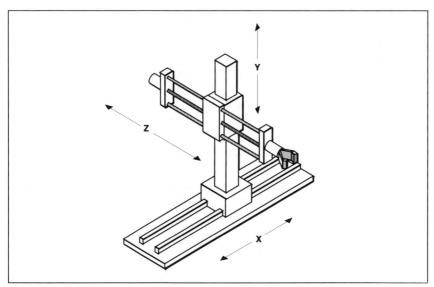

Figure 11-7. *Cartesian coordinate robot.*

PRACTICAL ROBOTIC INTEGRATION

Identifying the most appropriate type and size of robot and its ancillary equipment is contingent on the tasks to be performed in the cell. General questions that must be addressed at this point include:

1. What type of robot is most appropriate? How simple can it be?
2. What kind(s) of end effectors are necessary to manipulate the anticipated range of parts or tools accurately and reliably?
3. What sensors might be needed to guide the robot and accommodate process/part consistency?
4. What material presentation and auxiliary handling equipment may be necessary (i.e., vibratory bowls, racks, containers, etc.)?
5. How must the machine or assembly fixture be designed to facilitate robotic application?
6. What type of control is appropriate for the robot? (See the section on control in this chapter.)
7. What safety features and precautions are necessary to protect the work force from the robot and the robot from the work force?
8. How should the robot be configured to allow for automatic as well as manual workstation/cell operation, and preventive and unplanned maintenance?

In addition, many specific factors influence the selection of a robot appropriate to perform a particular task:

1. Accessibility to the workstation or machine to be tended by the robot— axes of access, clearance/interference distances, doors that open and close, etc.
2. Part or tool size, weight, and moments of inertia (the ratio of the torque applied to a rigid body free to rotate around an axis to the angular acceleration produced about that access).
3. Process speed, and time available for part movement, loading and unloading, gaging, etc.
4. Part or tool positioning accuracy and repeatability of positioning.
5. Variety of tasks or sequences of operations to be performed by the same robot—the need for programming ease and editing as required.
6. Workspace or cell area to be accessed by the robot—the robot's envelope or ''world.''
7. Complexity of tool or end effector motion necessary to perform the task.

ANCILLARY EQUIPMENT

Drive System Selection

Any of the robot types can be powered pneumatically, electrically, or hydraulically. The most appropriate drive system will be defined by the weight of the end effector and the components to be handled, the necessary/desired speed, and the desired positioning accuracy/repeatability. Also, the ''settling time" (how long it takes the robot to decide it has positioned the part or tool) will assist in drive selection.

Pneumatic drives (operated by air or another gas) are the simplest and least expensive, but they cannot accept complex programming and have the lowest payload capability. Maintenance skills/requirements and downtime are typically lowest with this type of drive system. However, cleanliness of the air supply is critical to the reliability of the robot. Hydraulic drive systems (operated by the action of water or another low-viscosity fluid) are often the most complex and slowest, while requiring the most maintenance. These systems, however, are compact and provide the greatest payload capacity. Electric drives are typically the most accurate, flexible, and fastest drives.

End Effector Selection and Design Criteria

Design, availability, and selection of the proper end effectors are important. The successful application of end effectors can be as integral to the robot's performance as the selection of the robot itself. End effectors come in all shapes and sizes *(Figure 11-8)*. They range from simple, generic grippers that simply grasp the part/tool like a vise, to very complex, customized tools dedicated to manipulating a single part.

Unfortunately, there are no specific rules for end effector design. The following criteria are simply guidelines to review when attempting to select end effectors from those available or when designing custom end effectors:

- Keep the end effector motions and force requirements as simple as possible.
- Design the end effector to accommodate any potential variations in part geometry and tolerances. If a single end effector will not fully accommodate the expected range of variation, mechanisms for changing end effectors can be built into the robot arm. These end effector exchange mechanisms substantially increase the cost of the robot and end effectors, while degrading the payload capacity and reliability of the robotic system.
- Select/design the lightest, smallest end effector that is practical. End effector weight directly impacts robot payload, as well as accuracy and settling time. End effector size also can restrict machine/workstation access.
- Design the end effector to accommodate slight variations in location accuracy of the part/tool. This design should allow the part/tool to be grasped and positioned accurately for processing or assembly. RCC mechanisms allow robots with medium positioning accuracy to perform high-accuracy assembly of certain part configurations and to accommodate small shifts in axial alignment during assembly.
- Select "finger" or gripper configurations and materials that prevent damage to the component and its finish. Remember to match the gripping force in the end effector to the task to be accomplished.

171

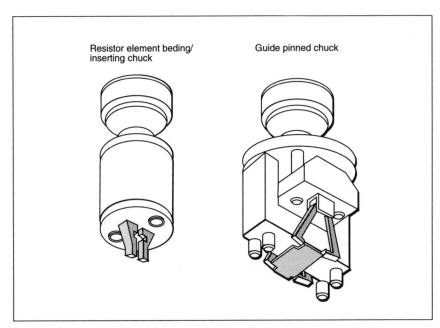

Resistor element beding/
inserting chuck

Guide pinned chuck

Figure 11-8. *Types of end effectors.*

Sensing Devices

Although a robot is simply a machine, it need not be blind or "unfeeling." Numerous types of sensors can be integrated into the robot arm and end effector. These sensors help guide its motion, acknowledge that it has grasped or let go of an object, control gripping or tightening force, recognize a part or tooling location, sense locations of moving objects in space, etc. However, all sensors add significant complexity and expense to the robotic application, make programming more difficult, and degrade the reliability of the robot and controller package. As with any "add-on" technology, sensor technology should *not* be substituted for improved component quality, accurate tooling and part presentation equipment, or products designed for ease of assembly.

When a sensor is truly necessary, there are three basic categories from which to select. The criteria depend on the complexity of the task to be accomplished by the robot and are listed in ascending order of complexity, expense, and support requirements:

1. *Tactile sensors, such as limit or mechanical proximity sensors, touch probes, and force sensors/pressure pads.* Tactile sensors can be integrated

into end effectors, tooling, or the robot arm itself at low cost and with minimal additional control requirements. These sensors can detect missing or poorly aligned/presented components, minor (as well as major) component size or shape variations, and the presence of a particular feature or shape in a part (such as the presence of a hole or machined pad). Torque controls and pressure regulation sensors are variants of tactile sensors, as they also require contact with the part to be effective. Tactile sensors require little or no maintenance and minimal operator programming skills.

2. *Noncontact optical pressure sensors, such as photocells and infrared sensors, and sonic sensors, such as primitive sonar devices and ultrasonic ranging devices.* Optical arm and sonic sensors are also used to sense part presence, size, shape, etc., and in situations where contact is not permissible, such as with freshly painted components or static-sensitive components. Sonic sensors are more complex to program and maintain than simple optical sensors, and they are usually only used where optical or tactile sensors are inappropriate.

3. *Vision sensors, such as "machine vision" video cameras, laser triangulation and ranging devices, and laser/video camera combinations.* Vision sensors are more complex. Machine vision is an engineering science of its own. And while robotic applications have been driving the improvement of video and laser sensing systems, the advances have been relatively slow to mature from technical and support viewpoints. For these reasons, all other means of improving the robot application should be exhausted before utilizing machine vision.

Both video and laser triangulation require complex programming and control capability, and they can be seriously affected by ambient light in the workspace, component shape and color, speed of movement of the robot arm or part, reflectivity of the part or tool surface finish, etc. Machine vision is most suited to situations where there are no alternative methods for controlling the attitude of part presentation, for differentiating parts or tools, or for guiding the robot arm and end effector through the completion of the desired task.

Ancillary Mechanisms for Material Presentation

Robots, even those equipped with complex sensors, require a degree of consistency in their workspace—parts and tools should be located in specific locations, be within a specific expected size range, and be presented in a specific attitude so the robot can be programmed to grasp and position the part successfully. Material presentation is most often accomplished by manually or automatically loading a tray, rack, or similar presentation device so components are accurately positioned in the correct attitude for the robot. The more complex the shape of the part or tool, the more specific the presentation mechanism or fixture must be.

More specifically, small components for robotic assembly may require orientation tooling when being fed from vibratory bowls; larger components may be gravity fed to a stop that locates the part on all three axes. More efficient component presentations eliminate the need for complexity and adaptability in the robot itself. Additional mechanisms, such as powered rotary and multi-axis worktables, or other simple robots, may be required when the parts are large and complex *(Figure 11-9)*. These mechanisms add versatility and should always be considered when part manipulation appears to require a complex, five or six-axis robot; quite often a simpler robot coupled through its controller to a standard mechanical positioning device is a reliable and less expensive alternative to a complex robot.

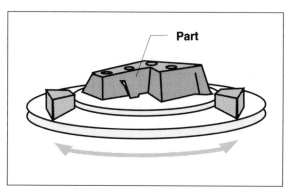

Figure 11-9. *Orientation tooling—powered rotary worktable.*

FIXTURE AND MACHINE REQUIREMENTS

Fixtures and machine tooling, such as lathe chucks, must be designed to accommodate the positioning accuracy of the robot selected and the tolerances of the components to be manipulated. Tooling should be designed to allow self-centering of components. It must be designed to provide gripping access around the component for placement and removal of the component. Machine guarding must be under machine or robot control and must provide for automatic opening and closing (doors, chucks, etc.) to allow access for the robot. The robot arm and end effector must be protected and insulated from any coolants and chemicals used in the manufacturing process. These components must also be capable of handling the temperature extremes likely to be encountered during operation (the controller is temperature-sensitive too).

Fixtures used in cells producing families of parts must accommodate the range of components produced by the cell. Designs should have simple part-related components that the robot can be programmed to exchange during setup for the next series of parts. Any tooling the robot must exchange must be stored within the robot's workspace for automatic exchange. This often consumes valuable workspace from other operations within the robot's reach. The size of the robot and storage of the tooling, fixtures, and end effectors must be considered as the floor space requirement is determined.

INTEGRATING THE ROBOT INTO THE CELL

At the design stage, workspace, axis of movement, and payload capacity should receive primary attention, followed by the requirements for material handling and machine access in the cell which define what types of robots are practical for integration into the specific cell or workstation. Despite this, four criteria are often overlooked during the design effort:

1. How will cell/machine maintenance be performed on the equipment in the cell, and how can production be performed if the robot fails?
2. Where will the controller be located in the cell (it often has temperature and electrical interference limitations)?
3. Is there enough space for part presentation (especially if different assemblies or components are used at different times)? Are tool/end effector storage and ancillary processes in the cell within the robot's reach?
4. If the robot is controlled primarily through the teaching mode, is there enough room to manipulate the arm through its various paths and patterns? (Typical paths and patterns are shown in *Figure 11-10, Figure 11-11,* and *Figure 11-12 .*)

Dealing with these issues during the design effort can alleviate many embarrassing situations after installing the robot.

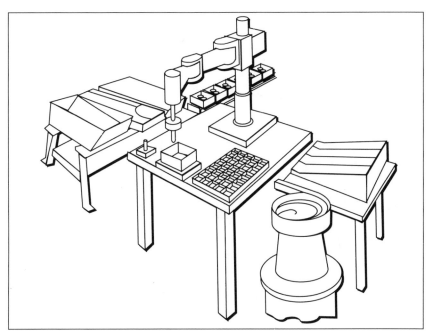

Figure 11-10. *Single robot pick and place assembly station.*

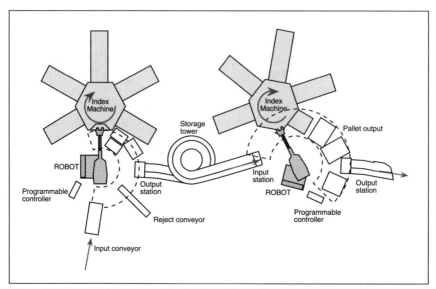

Figure 11-11. *FMC layout—machine loading robot cell.*

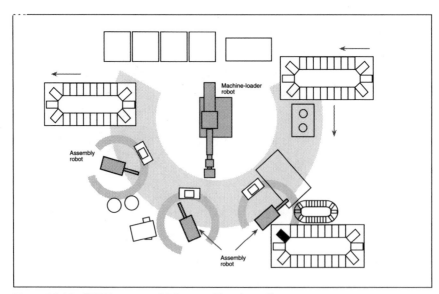

Figure 11-12. *FMC layout—multiple assembly robot cell.*

SAFETY—NOT AN ADD-ON

As with any application of technology in the workplace, safety features are not an option; they must be considered an integral part of the installation. There are two important aspects of robotic safety for the cell designer to consider:
1. How can the workers be protected from the robot during normal and abnormal operations?
2. How can the robot be protected from the work force and human error?

The most effective and obvious means of protecting workers from the robot (and vice versa) is to enclose the robot in its own room or shell. This is not typically done for access and "showcase" reasons. Other options to prevent employees from accidentally walking into the path of a moving robot include erecting physical barriers such as fixed railings and fences.

Light curtains with fixed barriers also are recommended to protect maintenance employees, employees restocking the material presentation devices, and other workers. A visual reference such as a blinking or revolving light should be integrated into the robot control so it functions whenever the robot is in cycle (this includes when the robot is waiting for parts to arrive or machining operations to complete, and other possible delays—not just when the robot is in motion). Whenever possible, workstations attended by employees should be separated from those tended by robots. Attempting to have humans and robots work around the same tooling is not recommended. The cell designer should review all state and federal OSHA regulations with respect to automated equipment and robot safety when designing the cell.

Physical barriers are also the best way to protect the robot from human abuse. The cell design must consider material handling options carefully. Forklifts and other mechanical material handling equipment used for tending the cell often conflict with these barriers. Robot controllers are easy targets for overhead cranes, disgruntled employees, etc., and should be protected from the "real" world as much as possible. Typically, material is provided via a conveyor into the work area and complete parts are also conveyed from the cell.

CONTROLS FOR ROBOTS AND CELLS

Three factors influence the definition and complexity of the controller to be specified with the robot:
- How complex are the robotic motions and speed profiles to be controlled, and how complex are the sensors to be integrated with the robot?
- How will the robot be programmed—will it be physically taught, or programmed off-line through the controller or other programming device?
- Should the robot controller also provide cell control, or should it respond to cell-level commands from an independent cell controller? (See Chapter 13 for a discussion of independent cell controllers.)

Obviously, the more complex the path the robot arm must traverse to manipulate a part or tool and the faster it must move through this path, the more

177

complex the control computer must be. Additionally, the greater the number of axes under simultaneous control, the more complex and expensive the controller. Sensor integration adds yet another layer of control complexity, and the number of sensors to be continuously monitored can also affect the cost and complexity of the controller. Robot manufacturers may be a resource in selecting the proper control for the tasks to be automated; they should not, however, be relied on to specify what controllers are required, only to define the operation characteristics required for specific configurations of equipment.

Programming the robot emphasizes a different capability of the control computer. There are basically two standard methods of describing the robotic task to the control:

1. Point-to-point programming involves physically moving the robot arm through the motions necessary to accomplish the task and "teaching" the control the specific points in that motion where arm direction or tool attitude change.

2. Off-line programming involves preparing a program for the robot at the controller or another computer using the programming language designed for the robot.

The simplest controllers require that an operator direct the motion of the robot with a joy stick or similar hand-held device (such as a spray paint end effector). In this scenario, the operator guides the robotic arm, pressing a teach button at the beginning and end of a specific path. This is typical in spray painting applications, where it is easier to capture the skills of workers as they perform their task than to try to program the individual positioning commands manually from the control console.

Advanced controls allow an operator to teach the robot by entering control commands from the operator console of the controller. This approach is usually faster if all movements are along the primary axes of the robot. Finally, the most flexible, complex, and expensive approach to programming is to assemble the program from a series of standard command statements. This is done at an off-line workstation such as a robotic control programming system, for example, often using a personal computer. The program is then transferred to the robot controller.

The off-line approach is the only way to maintain robot productivity while preparing the program (though some debugging and tryout time should be allowed to test the program in the workstation before committing to production). Significant operator/programmer training and equipment are necessary to support off-line programming, and the cell designer must trade off these costs with the cost of nonproductive time to teach the robot its next task. In a majority of tasks, the off-line programming approach is the most cost-effective.

Finally, if the robot is truly integrating cell operation through material handling and machine tending, it may be practical and cost-effective to use the robot control computer as the cell control. The robot controller can sense signals from the various workstations and material presentation devices in the cell to know when to move material, change tools, or perform an assembly or machining operation. Controls can signal these devices when they have completed their tasks. Most robot controllers available today can either control several independent workstations simultaneously or can be coupled with a programmable logic controller (PLC) for cell control, avoiding the cost, programming, and maintenance of separate cell controllers.

PERFORMANCE EVALUATION

Robots will have an impact on: productivity, throughput, quality, and financial performance.

The physical performance capability of the robotic workstation, typically measured by productivity, throughput, and/or flow time, must be determined *before* assessing the financial performance of the robot. Productivity is determined by measuring the time required by the robot to perform the necessary tasks as well as the throughput of the machines tended or of the robotic workstation, and comparing that time and throughput with the same productivity measures for the cell if manually operated.

Improvements in quality (less scrap and rework, fewer misassembled products, more consistency in the paint, more accurate hole locations and part positioning, etc.) are as important to quantify as time savings or increases in throughput. Accurate comparisons to actual employee performance on second and third shifts should be defined, rather than simply extrapolated from single-shift data (manual operations on second and third shifts tend to be less productive than first-shift operations.) Remember to include tool and fixture change/exchange times as well as debug and tryout times when estimating robot efficiency and anticipated uptime.

Once the physical performance parameters for the robotic application have been determined, the financial justification process can begin. Any machine tool justification should be performed on a total cost basis, and the justification of a robotic workstation should also be based on total cost. Most of the cost/savings categories in which the robot will have some impact are listed in the following paragraph. Almost any application should result in a savings or cost in these categories, and certainly there will be more categories in specific applications.

Costs

Robotic costs include:
- Robot cost:
  - Robotic equipment;
  - Controller, including PLC if necessary;
  - Sensors, including mounting and wiring;
  - End effector(s) design, fabrication, and tryout.
- Part presentation device (fixtures/tooling) costs, for both the robot as well as manual or ancillary operations, and material preparation costs, such as the preloading of trays, racks, etc., to feed the robot that would be unnecessary with human operators.

179

- Support costs:
    - Programming system(s)—usually for off-line programming;
    - Programming and debugging costs, as well as unproductive time necessary to either teach the robot or debug new programs;
    - Maintenance and spare parts;
    - Utilities and supplies, such as hydraulic fluid;
    - Service contracts, educational classes for maintenance, programming, orientation, etc.
- Installation and start-up costs, including productive time lost on cell equipment while interfacing them with the robot and debugging robotic operations, as well as any inventory ramp-up necessary to accommodate the downtime during robotic implementation.

Savings

Savings from making use of robotics include:
- The cost of one or more operators necessary to tend the workstation, machines in the cell, etc. that the robot may displace, including health care costs and worker's compensation resulting from current manual operation. As robotics are applied, operators are often relocated or retrained for other areas within cellular manufacturing.
- Costs of additional equipment avoided by implementing a robot. Improvements in equipment utilization due to the use of a robot can defer the purchase of new equipment that would have been needed to produce the targeted number of components had the cell been manually operated on one or more shifts. The salvage value of current equipment that can be eliminated due to increased operating time or throughput also can be credited to the robot. Quite often the displaced equipment can be used to avoid the purchase of equipment for other, unrelated component manufacturing in other areas of the factory. This cost avoidance should also be quantified and considered a benefit due to the robot.
- Reductions in the costs of quality, both internal (scrap, rework, lost productive time, detection, and reconciliation) and external (warranty costs due to improper assembly, damaged components, etc.) due to robotic operation.

There are certainly more cost categories and potential benefits that must be quantified to accurately estimate the value of implementing a robot within a cell, as well as measure that value after implementation. The more shifts the operation will function, and the more machines a single robot will tend, the easier it is to justify robotic application within manufacturing cells. Chapter Two provides more details on justification.

SUMMARY

Sophisticated robotics have received particular interest in the U.S. within automated, small batch production and assembly cells. Cells with standard products are often candidates for robots, while cells producing mixed standard and made-to-order products are unlikely robotic candidates. Again, robots must be applied as an integral part of the manufacturing cell, with user and management training completed before robotic implementation. Additionally, safety considerations for the operators and the robot must be paramount in the robotic installation. Close attention should be given to the underlying social and economic questions. Successful application of robotic technology takes common sense, enthusiasm, and a willingness to make it work.

QUALITY

"Quality is never an accident; it is always the result of intelligent effort."
John Ruskin, English essayist, c. 1850

In the past decade, quality has come into vogue for United States manufacturers. Traditionally, "quality control" focused on identifying (and sometimes preventing) defects in manufacturing. More recently, the emphasis has shifted from the "control" part of the definition to "quality," and manufacturers now work to eliminate waste in all functions of the organization. Popular terms like quality control (QC), quality assurance (QA), and total quality control or management (TQC, TQM) are all used and misused in advocating quality improvement programs in the manufacturing community.

Beyond just product quality, TQC encompasses aspects of quality assurance, process control, *and* quality improvement. It can therefore be defined as *the control of all transformation processes of an organization to best satisfy customer needs in the most economical way.*[1] Above all, the emphasis is on the customer and revolves around the concept of "customer satisfaction." The following criteria for evaluating customer satisfaction have been offered:

- The subjective comparison between customers' expectations before they receive the service and their actual experience with the service;
- Quality evaluations derived from the service process, as well as from the service outcome;
- The quality level at which regular service is delivered, and the quality level at which exceptions or problems are handled.

THE QUALITY IMPERATIVE—COST

Quality, however it is defined, is not superfluous; most U.S. executives "think the cost of bad quality is 5% of sales or less, although specialists place

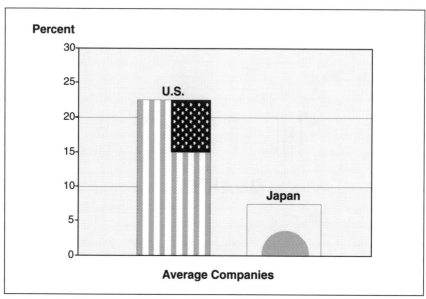

Figure 12-1. *Quality as part of manufacturing cost.*

it at 20% to 30%."[2] As depicted in *Figure 12-1*, the cost of quality in the U.S. appears to be far greater than estimates for Japanese companies.

Most conventional "cost of quality" calculations break down the components of quality into three "buckets"—prevention, appraisal, and failure—as shown in *Figure 12-2*. Traditionally, U.S. and other Western manufacturers incur the most costs within the failure portion of this quality continuum, by two to three times the next highest category. This is usually manifested in terms of scrap, rework, returns, and warranty (including initial installation) problems. The appraisal costs are usually associated with inspection and internal testing. Prevention costs, which should account for a much larger portion than they currently do in the U.S., typically come from quality engineering, process planning, training, and problem prevention. Besides tangible quality costs, there are numerous less manageable or intangible costs of quality. These types of costs are shown in *Figure 12-3*.

TRADITIONAL APPROACHES

Using the "quality control" mindset, manufacturers understand the quality process reactively and often try to inspect quality into the product; quality typically is the responsibility of a specialist who inspects the product before it goes out the door. In this environment, the emphasis is on technical issues and techniques to detect defects, like specifications compliance, efficiency ratings, numerical goals, and output quotas. Quality efforts in the U.S. have gone

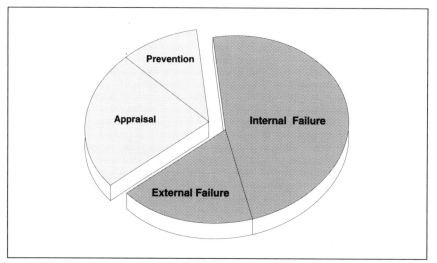

Figure 12-2. *Failure is the largest component of the cost of quality.*

through numerous phases during the 20th century (*Figure 12-4*). Surprisingly, most of these concepts began in the U.S. Their applications were expanded and perfected in Oriental cultures, however.

In the traditional quality control department, inspectors act as policemen or referees who coordinate a multifunctional process:

- Deviation requests and approvals;
- Material review board (MRB);
- Engineering "negotiation;"
- Disposition of rejected material or product, and
- "Beating on" vendors.

These traditional approaches to quality depend on inspection of the final product to cull out those not meeting customer specifications. This applies to clerical tasks as well; work (typically paper) is produced and then checked and rechecked for errors. Detecting errors in this fashion wastes resources. Time and material are invested in products or services that are not always usable. Costs incurred in scrap and rework occur too late to be recoverable.

THE NEW QUALITY

Today, especially where cells are being used, quality goes beyond piece/part inspection. Wallace Company chairman John Wallace, whose company won the 1990 National Malcolm Baldrige Quality Award in the small-business category,

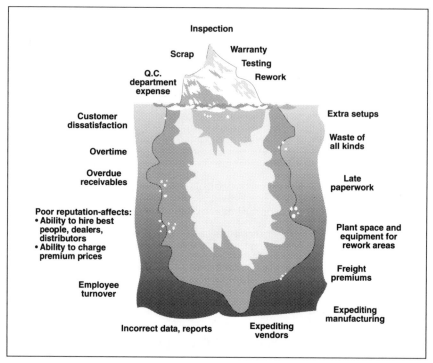

Inspection

Scrap

Warranty

Testing

Q.C. department expense

Rework

Customer dissatisfaction

Extra setups

Waste of all kinds

Overtime

Overdue receivables

Late paperwork

Poor reputation-affects:
• Ability to hire best people, dealers, distributors
• Ability to charge premium prices

Plant space and equipment for rework areas

Freight premiums

Employee turnover

Incorrect data, reports

Expediting vendors

Expediting manufacturing

Figure 12-3. *Intangible and less manageable costs of quality. The tangible costs can be just the tip of the proverbial iceberg.*

sums up this perspective: "There's not a process in this company that hasn't been redefined. . .It's a different way of doing business because quality has become all-encompassing here."[3]

This all-encompassing approach to quality is more effective because it encourages long-term solutions, not quick-fix, symptomatic relief. Taking responsibility for meeting quality goals out of the hands of the inspection or quality control department and giving it to the person who actually does the work fosters ownership of the process, which, in turn, encourages operators to embrace quality goals. Developing this environment of self-inspection and *feedback* to continually improve performance, on a company-wide basis, is the foundation of the new quality *culture* in the U.S. today.

QUALITY IN A CELLULAR ENVIRONMENT

Quality should be a primary consideration when investigating the use of cellular manufacturing. The traditional approach to quality is no longer appropriate since the objective of cellular manufacturing is to eliminate waste of

PHASE	ERA
Detection, then correction	1920s
Statistical process control (SPC)	1970s
Quality circles, zero defects	1980s
TQC/TQM, six sigma	1990s

Figure 12-4. *U.S. quality efforts this century.*

space, time, and other resources. The three steps to establishing quality in flexible manufacturing cells include:

- Define the quality culture;
- Define quality for the cell and product;
- Achieve FMC quality objectives.

Defining A Quality Culture

When FMCs are applied to manufacturing, it's essential for everyone in the organization to realize that responsibility for quality must be shared to ensure successful FMC implementation. A quality culture is not so much a set of rules or regulations as it is a mindset for the entire organization, focusing on quality in *all* aspects of running the business—from processing orders to ultimate customer satisfaction (*Figure 12-5*). A quality culture involves management, design or product engineers, purchasing personnel, manufacturing and process control personnel, cell team members, and support personnel. For example, some elements of a quality culture might be:

- A plan to keep improving all operations continuously;
- A system for accurately measuring these improvements;
- A strategic plan based on performance benchmarks;
- A close partnership with suppliers and customers;
- A deep understanding of the customers;
- A long-lasting relationship with customers;

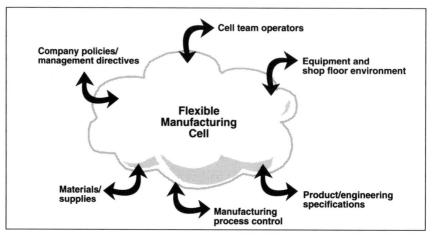

Figure 12-5. *A "quality culture."*

- A focus on preventing mistakes rather than merely correcting them;
- A commitment to improving quality that runs from the top of the organization to the bottom, and from the bottom to the top.

Some of the cultural transformations that have occurred during the past two decades to make this new quality focus possible are shown in *Figure 12-6.* The four key issues that must be addressed to establish this new culture are communication, information, training and education, and ownership.

Communication. Upper management, as the policy and standard-setting arm of the organization, is obligated to communicate the goals and directions they've set to the rest of the company. For continuous improvement, management must:

- Explore methods for *exceeding* customer needs and expectations;
- Expand participation to all employees, customers, and suppliers;
- Build suppliers into the system;
- Focus on continuous improvement of products and services, and
- Consistently relate the objectives of the firm in terms of quality.

At other levels of the organization, quality should become an everyday consideration. Design or product engineers share the responsibility for setting product specifications. Products should be reviewed specifically for quality as well as design for manufacture/design for assembly (DFM/DFA) objectives; these techniques can be a valuable tool in this review process. Raw material should be purchased to support the cell's quality objectives; vendor certification (discussed later) is one means toward that end. Methods, equipment, and manufacturing processes for process control should be specified to meet the quality objectives; identifying and maintaining a capable process is a key component in this area.

Traditional Focus	Progressive Culture
Doers	Thinkers
Single, repetitive functions	Group problem-solving
Individual piecework	Team-centered
Single job for a lifetime	Flexible learners
Familiar with simple machines	Know the technology
Single-task orientation	Information processors
Carry out the decisions of others	Decision-making by those closest to the product

Figure 12-6. *Recent cultural transformation in the manufacturing environment has contributed to a new focus on quality.*

Finally, at the cell level, the operators who are responsible for making parts to the prescribed specifications and who perform in-process inspection must understand the nuances of the quality culture. The most difficult aspect of this is usually making two-way communication viable. In the traditional environment, operators have been told what to do and when to do it; now they're being asked for opinions and information to make the process more effective. This is a radical shift in roles and will require encouragement and patience, but the results (surprising answers from the people who have been doing the work for years) are well worth the effort.

Information. The data and information coming into the cell must meet the same standards of quality as the actual material used to manufacture the part or components. This information must be clearly defined, accurate, and understandable, and the cell operators must be prepared to use it. To be sure, it is just as important that the information leaving the cell meet the same criteria.

The functional areas—engineering, scheduling, and maintenance—must provide the information in a reasonable, practical, and usable format. For example, engineering drawings should be current, specifications for quality standards must be definable, schedules should be made available in a timely manner, and tooling and setup instructions should be accurate and understandable to all who will use them.

Training and education. One of the key concerns with cell implementation will be the training required for quality. Initially, brainstorming sessions can be an effective training tool. With this technique, a typical problem is introduced to a group of cell operators who investigate alternative approaches to dealing with that problem. When designing the cell, this can be a powerful method for improving quality. More specifically, training will be required in three areas concerned with quality, as shown in *Figure 12-7*.

Engineering	Managing	Controlling
Design and failure mode and effect analysis	Quality policies	System audit program
Process failure mode and effect analysis	Process capability	Quality indicator system
Competitive product analysis	Targeted quality improvement program	Statistical methods and design of experiment
Design for manufacturability: value analysis, structured product/process analysis	Training: job skills, problem solving	Supplier development and supplier quality improvement
	Manufacturing teams (cells)	
	Inspection planning	
	Preventive maintenance	

Figure 12-7. *Training is required in three areas that affect quality.*

Training (in quality and other areas) cannot be overlooked; it is a substantial part of FMC implementation. Organizations preparing for FMCs should explore their educational resources. Technical schools or community colleges may offer courses in-house, or help your personnel develop education and training programs. Many education services and professional groups offer seminars that will provide the necessary information. The cost of training can typically reach 10% to 20% of the total investment in cell implementation.

Ownership. Involving the people who will work in the FMC in its development and design is essential for success. This involvement naturally allows them to help shape and take ownership of the new environment. The impact of this involvement or ownership on quality cannot be overstated—it is as essential as a properly tooled machine or an accurate schematic drawing. Since FMCs involve organizational changes and philosophies that are often radically different from what was done, people will require a more comprehensive level of understanding of their individual roles to act effectively in the FMC environment. This, coupled with an understanding of the "big picture," provides this sense of ownership.

Lack of ownership on the part of the people who are surrendering the quality function (typically quality control management and inspectors) can be overcome by the education and training process. These people must be assured that the cell operator possesses the information and ability to identify and correct quality problems. Roles must be redefined for *everyone* so that supervisors and quality personnel take more of a preventive, troubleshooter, or auditing capacity.

Defining Quality for the Cell and Product

From a manufacturing perspective, quality is usually thought of as "conformance to specifications." These specifications are typically defined by the firm's marketers as they relate to customers, and they will infrequently relate them to manufacturing process capability. As shown in *Figure 12-8*, defining quality requires balancing customer perceptions against the design and manufacturing specifications; the customer's expectations must be supported by the manufacturing objectives from the onset, not attempted to be built into the process later.

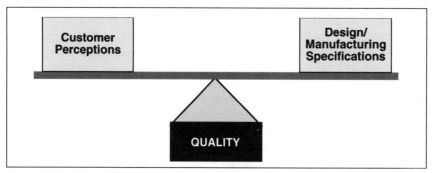

Figure 12-8. *The quality balance.*

At this point, an understanding of what quality is and what it will be must be established from two perspectives:
- *Internally*, it is important to know how the shop floor workers, supervisors, managers, and other support personnel perceive quality. They will provide valuable information on the present state of quality, as well as specific areas that may be causing shortcomings.
- *Externally* (which might also be the next cell or assembly station). Improvement alternatives that may not be immediately obvious can be brought to light using customer feedback mechanisms (product reply cards, warranties, complaints, etc.) supplemented by customer interviews. The customer's opinions should be given careful consideration; achieving true customer satisfaction is the most basic form of product quality.

These two sources of information should provide a basis for planning quality into the cell and a benchmark for measuring the continual improvement of products and services. Designing quality into an FMC requires significant emphasis on five elements: fixtures and tooling, preventive maintenance, vendor certification, in-process verification, and statistical process control (SPC).

Fixtures and tooling. Within the FMC, fixtures and tooling should be designed and operated in such a way that they will not accept defective parts; they should automatically sense the presence and correctness of incoming material or parts. One of the most effective fixture design techniques involves poka-yokes, which often incorporate simple devices (locator pins, notches, etc.) into the fixture. This method must be initiated at the design stage and will reduce, if not eliminate, human error. This mistake-proofing process, shown in *Figure 12-9*, provides immediate visibility for corrective action.

Other simple ways of mistake-proofing manufacturing hardware include:

- Material handling—reduce the number of handlings and improve the lifting and handling points on the parts, tools, and fixtures.
- Instruction manuals—verify that the directions for machine operation are easy to read and include basic problem-solving steps.
- Part tracking—stamp the part and operation number on the fixtures as well as on the tooling.

Preventive maintenance. Another simple way to ensure the consistent production of quality products within the cell is comprehensive preventive maintenance. While machine breakdowns are inevitable, equipment should be kept within the allowable operation limits. A preventive maintenance program should consider basic issues like verifying machine alignment and adjustment, and keeping spare parts and diagnostic equipment available for unplanned maintenance. In addition, a record of preventive and corrective maintenance should be kept with each machine. Within the FMC, the goal of maintenance should be to eliminate downtime (such as waiting for repair personnel or spare parts). Once a process is made capable, the preventive maintenance goal should be to support the continuation of that process—no matter what it takes.

Vendor certification. Vendor certification is the bond between the customer and supplier that is based on performance. Obviously, a product that exceeds quality expectations is one of the keys to satisfying the customer, and it all starts with the raw material. In the traditional approach, material goes through a receiving inspection or a sample check to ensure that it meets specifications. While high-quality material is still essential in the new environment, material and vendor certification programs are eliminating the need for these preliminary inspections. The benefits of vendor certification directly parallel the benefits of cellular manufacturing:

- Eliminates waste:
  - Checking and rechecking material;
  - Material handling;
  - Nonvalue-adding activity;
  - Redundant activity.
- Improves flow of cell (because material is immediately available and deliverable directly to point of use).

192

Figure 12-9. *"Mistake-proofing."*

- Facilitates better communication and feedback (from both supplier to cell and cell to supplier).

A typical vendor certification program requires a supplier to comply with certain quality levels (usually measured by the vendor), demonstrate that quality standards and techniques are understood and used (regular tool gaging), and provide standardized reports to demonstrate compliance. Some companies suggest that vendors working toward certification should attend quality seminars sponsored by the certifying company or a designated third party.

Upon certification, a supplier is usually offered a long-term or blanket order to supply material needed over a period of time, or guaranteed a specific volume of work. Material will not be subject to receiving inspection, and warehousing may be eliminated by point-of-use delivery. Some companies are using electronic data interchange (EDI) systems to exchange material requirements, shipping data, and order information electronically (this topic is covered in some detail in Chapter Nine). This relieves both the vendor and the manufacturer of a great deal of paperwork and manual data entry while making the order process virtually automatic.

These programs are intended to benefit both the vendor and the manufacturer; they must, however, be judiciously administered to ensure that they are more than just a "cosmetic process" that fails to solve any real problems. The potential drawbacks of using vendor certification to mandate quality include:

- There must be equal interest in quality goals coupled with equal power to accomplish them; the goals, however, are usually developed by the manufacturer with little or no supplier input.
- Mandated quality improvements usually "exist only cosmetically at OEM locations;"[4] benchmarks to measure improvement are not fully established.
- The process assumes that the supplier is ignorant about quality by looking "at conformance to specified quality procedures, rather than suppliers' actual success at improving product quality."[5]

When vendor certification is used to improve quality, it transforms an attitude of defensiveness into one of constructive cooperation. It can be particularly effective when it is teamed with the FMC concept.

In-process verification. During production, the cell operators are responsible for maintaining the level of quality established for their part or component. In most FMCs, in-process verification will ensure quality, and the inspection method must be planned into the cellular manufacturing process. Recall earlier suggestions for mistake-proofing as much of the process as possible. The equipment required to check the parts or tools should be provided as part of the cell and be simple to use.

In the FMC, verification equipment should also be consistent with whatever equipment is used by the next customer. Using different methods or equipment to measure the product will only result in conflict between the supplier and customer. Equipment that may be used in the cell include fixed and indicating gages (plug, ring, micrometer), comparators, and coordinate measuring machines (CMMs). Frequent calibration and gage certification is required to make these tools effective. These maintenance activities should be considered during the planning process.

Statistical process control. Another technique supporting FMCs is SPC. Based on the laws of probability, it is a method of evaluating the quality of production and controlling it based on specified limits. SPC makes it possible to determine whether a manufacturing process is producing the parts within these limits and to predict trends that may indicate future violations of the limits.

With SPC, established characteristics or features (hole location and size, surface finish, etc.) are measured and the averages and ranges are graphed to determine if they fall between the upper and lower control limits prescribed for the process (an example is shown in *Figure 12-10*). If the sample is unacceptable, corrections in the production process are made until the specifications are met. The advantages of SPC support the goals of the FMC by:

- Ensuring product uniformity;
- Detecting errors at discrete manufacturing points;
- Providing quick feedback and dynamic inspection information directly from the value-adding operation;
- Reporting different types of errors;
- Reducing the cost of quality;
- Establishing a basis for cell output capability and providing an attainable specification;
- Providing a direct input for preventive maintenance.

The most significant advantage of SPC is its ability to verify that a *given manufacturing process* is capable of achieving a desired result; it does not verify the part itself. SPC can, therefore, be applied to processes producing lot sizes of one with predictable results. However, statistical sampling is still a sampling plan to reduce and to maintain an accepted defect rate; while it identifies the source of defects, it does not eliminate them.

Achieving FMC Quality Objectives

Establishing and maintaining a quality-oriented manufacturing environment requires specific elements:

- Commitment to improving quality that extends from the top of the organization to the bottom;
- Comprehensive understanding of the customer;
- Emphasis on preventing mistakes rather than merely correcting them;
- A close partnership with suppliers and customers, built on a long-term relationship;
- A plan to keep improving all operations continuously using established performance benchmarks;
- A system for accurately measuring those improvements.

In the FMC environment, it is essential to focus on customer satisfaction—satisfying the *next* customer, internal or external, as well as the final customer who gets the finished product. The definition of quality must be exploded to include all activities—entering the order, creating the schedule, and giving instructions—not just running the machine. This quality culture can be effective in the FMC because activities are directly related.

195

Work center number ___365-2___
Quality characteristics ___Dies___   Date ___3/6/91___

Time	8:30 AM	9:30 AM	10:40 AM	11:50 AM	1:30 PM			
Subgroup	1	2	3	4	5	6	7	8
X_1	55	51	48	45	53			
X_2	52	52	49	43	50			
X_3	51	57	50	45	48			
X_4	53	50	49	43	50			
Sum	211	210	196	176	201			
$\overline{X}$	52.8	52.5	49	44	50.2			
R	4	7	2	2	5			

Durometer measurement readings

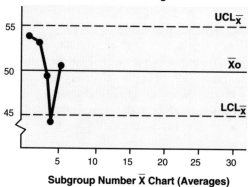

Subgroup Number $\overline{X}$ Chart (Averages)

Durometer measurement readings

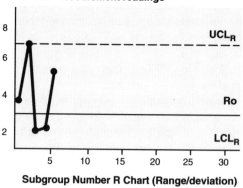

Subgroup Number R Chart (Range/deviation)

Figure 12-10. *SPC measures and graphs. Ongoing measurement tracking will assist in identifying trends going out of specification.*

SUMMARY

Total quality control and the next step, total quality management, are based on three main steps relative to the FMC environment:

1. Eliminate the communication barriers between different functional disciplines, such as engineering and manufacturing. This can be implemented through cross-functional quality teams centered on products or product lines.
2. Measure these teams against quantifiable measurements. *All* members of the work force should be included and monitored against nontraditional measurements other than defects per million.
3. Motivate teams to strive for continuous improvement. Quality is not a goal to be achieved, but a never-ending process.

Effective quality programs require full top management support. This means going beyond the annual quality "pick me up" speech and award banquet. Every quality program has the same fundamental elements; the key characteristics include an expression of the company's quality philosophy, an assessment of process capability, a written quality plan, corrective action plans and procedures, a recognition plan, and several management reviews to keep all parties acting, involved, and informed (as illustrated in *Figure 12-11*).

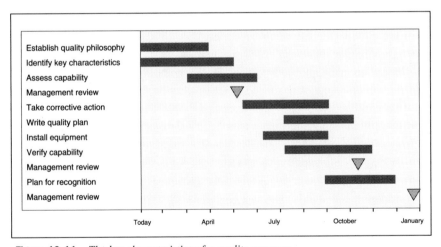

Figure 12-11. *The key characteristics of a quality program.*

In short, individual commitment to quality produces quality. When processes, fixtures, tools, procedures, and vendors break down or fail, only people can make them right again. An FMC, with its collection of machines to finish a part or a series of work steps, provides an ideal environment for *people* to actually take control of quality.

ENDNOTES

1. Vansina, Leopold S., "Total Quality Control: An Overall Organizational Improvement Strategy," *National Productivity Review*, Vol. 9, No. 1, Winter 1989/1990, p. 63.
2. Rohan, Thomas M., "Sermons Fall on Deaf Ears," *Industry Week*, November 20, 1989, pp. 35-36.
3. Rohan, Thomas M., "Services Win Baldrige," *Industry Week*, November 5, 1990, pp. 52-53.
4. Bajaria, Hans J., "A Win-Win Strategy for Supplier Quality," *Manufacturing Engineering*, April 1990, p. 10.
5. Ibid.

IIIII 13 IIIII

CONTROL SYSTEMS

This chapter discusses the role of systems in flexible manufacturing cells. It focuses on the three levels of systems support:

- Business and engineering systems;
- Cell control systems;
- Machine controls.

Emphasis is placed on the design and integration of these levels in support of cells. A schematic of the three levels is shown in *Figure 13-1*.

Manufacturing cell systems are an important part of every cell design and implementation project. If the systems are not emphasized and not clearly specified, the result is the installation of systems which do not support the FMC. When the wrong systems are installed, many of the benefits of cells are negated, and the projected benefits, especially in overhead reductions, will never be realized.

When elaborate systems are installed in places where simple, inexpensive techniques are more appropriate, time and money will be spent administering a system that is beyond the cell's needs. Symptoms of this include:

- Additional shop expediters and schedulers to manage rapidly changing schedules;
- Extra inventory to compensate for system complexities in scheduling or planning;
- Added data processing costs for unnecessary reports and programs;
- The inability to operate over an extended period of time without having data processing staff intervention.

If controls are too simple, however, the cell operators must create or adapt informal techniques to offset the lack of "formal" tools that should have been put in place. This situation is evident by:

- Manually created shortages lists identifying assembly needs and component availability;

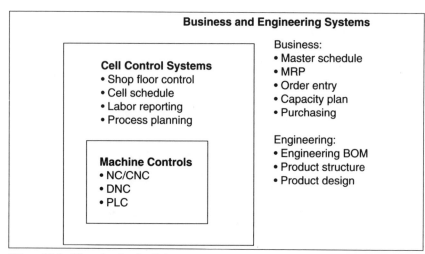

Figure 13-1. *Three levels of cellular support.*

- "Hot lists" which supplement formal dispatch lists;
- Handwritten requisitions required for inaccurate bills of material;
- Individuals' notebooks filled with personal notes on setup, routings, locations, tools, etc.

To avoid developing a system which does not fit cellular manufacturing, clear functional criteria must be established. The systems must be designed to meet those criteria. The ranking and evaluation of these criteria will lead not only to the proper level of automation, but also to the necessary level of integration between the three levels. The result can range from the levels being tied together with simple, manual methods, to the use of complex mainframe computer systems. Other scenarios include combining computer links for some functions, with manual links for others. For example, the sequence of jobs to be run in a cell may be developed manually, but downloaded to the machine through DNC links.

In some instances, cell control systems and machine controls might be tied together directly or use common hardware and software, as in the case of Mazak controls or DEC VAX computers. There are systems combining the requirements of business systems and cell control systems by implementing shop-wide scheduling via MRP modules. Determining the level at which controls reside, what data and information is appropriate at each, and how they are divided and integrated among functions comes out of the team's task of designing the system. The first step, however, is to identify what must be controlled.

WHAT NEEDS TO BE CONTROLLED?

The first question answered by the multifunctional design team specifying the system to support cellular manufacturing should be: "What needs to be controlled?'' The starting point *must* be the cell operating parameters. The philosophies of cell design and the technical makeup of the cell must be understood to select effective support systems. A cell for high-volume, low-variety production with manually controlled machines will need different support systems than one constructed with CNC machines for higher variety at lower volumes. An example for each level follows:

- Business and engineering systems—Control of raw materials and finished product can be accomplished through inventory management and material planning systems that are a part of MRP-based business software. Manufacturing planning and procurement policies must be known by the design team before the parameters can be set up within this software. If, for example, the plan is to stock raw material and finished product in the cell, the material procurement quantities and finished goods quantities must be known. Otherwise, they will not complement the physical design of the cell.
- Cell control systems—Cell control may simply be work sequencing or daily production rates. For example, in cells producing many products, sequencing the work can be very important to minimizing setup time when changing over.
- Machine control—Machines will have either manual controls or numeric controls. Generally, the cell control systems designer will accept the existing machine controls as a ''given,'' unless there is a compelling reason to consider retrofitting machines for the cell, such as substantial anticipated benefits. Upgrading machine controls is usually a costly option.

What follows are the details an FMC planner should consider during the design stages.

BUSINESS AND ENGINEERING SYSTEMS

Today's fully integrated management information systems contain modules for many applications (*Figure 13-2*), including support functions such as human resources and marketing. With the advent of computer-integrated manufacturing (CIM), the need for integrating the management information system to the engineering system (and the cell and machine levels) becomes apparent. This section discusses the design of these business systems, their applicability and adaptability to flexible manufacturing cells, and how they lead to the design of the lower level systems. These systems can be set up on all sizes of computer hardware, from personal computers to the largest of mainframes. The systems arrangement is dependent upon the complexity and size of the business.

The discussion about the business system side of CIM will focus only on those modules falling under ''MRP-based manufacturing logistics.'' Central to this is an MRP component inventory planning module which goes through a requirement ''netting'' process. It is closely associated with and integrated with the other modules of order entry, forecasting, master scheduling, rough-cut capacity

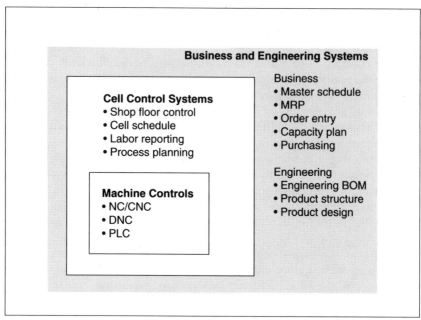

Figure 13-2. *Business and engineering systems contain numerous modules for design, process, and information management that can be applied to the company, including its support functions.*

planning, and purchasing. The discussion about the engineering input focuses on the bill of materials, which is a primary input to the MRP logic.

Basic Structure Of A Business System

The logic built into a typical MRP-type system starts with a set of forecasted orders and the real customer orders. From these inputs, a master production schedule is created. The master scheduling task is often aided by a rough-cut capacity planning module. The MRP "explosion" affects the timing of the release of purchase orders for procurement of raw material and components. It also impacts production orders based on planned lead times and predetermined lot sizes. Planned order releases from the MRP inventory system, combined with process and routing data, are used to produce machine load information. Orders are released down to the cell control system. The factory scheduling module uses the MRP due date as a means for providing priority to orders sequenced through the factory in competition for limited resources. This order process is illustrated in *Figure 13-3*.

With this basic structure in mind, it is now possible to look at the design of a system for cellular manufacturing and how well it can be adapted to the MRP logic. The designers should be part of the design team providing the final cell layouts, since the system design will be heavily impacted by the physical cell layout.

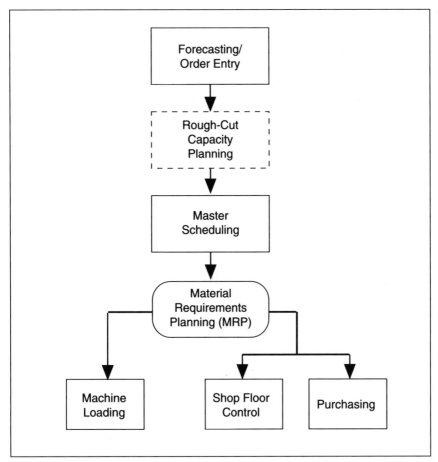

Figure 13-3. *Ordering processing sequence.*

Elements Of The Business And Engineering Systems

Demand: order entry and forecasting. The driving input to the business system is demand. It is entered into the system either as firm demand via the order entry module, or as forecasted demand via the forecasting module. Forecasted demand is based on expected orders.

The reason forecasted demand is an input is because material procurement lead times or manufacturing lead times are often longer than the quoted customer delivery lead times. Therefore, material procurement and/or manufacturing operations must be in process prior to receiving a firm order. Unfortunately, this introduces a level of uncertainty. It is seldom possible to know exactly what customers will want prior to their orders.

One of the reasons for going to cellular manufacturing is to shorten manufacturing lead times. This provides a manufacturer with the ability to "make to order." When this is achieved, the forecasted demand element of the system disappears and this part of the business system becomes much easier to manage. When designing the system to support the cells, forecasting should be downplayed and reliance on actual orders emphasized.

The output from this activity is a delivery plan which reflects the delivery dates promised to the customers. Next, a "preliminary" master schedule is produced which reflects the exact delivery plan. This preliminary master schedule is tested against capacity constraints. If the schedule is unachievable due to lack of capacity, a "revised" master schedule is created and tested against capacity. These iterations continue until a "final" master schedule is set which meets capacity considerations. Master scheduling and capacity planning are closely linked.

Master scheduling. The order entry system provides to the master scheduling system the delivery plan based on customer orders received, and if necessary, forecasted orders. When the flexible manufacturing cells are designed to take a product through to its completion, this delivery plan should be organized to include such data as order numbers, part numbers, quantities ordered, promise dates, and sales values.

The master scheduling system will backward schedule from an order's promised delivery date to the primary activities. The master schedule is the material, manufacturing, and assembly plan. It determines when material and tooling must be available, and when components must be completed to be assembled and packaged. The master schedule must be frozen at a specific advance time and changed only in special circumstances (when customer demand changes).

Based on the master schedule, the system can provide information on the expected value of shipments per week. It will also be the source of measuring the performance of the materials and operations functions, based on how well they executed the master schedule.

The production function must not dictate when something will be made. Work will not be released to a flexible manufacturing cell in order to keep producing. (This work will only get in the way and cost money to hold in inventory.) If a master schedule activity date is missed, the reason can be determined by looking for a problem in production, material, etc., and taking steps to solve the problem.

The master scheduling module releases the master schedule to the cell control system. It becomes the responsibility of the cell to sequence work to meet the master schedule. Performance measures must be in place to discourage the cell from working behind or ahead of schedule.

To better describe how the master scheduling module should be designed to fit into cellular manufacturing, let's follow a hypothetical order through the system:

1. An order is received by sales from a customer and contains the part numbers, quantities ordered, due dates, and sales value.
2. Sales adds the order information to the order entry module, creating a delivery plan integrated with the master scheduling module.

3. The master scheduling module backward schedules from the delivery plan date the functions needed to complete the order—for example, design engineering, material procurement, and manufacturing. This backward scheduling can also take current cell and support staff loads into account. If the functions do not have the resources to meet the schedule, alternatives can be explored to increase resources (adding machine capacity or staff) to accommodate the capacity constraint.

4. This master schedule is frozen. (With cellular manufacturing and reduced lead times, there will be fewer order changes because customers will not have to plan so far in advance. They can place their orders nearer to the time the product is needed and be assured that their needs will not change.)

5. This firm master schedule then gives the timing of general activities to be performed by the various functions. The detailed planning of the order can then take place. For example, with the definition of the manufacturing time frame for an order, the work can be released to the cell control system, which can detail schedule and sequence jobs based on minimum setup time and other criteria.

6. Through integration (either manual or automated) with the cell control system, the business system can measure (after the execution of the master schedule) what was actually done versus what was planned.

The master scheduling module also can provide the sales function with time periods in which cell loads are planned to be at maximum capacity. Therefore, salespeople should not promise to the customers any additional product delivery in those weeks.

Rough-cut capacity planning. Capacity planning supports the master scheduling module. A cell has a finite amount of time available, so when the load on the cell exceeds the time available, a capacity constraint exists. Often, there are capacity constraints (bottleneck machines or people) that do not allow an order to be completed just prior to the customer's promised date. Capacity planning will provide the information that a capacity problem exists, and suggest the means to alleviate the constraint. For example, a decision might be made to schedule an order early on the master schedule if there is capacity available in an earlier period. Also, it may be possible to work overtime, hire people, buy equipment, or add a shift to increase capacity either temporarily or permanently. The capacity planning module will allow the constraint to be identified and managed early in the process, before the creation of past-due work.

Macro-routings can serve as the base data for capacity planning. They provide, by part number (or family of parts), the estimated times required to set up and produce a part. For example, in the shop there can be groups of macro-routings by cell which include average operation times for just bottleneck machines.

A simple example of a macro-routing will help illustrate its usefulness. Suppose there is a manufacturing cell designed to make a family of widgets (*Figure 13-4*). It operates on a one-shift, 40 hour per week schedule. The cell has three work centers, A, B, and C. Every widget goes through machine B.

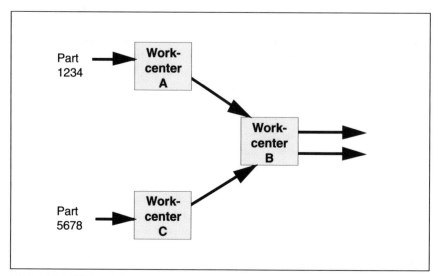

Figure 13-4. *Widget cell part sequencing and flow. Operation B is the bottleneck.*

Machine B is the slowest operation for both setup and run. Therefore, B is the bottleneck operation in the cell—that is, it controls the output of the cell. A simple, manageable way of handling rough-cut capacity planning would be to create a macro-routing for each widget in the cell, but include only information about the bottleneck work center, B, on the macro-routing (*Figure 13-5*).

Now suppose for a particular week there were orders for five of part number 1234 and 10 of part number 5678. The rough-cut capacity planning module would refer to the macro-routings and make the calculations shown in *Figure 13-6*. The cell would be at 30 of the available 40 hours, or be at 75% of capacity; this does not create a master scheduling problem and enables sales to continue to make promises for widget manufacturing in this week (up to 10 hours' additional capacity).

Put another way, the rough-cut capacity planning activity uses the master-scheduled dates and quantities of items, and macro-routing with activity times to output a capacity plan showing constrained resources. Therefore, the capacity planning module must maintain available capacity per cell (as opposed to capacity per work center, which the MRP-based systems are typically geared to use), part numbers assigned to a cell, macro-routings, load within a cell by period, and cells with maximum load.

Bills of material. Once the master schedule is in place based on the demand and the capacity to meet demand, the concern is providing the material required to meet the schedule. The first step is for the system to take the master-scheduled due dates for the products along with the product quantities, and "explode" them through the bills of material. The process of exploding through the bill of materials is where the MRP component inventory planning module creates

MACRO-ROUTING

Part Number: 1234   Cell: Widget

Control Operation(s)	Operation Number	Work-center	Average Operation Time	Average Setup Time
Threading	20	B	1 hour	2 hours

MACRO-ROUTING

Part Number: 5678   Cell: Widget

Control Operation(s)	Operation Number	Work-center	Average Operation Time	Average Setup Time
Threading	30	B	2 hours	3 hours

Figure 13-5. *An example of macro-routing, where both parts need only reflect routing through the bottleneck work center, B.*

Part number 1234: (5 parts x 1 hour) + 2 hours setup	7 hours
Part number 5678: (10 parts x 2 hours) + 3 hours setup	23 hours
TOTAL	**30 hours**

Figure 13-6. *Macro-routing calculation example.*

requirements. This is done by accessing the bill of material for the master-scheduled demand, taking each line item on the bill, and multiplying the master-scheduled quantities by the "quantity per build" on the bill.

Typically, a bill of material is developed as a part of the product design on the engineering system. Some type of integration must take place to pass the information to the business system. Most computer-aided design software packages provide modules for passing bill of material data directly to the MRP system.

On a bill of material, each line (part number) can be associated with numerous fields, such as part description and quantity per build, or the total number of components of a specific part number required. There also can be a field specifying the type of material control. For example, material may be controlled with a Kanban system, it may be a purchased component handled through purchasing, or it may be raw material requiring value-adding operations. (Kanban is discussed in the cell control section of this chapter.)

Kanban parts are controlled via a two-bin system. Controlling these parts as part of an MRP system defeats the benefits of managing them as Kanban parts. The types of parts which should be controlled by MRP are:

- Purchased finished components that are not Kanban-controlled, and purchased raw material;
- Work internal to the company but coming from outside of the cell, such as work from within the same plant or from a different plant.

An important step toward integrating cellular manufacturing with the MRP system takes place during the design. This is when every line item on the bill of materials is evaluated to determine the appropriate type of material control. When executing MRP, the system will not be concerned with requirements for Kanban parts; it will track the requirements of those parts designated on the bill of materials as needing control.

Netting: MRP component inventory planning. Once the line items of controlled materials for all master-scheduled products have been identified from the bills of material, and the MRP component inventory planning module has multiplied the master-scheduled demand quantities by the "quantity per build" on the bills of material to create requirements, the requirements across all products are combined. This is called "netting." It is necessary because there are common parts between different types of products. (If there were no part commonality, the use of MRP would be in question.)

The result is a material requirement by part number for all products currently demanded. (This requirement does not include those parts designated as Kanban.) MRP looks out into the future, comparing the requirements to the current inventory plus the parts already on order. MRP determines which demand is not covered—that is, *when* the next lot of material needs to be ordered and *how much* material is needed.

Purchasing. There are two ways the business system can approach purchasing activities. The design team will have to decide the most appropriate approach for the business and specify it. After MRP explosion identifies a need for material and determines that the need is not covered by current inventory, purchase requisitions can be prepared.

In the first approach, the purchase requisitions can provide the exact quantity needed to cover the requirement. Then the requisition goes to purchasing. It is possible to design this task to be done on-line. Purchasing determines the lot quantity and specifics of the order, then releases the purchase order to the supplier.

Alternatively, prior to producing the purchase requisition, the MRP system can access the "item master" or "order rules" to determine the procurement and delivery procedures. These procedures can include the lot quantity, point-of-use delivery, and other specifics of the order. Then, the system could provide the purchase order, and all the purchasing department would have to do is mail it.

The second approach is more efficient, requiring less manual intervention. It is well adapted for Just In Time material policies where all the particulars of an order have been established before the material needed has actually been identified. However, manufacturers must beware using this in a non-JIT situation. Businesses have adopted this approach carelessly too often. For example, the item master may specify ordering a particular high-volume part in quantities of perhaps 100 pieces. Then demand for this part may have dropped significantly and the MRP system produces a requirement of two pieces over one month. If the system automatically orders 100 pieces, then obviously inventory goes up to 50 months of usage, and high inventory costs follow. When moving to flexible cells, purchase requisitions are generally handled both ways, with the intent of moving toward their automated release.

In relation to purchasing, the idea of lot quantities requires some explanation. Traditionally, when talking about lot quantities in the context of purchasing, economic order quantity theory was used to determine how much should be ordered. The basic premise of the theory is that there is a cost associated with producing a purchase order; there is also a cost associated with holding inventory. Economic lot size calculations balanced these two costs to produce a quantity that would provide the lowest cost. It sounds good in theory. In practice, however, inventories frequently grew. This was because:

- The economic order quantities were often larger than the known demand. If demand decreased or dried up, the company was left holding the inventory;
- It was difficult to know the real cost of processing a purchase order;
- The cost of holding inventory was usually understated by more than 50%, forcing the lot quantities to be too large.

To get away from these problems, businesses need to concentrate only on procuring material for which there is a firm demand. The material should arrive just before it is needed—that is, Just In Time. But what about the cost to process a purchase order? Much attention has been put on reducing this cost to the point where it is of no concern. This is done through prearranged partnerships with vendors, where the same vendor is always used and agreements have been made for small quantities and frequent deliveries. Companies have also set up electronic data interfaces (EDI) with suppliers, where orders and material supply information are electronically transferred between supplier and customer. Also, purchasing information systems are continually making it easier and less costly to produce purchase orders.

CELL CONTROL SYSTEMS

The objectives of cell control are:

- Sequencing the work through a cell;
- Reducing the material between production nodes;
- Reducing manufacturing lead times;
- Increasing the ability to respond to changes in demand;
- Increasing labor productivity and quality levels;
- Tracking part location.

Collecting cell data, scheduling cells, and managing FMCs use many techniques, ranging from the most sophisticated automated collection devices located on machines, to manual sheets, calculations, and spreadsheets. The key to successfully using either method is to initially specify the need, determine the volume of activity, and then determine the appropriate level of integration and automation.

Cell control systems schedule machines or cells and coordinate the machine controllers within the cells, as shown in *Figure 13-7*. The primary interface between a cell control system and a business system at the highest level of support is via a shop floor control module. This module can be supplied as an integrated part of the management information system, or bought and developed separately. If it is separate, the procedures for integration to the business system must be designed to be handled automatically or manually. This section discusses the design of cell control systems and their integration with business systems:

- Shop floor control—using business system information to run cells and report back to the business system;
- Determining the requirements—the information (and its timing and format) that is needed to make cells run (process plans, tooling requirements, etc.);
- Control points—where to gather the information that management needs fed back to the business system level about the cells' activities;
- Labor reporting and scheduling—control points to capture labor activities specifically;
- Kanban and manually controlled cell support systems—simple techniques to control manufacturing;
- Systems developed with personal computers—custom options more easily available;
- Simulation to help with the design—modelling cells to determine schedules.

Shop Floor Control

From a systems perspective, shop floor control is that activity which takes the business information and translates it into the information used to run a cell. It also takes information that is created at the cell level and passes it back up to the business system. For example, one of the functions of shop floor control is to

```
                Business and Engineering Systems

  ┌─────────────────────────────────┐   Business
  │                                 │   • Master schedule
  │      Cell Control Systems       │   • MRP
  │      • Shop floor control       │   • Order entry
  │      • Cell schedule            │   • Capacity plan
  │      • Labor reporting          │   • Purchasing
  │      • Process planning         │
  │                                 │   Engineering
  │   ┌─────────────────────────┐   │   • Engineering BOM
  │   │                         │   │   • Product structure
  │   │   Machine Controls      │   │   • Product design
  │   │   • NC/CNC              │   │
  │   │   • DNC                │   │
  │   │   • PLC                │   │
  │   │                         │   │
  │   └─────────────────────────┘   │
  │                                 │
  └─────────────────────────────────┘
```

Figure 13-7. Cell control systems schedule machines or cells and coordinate the machine controllers within the cells.

provide trend and performance information to management relating to daily labor and cost status, performance to schedule, dispatching effectiveness, and quality performance:

- Labor and cost reporting is the process of collecting information on the amount of time spent to produce products. This is utilized for financial valuation of inventory and orders, payment of factory employees, and distribution of overhead costs (if absorption accounting practices are being used).
- Sequencing and dispatching are required to manage and control the shop floor while ensuring customer demands are met. These can be part of a stand-alone system or part of a mainframe MRP-based package. Master-scheduled orders are assigned to cells, then sequenced within the shop floor control module based on processing times and setup changeovers. Work center loads can be calculated based on standard or projected hours.
- Quality data can be accumulated using automated devices located at machines or at stand-alone stations, or it can be manually collected and input through terminals. Some typical techniques of data accumulation include bar coding, and visual and mechanical sensing devices. Each can be hooked directly to numerous types of computers that can store data and pass it to other systems. For example, once collected, quality data can be compared to print specifications in a CAD system or be analyzed as part of a quality control program.

211

When designing the shop floor control requirements and uses, commercial vendors can help by providing specifications and valuable insight to appropriate use of their products; vendors who have experience in this process can be a key resource. Generally, this software is on the low end of the cost scale.

Determining the Requirements

The cell control system informational requirements will vary from cell to cell, even within the same factory. Each cell must be individually evaluated to determine what is required. The starting point for the design of cell controls is the cell's informational needs:

- What objectives are to be met:
  - Cost;
  - Quality;
  - Timeliness.
- What needs to be produced?
- When is it needed?
- What materials are required and when do they need to be available?
- What tools, fixtures, and other manufacturing devices are required?

After the identification of the informational needs, a design specification can be produced for integration with the business systems. Typically, it specifies the integration of schedules—that is, taking information from programs that generate schedules (such as the master schedule) from MRP or other business-related systems. A planning module which includes master scheduling and rough-cut capacity planning should be employed for order release to the FMC. The FMC is responsible for sequencing orders, and taking into account tooling and material availability and the time required to go from setup to setup. Also, some leeway can be built into the schedule, which permits flexibility for handling periodic disruptions at the cell level.

This planning module will communicate with its counterpart at the higher (business) and lower (cell or machine) level. Initially, this will require some communication interfacing. The key to building a global system is deciding *what* information should be passed between different levels. Close attention to information needs will limit performance problems and provide greater accuracies in planning, scheduling, and reaction time.

Information about what needs to be produced and when is usually provided through production control or dispatched from the shop floor control system. In dedicated part cells, schedules can be as simple as daily production rates. In cells that make many different parts, schedules may be published weekly, daily, hourly, or even be presented in real-time on a computer terminal. With a pull system, the schedule can even be determined by receipt of a tag, or an empty space on the floor or in a storage stacker.

Determining the frequency and makeup of schedules depends on the characteristics of the cell:

- Cells making complicated parts with long manufacturing cycles will need

212

schedules less frequently than cells making simple, high-variety parts in short manufacturing cycles.

- Dedicated cells may only need to know daily production rates. Those that make different engineered parts will need to know part numbers, quantities, revision levels, tooling lists, testing requirements, and material requirements.

Meeting a schedule is contingent on the availability of raw materials such as steel, bar stock, or castings. Besides raw materials, there might be supplies required, like paint, chemicals, welding materials, etc., that must be manipulated or preordered. Just as important as the actual production equipment involved are the peripheral manufacturing devices like holding fixtures, weld jigs, expendable tooling, dies, or handling mechanisms. The cell system must provide data relative to the location, availability, condition, and capacity of all of these peripherals. The current documentation (drawings and specifications) at the correct engineering change level is also a requirement for meeting the schedule.

Also integral to design are control considerations such as transportation restrictions and actual physical handling mechanisms. Developing all of these features during the design of the cell system will provide the selection criteria for the appropriate type of cell control and control points. Systems rely on control points to gather data for management. This includes cell-level data and overall business information such as cost, schedule, inventory, and order status data. Cell scheduling requires:

- Definition of scheduling control points;
- Appropriate labor reporting.

Control Points

A material or labor control point is defined as any point or location within manufacturing operations where management needs to gather data, manage inventory, or schedule or control a process. As part of a cell design project, it is essential that thought be given to the locations of control points; they dictate where communication links and reporting procedures are required. Examples of control points include:

- The entry point of a cell where raw materials are either purchased from the outside, requisitioned from stores, or received from another cell or part of the factory.
- Any point in the cell where labor reporting is required to accumulate product costs or manufacturing order status.
- Any point inside the cell where products go "off line" and need discrete identification, such as to outside plating or heat treat. In this case, another level of the bill of materials or routing operation may not be required if the control systems can effectively handle outside operation subcontracting.
- The exit point of a cell when identification is needed to store or ship the parts or products.

The number one criteria in determining where to put control points are the minimum information requirements. Additionally, when long distances between operations create the need to batch materials, a control point may be required. When operations are close together and there is complete physical control of parts (such as dedicated material movement devices or conveyors), control points should be minimized. When cells are introduced and routings become much simpler to maintain, most typical process manufacturing routings that delineate movement in and out of a department, workstation, or stores (usually for work center labor reporting) become unnecessary. An example of how the number of control points is reduced with cells is depicted in *Figure 13-8*.

Historically, most factories have emphasized control points because management wanted to know what was where and when. In complex batch manufacturing environments, where there are large amounts of material to be accounted

Old Batch Routing		New Cell Routing	
Operation	Description	Operation	Description
10	Receive	10	Receive/inspect
20	Move to store	20	Move to cell—manufacture complete
30	Move to inspection	30	Pack and ship from cell
40	Inspect		
50	Move to weld		
60	Weld		
70	Move to store		
80	Move to rivet		
90	Rivet		
100	Move to assembly		
110	Assemble		
120	Inspect		
130	Move to shipping		
140	Pack and ship		

Figure 13-8. *Routings become simpler in cellular manufacturing.*

for at any one time, this system is indispensable. In the FMC arena, however, material handling is limited to virtually one spot, and control points can be reduced by as much as 95%.

Labor Reporting

Labor reporting can be a difficult element to manage at the manufacturing cell control level in terms of locating the control points and determining how the data is to be collected. Labor reporting has far-reaching implications in cost accounting, labor management, human resources, planning, and customer service. Its information requirements must be understood by the cell control designer; involvement from other departments who will use this information and data is essential, as is taking a look at reducing reporting activities to those that are essential. The data typically includes:

- Inputs to cost of sales for cost accounting function;
- Time, attendance, and quantity of work (if using incentives) for payroll;
- Scrap and reject quantities for quality and replacement requirements;
- Orders shipped for accounts receivable;
- Part status and inventory levels for production control;
- Bottleneck, capacity, and product mix data for planning.

Minimizing work order reporting for order locations within FMCs means all parts should be considered as being in one location—the cell. Concentrating on simplicity by making the transition from individual worker reporting to team reporting, and from individual operation reporting to cell reporting means *one* "cell" work center control point.

Kanban and Manually Controlled Cell Support Systems

There are highly effective methods for controlling cellular manufacturing which are also simple to conceptualize and implement. In well-designed cells, simple techniques include manual, physical, or visual practices. The most widely known manual control technique is Kanban.

Kanban was developed in Japan during the 1950s for the control of parts and material between production nodes on the shop floor. The objectives of Kanban are to:

- Reduce the amount of material between production nodes;
- Reduce manufacturing lead times;
- Increase the ability to respond to changes in demand;
- Increase labor productivity.

There are several types of Kanban systems; one can be employed in almost all production situations. The common element of all Kanbans is a manual signal (usually an empty container) which is used to *pull* a specific amount of material to the point of consumption. Unlike MRP-based scheduling systems, which link material movement to a production order, Kanban systems move material from a point of supply to a point of production *only when* the material ahead of it is consumed. The consumption of a container of parts triggers further replenishment.

Kanban is a powerful tool for the control of the flow of goods through and between cells when certain conditions exist. Most of the preconditions are the same for Kanban as for cellular production, which is one of the reasons that Kanban is suitable for FMCs. Because most of these conditions have been explored in depth elsewhere in the book, they are only briefly mentioned here:

- Smooth, stable demand for the products and parts is essential if Kanban is going to be successfully employed. These are typically the parts with high usage.
- The layout of production facilities must be such that the physical distance of production (the consumers) from the suppliers is short.
- The process must be appropriate. Continuous process industries like steel and wire production, with no breaks in the introduction of raw material to completion of the product, obviously have no need for Kanban control.
- Kanban control is predicated on the desire and motivation of the production workers to make the system work. It requires training and education, team spirit, and a reward system that presents a payoff for using it as it was intended.
- The basic principle of Kanban—suppliers moving products directly to consumers—presumes that the quality of the product is impeccable. High quality is as important in internal supplier-customer relationships as is the qualification of external suppliers.
- The movement to Kanban is ineffective unless manufacturing lead times are reduced by reducing the setup times. Otherwise, the Kanban size is unmanageable. The basic rule of thumb is that there is no more than five days' supply in Kanban. This cannot be achieved without substantial reduction in setup time.

Kanban should be used where possible, but it is not appropriate or desirable for all parts, products, or industries. In most cases, material control and movement is best accomplished with some hybrid system utilizing Kanban control of high-demand parts through FMSs, and MRP control of less utilized parts, processes, and products.

Several other simple floor control systems, such as visual methods (sometimes called Andon Systems) can be used. These are applications where one manufacturing area signals its supplier with visual aids like colored signal lights when it needs the next supply of material. Some incremental station assembly lines will have confirmation signals; when all stations confirm their status as complete, the entire assembly line moves one station forward.

Systems Developed with Personal Computers

Steadily improving personal computer technology has increased the options for systems to support cells. Coupled with these technical transformations is the availability of a new generation of machine-level controllers featuring connectivity and programmability by plant-floor operators and technicians. The connections are adaptable to nearly any machine (machine tools, robots, assembly equipment, special machines, etc.). As a result, cell control systems will become major sources of product tracking information and data collection. Other compatibility can be achieved with operating systems, languages, software tools, and communications. All of these are becoming more modular in nature, allowing them to be phased into the operating arena. This capability has lasting benefits, as factory technicians can configure software to accommodate hardware modifications, such as adding a next-generation controller at the machine level.

Simulation to Help with the Design

Modeling cells using simulation software (e.g., GPSS, WITNESS, SIMON) will assist in designing the scheduling module. The base data used in the simulation must be accurate and complete. Experience has shown that much existing data in manufacturing facilities is out of date, inaccurate, or not appropriate for cells. To use simulation correctly, process planning data (using the planned cell process, not the existing process) must be provided or developed. Various product mix and production level scenarios can then be simulated to show where bottlenecks exist; the schedule must address these constraints. Other key information provided by simulation includes average throughput time, inventory levels, and minimum, maximum, and average sizes of buffers between operations.

MACHINE CONTROL

The primary interface between a cell system and machine control at the lowest level (except for manual control) is numerical control, as shown in *Figure 13-9*. This section defines numerical control and other types of machine control for FMCs, and discusses the design of machine controls and their integration to cell systems. Included are:

- Numerical control (NC);
- Computer numerical control (CNC);
- Distributive numerical control (DNC),
- Programmable logic control (PLC).

Numerical Control

NC is the interface between programmed manufacturing instructions and the operation or functions of a machine tool. In the manufacturing environment, NC evolved from simple two-axis motion machine tools with controls that required tape and tape readers. As a tape was "read," each instruction or block of data was executed. These simple controllers did only point-to-point movement, were slow, and allowed no flexibility at the machine tool. High programming skills were important for smooth operation of the NC machine. Initially, these control units were "hard-wired" numerical controllers. Data was usually read from punched tape by the control unit, with signals sent to the machine for execution.

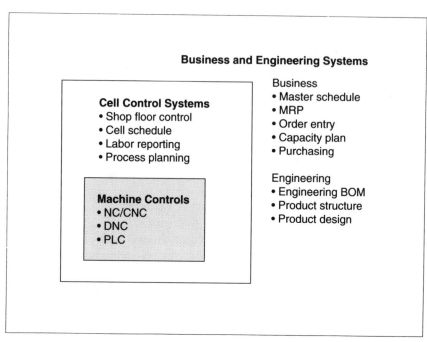

Business and Engineering Systems

Cell Control Systems
- Shop floor control
- Cell schedule
- Labor reporting
- Process planning

Machine Controls
- NC/CNC
- DNC
- PLC

Business
- Master schedule
- MRP
- Order entry
- Capacity plan
- Purchasing

Engineering
- Engineering BOM
- Product structure
- Product design

Figure 13-9. Machine controls are the local managers of programming instructions.

Older NC machines can create problems in cell applications because often they lack the ability to communicate with other computer sources. Also, they cannot be retrofitted for DNC applications. They are dependent upon manual tapes, including the cost to punch, carefully store, and maintain these tapes. Typically, these 1960s machines are expensive to maintain due to lack of parts and experienced maintenance personnel. Machines may be considered for a retrofit if:

- There is a need to standardize on machine control units for efficiencies in NC programming or maintenance.
- Obsolete controllers (hard-wired with relay logic) need replacing or are no longer supported.
- Controllers are not compatible or are incapable of sending and receiving data as simple as "cycle on" or "cycle complete."

Computer Numerical Control

Today's machine controllers are computer operated, and have the ability to store and activate programs from computer memory. In applications within FMCs, these CNCs have bidirectional communication capability to report alarms and provide status to the cell controller. Since the early 1970s, controllers have become much more sophisticated. Programs are input, accessed, and executed from memory quickly, not block by block. These controllers have evolved from 16-bit to 32-bit processors with much faster executable speed. This allows the control of multiaxes movement in a continuous mode not previously possible with NC. This continuous motion allows cuts (linear, circular, and rotational) to be generated accurately with controlled synchronization of axes.

In addition, editing functions eliminate punching new tapes and re-inputting them into controls. Today's CNCs allow storage of executive programs—that is, operating parameters and instructions for the specific machine tool and part programs, as well as editing and tool offset data. Often, these controllers contain maintenance and self-diagnostic data, stored and executed by both operating and maintenance personnel.

As industry moves toward cellular manufacturing, the use of CNC is necessary. CNCs outperform manual equipment through a significant reduction in run time; CNC equipment is typically three to six times faster. Probably even more important is the quickness of part-to-part setup time—flexibility is gained with the faster exchange of part and tooling information. Therefore, it is important to evaluate those CNC features that assist in achieving optimal, efficient performance. Here are eight functions to consider during design:

- Program storage—storing of all NC part programs, executive programs, tooling data, and editing and diagnostics programs. If size becomes exorbitant, consider DNC options for electronic data transfer: RS-232 or RS-422 for distances over 100 feet.
- Tool probing and quality applications—probing, then signaling machine or cell controllers about out-of-tolerance conditions, selection of a redundant tool, verification of a part in a fixture, etc.
- Tool management—indicating for tool/pallet coding, tool availability and location, tool setup, tool life/status, and pallet/part identification.
- Production status—monitoring on/off, production, prove-out, setup, training, maintenance, or nonproduction status.
- Start/stop and operator messages—signaling instructional data to operators for setup, prove-out, and inspection requirements.
- Parametric and decision logic—using "if-then" logic such as stock check to remove more stock or to eliminate cuts as required, verifying program/ pallet mismatch that would abort run, signaling operators if parts are not secured properly, checking if all parts are mounted in multipart fixtures (omit machining, eliminate cutting air), and using family part programming capability.
- Adaptive control—monitoring, then adjusting speed and feed control due to material/cutter condition.
- Machine/system diagnostics—troubleshooting for maintenance.

Direct and Distributive Numerical Control

The first generation of DNC—*direct* numerical control—was a forerunner to CNC, where controllers at the machine tool received their NC data electronically from a centralized computer. This computer had "direct links" to the shop floor-based machines. Today it is hard to find DNC operating as it was introduced in the early 1970s. CNC caused the demise of direct control, as costs for both hardware and software decreased and availability of "hardened" computers became more prevalent in the shop.

The second generation of DNC—now *distributive* numerical control—utilizes electronic transfer of manufacturing data (NC programs, tooling and scheduling information, etc.). Data can be bidirectional, from a business system computer or cell control systems, or individual PCs using modems. This system only distributes data, usually to CNC machines. It does not *control* them. This distributive system is shown in *Figure 13-10*.

Distributive transfer replaces the NC tape, the need for the NC punch, and the loading of punched tape via a tape reader into an NC or CNC unit. It potentially eliminates the need to manually input tool identifications, locations in the specific machine tool, offsets (both fixture and tools), and parameters for adaptive control. For example, file transfer software for network installations such as DECNET or protocol converter software such as KERMIT and BLAST are often used in the application of electronic transfer (uploading and down-loading) of data. Additionally, a cell controller which has process translation capability can be utilized to "distribute" data to or from other machine controllers. DNC systems should be considered for cell applications to eliminate or reduce the time lost due to errors in loading, debugging, storing, and retrieving NC data.

Programmable Logic Controllers

A programmable logic controller (PLC) is a general-purpose system that will accept and react to input from a variety of sensors such as limit switches or push buttons. PLCs can be integrated as a part of a machine control system or used as a full-function machine tool controller. Most PLC vendors provide axis control modules that plug into the input/output back plane of the PLC controller. Some controls can handle up to eight axes. Axis control is a part of the standard ladder logic functions. This can allow the PLC to control all functions of a machine tool, including auxiliary functions such as load/unload commands, pallet tracking, and coolant on/off. Examples of PLC applications within cells are:

- Material handling control from conveyors to robots to machines within the cell;
- Replacements for electromechanical relay systems now used on older machines;
- Ancillary functions of CNC machines such as tool changer motions, coolant and chip removal systems, or part pallet exchanges.

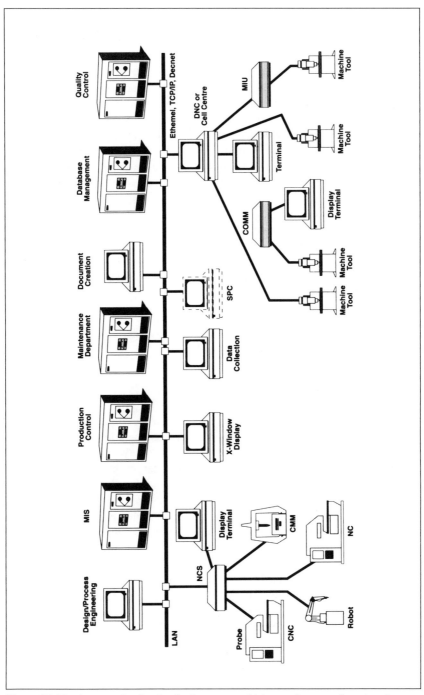

Figure 13-10. *An example of a distributive system which transfers manufacturing data electronically.*

221

SUMMARY

Cell implementation requires more than just the physical movement of machines and products. It has a major effect on office operations and corporate systems. Interaction and changes required in these other areas of the business must be coordinated with the physical moves of machines. Therefore, as part of a cell design, an implementation plan that allows adequate time and effort for design of support systems is essential. It must include the time for systems concepts to be communicated, learned, and adopted throughout the company.

A good first step is to document the existing flow of information and reports among and between functions, whether the flow be automated or manual. Then flows in support of cellular manufacturing can be developed, specifying reports and data requirements. From this comes a functional specification and criteria by which to evaluate support systems.

Certain procedures may need reviewing and adjusting as new cells are being installed. Many procedural changes will focus on planning, inventory control, labor reporting, and cost accounting. However, all other areas should be reviewed so nothing is missed. Again, an understanding of the existing methods is required to identify and quantify changes.

As changes are identified, the project team must develop cost-effective solutions and consider:

- New hardware;
- Purchased software;
- Software modifications;
- Operational policy changes,
- Procedure changes.

Each alternative must be analyzed on its own merit, with heavy emphasis on how the solution may affect the project or business operations.

Hardware and software changes can be a major part of a cell project depending on the sophistication and communication being planned. Cell designers need to get business systems personnel involved early to avoid last-minute technical difficulties that can easily stop the project. Up-front planning will identify, prevent, and minimize unforeseen problems.

The common message within the three levels of control—business and engineering systems, cell control system, and machine control—is that no two environments will have the same packaged systems solution. Providing the best system requires extensive analysis, planning, and specification development. However, the rapid evolution of control software in compatibility and user-friendliness have dramatically improved the potential for finding a good systems solution and being able to implement it with ease.

IIII 14 IIII

CONDUCTING DETAIL DESIGN

Several previous chapters have set the stage for beginning the process of detail cell design. Since each of these covered some specific, detailed approaches on their own, it is worthwhile to review these methodologies for formation of cells. In Chapter Five, the emphasis was on overall macro planning of the facility. This was predicated on the assumption that more than just one cell would come into existence. The macro discipline was encouraged at this stage to provide not only the big picture, but to begin the rational formation of an overall direction in thinking. It must be remembered that employee teams include the hourly workforce, who by tradition (and practice) are encouraged to confine their views to the job at hand. Part of the training occurring in the team process breaks down these paradigms and encourages participants to broaden their approaches, being guided by, but reaching beyond, their everyday experiences.

Chapter Six introduced most of the tools and organization that need to take place at the next level of detail. In many organizations, this level of detail would have provided enough proof of the concept to get a comfortable ''go ahead.'' Ultimately though, to ensure success and minimize surprises, the work to create a cellular manufacturing plan must include finite detail. The development of this next level of detail required to accomplish installation is what this chapter is all about.

This chapter will expand on the tools and techniques already introduced. The overall process as pictured in *Figure 14-1* illustrates the evolutionary steps that must usually be followed from the macro view through an installation. In this chapter, the major topics will be:

- Documenting and organizing data. *Tools*: Representative parts (Chapter Five), bills of materials, specially formatted database inquiries, make-versus-buy analyses.
- Transforming volume, process, flow, and proximity into equipment requirements. *Tools*: Process volume charting, proximity analyses, process flow charting, simulation.

223

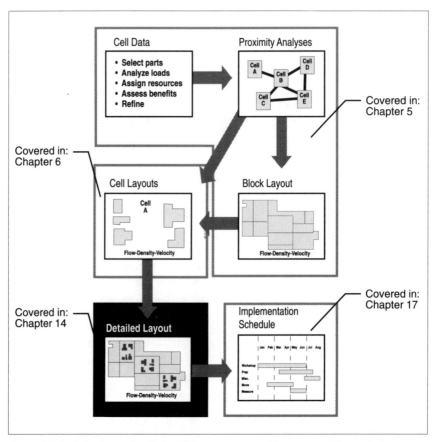

Figure 14-1. *Achieving a detailed layout requires several iterations of the analysis process. Chapter 14 focuses on the detailed layout process.*

- Validating all past assumptions. *Tools*: Assessment diagnostics, facilitated brainstorming sessions, strategic planning review and presentation (Chapters Three and Four).
- Adding the full and necessary complement of equipment to enable complete production. *Tools*: Equipment design and specifications (Chapter 15), technology assessment (Chapter Nine), machine/manning spreadsheets, operator involvement (Chapter Three).
- Developing operating procedures and supporting activities. *Tools*: Implementation plan (Chapter 17), team process, setup reduction, continuous improvement process, scheduling techniques.
- Developing a detailed layout illustrating flow, density, and velocity. *Tools*: Previous cell progressions (Chapters Five and Six), CAD software, part flow charting.

The section on the fundamental principles of *flow*, *density*, and *velocity*, which ultimately drive the proper formation of a production facility, is most important. Most plants and cells come up short of outstanding success when these concepts are not integrated into their design.

ORGANIZE THE RIGHT DATA

In Chapter Six, a number of analytic tools were used to help provide a broad perspective from a sample of parts that the cell was initially conceived to manufacture. This time-saving approach of using *representative parts* must now be expanded. At this level of detail planning, team members must look at all of the parts the cell will produce. Also, it was noted that more than one specific type of cell would probably come into being in an actual facility plan, but again, it is assumed that the cell configurations are product-based. Where specific applications are different, that detail will be highlighted.

The starting point then is to assemble bills of material for the top-level part numbers or assemblies the cell will be producing. Remember to include all parts or assemblies. Since this may constitute a massive undertaking, some form of discipline and checklist is advisable before the team pours through what can often be reams of data. The MIS (management information systems) group can often provide "sort" lists of active parts or assemblies from appropriate, reliable databases. Devising a sort list request is time well spent against the possibility that some part or assembly might be overlooked in the planning. Criteria for sorts might consider part number codes, material, or part descriptions (naming convention). Output lists could include purchased or finished information, last active usage data, and volume information. Whatever the case, a master listing should be established and posted as one of the main "war room" documents so everyone can track progress and be reassured that no component or assembly has been overlooked.

As bills of material are collected (and checked off against the master listing), the team should begin extracting manually or mechanically (again, MIS can be an agent in this process) all parts which have any kind of labor added. At this step it may also prove beneficial to have routed operations, setup times, run times, and operation numbers extracted also. Quite frequently, active organization databases aren't maintained accurately. As is often the case with downsizing of staff and elimination of departments, detail recordkeeping falls by the wayside. The team must beware of missing operations, combined operations, or undocumented routings which actually take place on the shop floor.

Procedures must be set up for team members to audit the documents collected for accurate information. If available database information proves extremely inaccurate or out of date, it may be necessary to create a database from scratch. For the purposes of this analysis, the assumption is that information is available in some form through which hard copies of the pertinent statistics can be collected. The hard copies will be more useful in the detail planning because a master file will be started. While on-line databases are important, many times the data will change and little or no record can be kept of the initial data extracted for the detailed planning. The hard copy master file will document starting points and primary data positions.

As requests are made for the bill of material extracts, a parallel activity of acquiring the blueprints for all of the parts should be initiated. This activity serves as a double check against the master list, while also serving as the hard copy documentation for process planning. As in the case of bills of material, an audit of the documents is in order since blueprint updates are subject to the same inaccuracies as other facility data.

There might be some concern about purchased parts being included in these requests, or that the volume of data can be reduced by excluding them. Given the new manufacturing environment being created, a review of all make-versus-buy decisions is in order (and the opportunity represents additional justification value). It should be recognized right off that some parts now purchased might fit into new cells without any major equipment or manpower additions. Secondly, from a quality perspective, when questions arise about mating part characteristics, either in practice or in the planning, all the documentation would be at hand. Finally, looking at purchased parts gives perspective on how material handling and storage must be accommodated for purchased components, whether they become vertically integrated or not.

MANIPULATE/CONVERT THE DATA

Note in *Figure 14-2* that the macro-level process volume chart developed in Chapter Five has now been expanded to include all the parts and their variants to be produced in the cell. For example, part number 12345 turns out to have three variants: A, B, and C. Note that the volume for all three variants totals 550,000, as was used in the macro approach. It is also evident from a careful examination of *Figure 14-2* that more process volume categories have shown up. Why wasn't this developed with the representative part information? The complete process routing and operation times reveal that what has been detailed is now internal or done simultaneously with the controlling operations that were developed in Chapter Six.

The proximity analysis done at the macro level showed 831,000 units going to assembly from wash (total incoming minus total outgoing to other operations). But the detail operation routings now show that all these parts need an oil dip. This is done internal to the wash cycle, but the detail makes it apparent that a dip tank must be included in the macro square footage, equipment list, and costs. Note that the same is true for deburring, which must be done after the milling and broaching operations, as well as the gaging, which must be done after grind.

Proximity Analysis

The construction of the proximity analysis uses the operation flow charts. A thorough examination of the technique of proximity diagramming was discussed in Chapter Five. That same approach is applied to the detailed planning.

PROCESS VOLUME CHART

PART NUMBER	VOLUME	WASH	OIL DIP	MILL	DEBURR	CENTER	TURN	BROACH	DEBURR	GRIND	GAGING	TOTAL MANNING
12345-A	275,000	550,000	275,000	275,000	275,000	275,000	275,000					
12345-B	183,000	366,000	183,000	183,000	183,000	183,000	183,000					
12345-C	92,000	184,000	92,000	92,000	92,000	92,000	92,000					
23456-A	188,000	188,000	188,000	188,000	188,000	188,000	188,000	188,000	188,000			
23456-B	47,000	47,000	47,000	47,000	47,000	47,000	47,000	47,000	47,000			
34567-A	178,000	356,000		178,000	178,000		178,000			178,000	178,000	
45678-A	23,500	23,500	23,500				23,500	23,500	23,500			
45678-B	19,500	19,500	19,500				19,500	19,500	19,500			
56789-A	3,000	3,000	3,000	3,000	3,000	3,000	3,000			3,000	3,000	
TOTAL	1,009,000	1,737,000	831,000	966,000	966,000	788,000	1,009,000	278,000	278,000	181,000	181,000	
MACHINE NUMBER		BT-1234	BT-8093	BT-5678	BT9011	BT-9123	BT-4567	BT-8912	BT9012	BT-3456	BT8883	
OPERATION TIMES:												
STD. MIN. PER PIECE		0.34	0.01	0.62	0.08	0.28	2.73	0.68	0.06	2.03	0.17	
STD. HRS. PER DAY		39.72	0.55	39.93	5.15	14.71	183.64	12.60	1.11	24.50	2.05	
% PERSON ATTENDING		25.00%	100.00%	60.00%	100.00%	20.00%	65.00%	60.00%	100.00%	50.00%	100.00%	
PEOPLE REQ'D./24 HRS		0.41	0.02	1.00	0.21	0.12	4.97	0.32	0.05	0.51	0.09	7.70
SETUP TIMES:												
SETUPS PER WEEK		0	0	8	0	7	8	2	0	2	2	
MINUTES PER SETUP		0	0	20.5	0	10.0	37.7	60.0	0	48.5	8.1	
SETUP HOURS PER DAY		0	0	0.55	0.00	0.23	1.01	0.40	0	0.32	0.05	
% PERSON ATTENDING		100.00%	100.00%	100.00%	100.00%	100.00%	100.00%	100.00%	100.00%	100.00%	100.00%	
PEOPLE REQ'D./24 HRS		0.00	0.00	0.07	0.00	0.03	0.13	0.05	0.00	0.04	0.01	0.32
TOTAL HOURS PER DAY		39.72	0.55	40.47	5.15	14.94	184.64	13.00	1.11	24.82	2.11	
MACHINES REQUIRED		1.7	0.0	1.7	0.2	0.6	7.7	0.5	0.0	1.0	0.1	
PEOPLE REQ'D/24 HRS		0.41	0.02	1.07	0.21	0.15	5.10	0.37	0.05	0.55	0.09	8.02
PEOPLE/SHIFT		0.14	0.01	0.36	0.07	0.05	1.70	0.12	0.02	0.18	0.03	2.67

Figure 14-2. *The process volume chart forms the basis for capacity, equipment, and staffing requirements. This cell produces each part number twice per week. (Note: Numbers are rounded to the nearest hundredth.)*

As shown in *Figure 14-3*, the "from-to" flow chart will identify specific flow patterns. Intuitively, the two different deburring operations will not be one work center (due to how deburring occurs in the overall flow). Therefore, an arbitrary starting point might be 10 boxes, as shown in *Figure 14-4*. The boxes should be labelled with the same names that appear on the flow chart: wash, turn, mill, broach, grind, etc. It will quickly become obvious that those operations receiving raw material should be kept on the periphery of the diagram. The proximity analysis is completed by using the operation flow chart data and drawing arrows and quantities going into each piece of equipment, with corresponding arrows and quantities flowing out to the next operations.

To establish some sense of proper relationship, the objective here is to rearrange the box positions to minimize the crossover of flow arrows (much as one might want material flow in the plant to be simple and in a straight line, with a minimum of crossing flows). This is done for clarity of viewing as well. Several different iterations should produce a lower number of crossing lines, until a reasonably less cluttered relationship diagram is achieved. This is *not* the layout. It is simply a tool to give a graphical representation of the many relationships that will exist within the cell. It does, however, suggest which operations should be close to each other, and which could tolerate some distance. Properly constructed, certain relationships will parallel cell layout alternatives. These should be developed from the criteria of the relationship diagram, not mirroring the appearance of it.

FROM	Wash	Turn	Center	Deburr	Mill	Broach	Grind	Oil dip	Gaging	Ass'y A	Ass'y B
Raw material	728	43			238						
Wash					728		178	831	831		
Turn	963					43	3				
Center		553			235						
Deburr	43	413	788								
Mill				966							
Broach				278							
Grind									181		
Oil dip									831		
Gaging	3										178

Figure 14-3. *"From-to" flow chart illustrating movement between work centers of quantities of parts.*

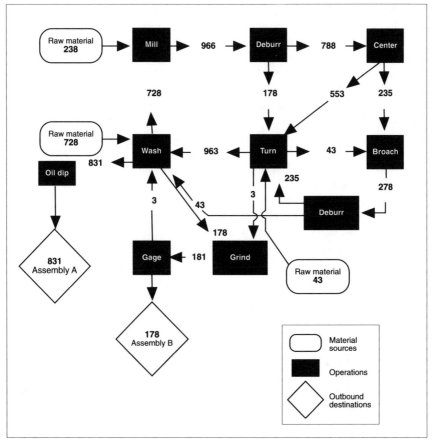

Figure 14-4. *Using the operation flow chart data, the proximity analysis is completed by drawing arrows to show where parts go into and flow out of each piece of equipment. The quantities involved are also detailed.*

Once all of the data has been put onto a single proximity analysis, it can be studied to understand volume relationships. Additional information can be added to the chart as required. Copies can be made, and color codes used for the flow lines representing parts of different weights, for example. This provides graphical insights where there might be special material handling implications.

The process volume chart has also changed slightly with the addition of some calculations to estimate the number of operators needed. In the case of the wash operations, note that 39.72 machine hours are required per 24-hour day. This means that 1.7 machines are required to process the annual volume. However, it is important to note that the cycle time for the wash requires the assistance of an operator for only 25% of the cycle hours. Hence, it can be shown that only .41 of an operator is required to load and unload the wash machine.

Preliminary cell operator calculations are important at this detailed stage because they affect how well cell operations will be balanced, and how well the operators will be utilized within the cell. This utilization is a function of machine placement and understanding the actual work being accomplished. If a layout were made that physically isolated the wash machine operator from other machines, it is evident on a theoretical (calculated) basis that 85% of that person's time would be nonvalue-adding. Frequent reality checks are a good habit for all theories and calculations. It is not hard to visualize placing the wash machine next to another machine to utilize some portion of the wash operator's wait time. This then begins to impact how the layout of the cell might appear.

Stepping further down into the detail, the process flow table from Chapter Six has also been revised in *Figure 14-5* to reflect the total operations for all parts now known for the cell. The number of operations columns used in the macro effort has expanded from five to 10. This begins to provide a more intimate feel for the complexity that must be planned. The next step, as developed in Chapter Six, is to convert the process flow table into operation flow charts.

Note how the number of operations has increased from the original six to now include the internal operations of oil dip, deburr, and gaging. At this point in

					PROCESS FLOW TABLE					
Part no.	Volume (000s)	Op. 10	Op. 15	Op. 20	Op. 25	Op. 30	Op. 35	Op. 40	Op. 45	Op. 50	Op. 55
12345-A	275	Wash		Mill	Deburr	Center		Turn		Wash	Oil dip
12345-B	183	Wash		Mill	Deburr	Center		Turn		Wash	Oil dip
12345-C	92	Wash		Mill	Deburr	Center		Turn		Wash	Oil dip
23456-A	188	Mill	Deburr	Center		Broach	Deburr	Turn		Wash	Oil dip
23456-B	47	Mill	Deburr	Center		Broach	Deburr	Turn		Wash	Oil dip
34567-A	178	Wash		Mill	Deburr	Turn		Wash		Grind	Gaging
45678-A	23.5	Turn		Broach	Deburr	Wash	Oil dip				
45678-B	19.5	Turn		Broach	Deburr	Wash	Oil dip				
56789-A	3	Mill	Deburr	Center		Turn		Grind	Gaging		Oil dip
TOTAL	1,009										

Sequence of operation flow ⟶

Figure 14-5. *The process flow table is now updated to include all nine cell part numbers and additional sequential detail operations.*

detailing, there should be sufficient data to be able to use a simulation package if desired. Many of the current packages available will do the job. Some offer a number of interesting features that can enhance the presentation of results. While the intent is not to go into the fine points of simulation packages, it is worthwhile to note the packages' end objective, which is to verify the throughput, machine utilization, and manning assumptions. More importantly, simulation can reveal situations that static modeling may not highlight.

Use of simulation software provides some additional disciplines as well, the foremost being that it requires a definition of operating parameters for the cell. It also allows for accelerated response to parameter changes, regenerating change impacts quickly, and demonstrating the effects of volume swings, process improvements, or changes in mix. However, many simulation packages are predicated on a batch "push" philosophy; this means that they may have difficulty in simulating a "pull" or Kanban production strategy.

VALIDATE THE ASSUMPTIONS

Before any significant amount of detailing is done on the cell design, it is wise to reconfirm base parameters such as volume and mix. This will prevent surprises later on and reaffirm the planning parameters with top management in case their thinking has changed. By establishing firm planning priorities, all combinations and permutations of the cell's output can be validated and significant impacts documented. Sales and marketing people specifically should play an active role in this assignment. They, along with product planners, will have the most insight about the products and each one's current state within product life cycles and market swings.

It is also critical that any service part or spare part markets be factored into annual volume assumptions. Sales and marketing are an important source of input for this information. In many industries, service parts are among the most profitable segments of the business. Overlooking these requirements will cause gaping holes in the overall plan. Typically, service parts requirements can increase overall volumes from 5 to 10%. For high-mortality products such as pump impellers or tool bits, volume impacts can be as high as 50 to 70% over regular production.

For this example, the assumption is that all volume and mix considerations have been verified. The next step is an overall scan of the data for any glaring or obvious problems. One of the first items to review is the estimated number of operations in the cell. Numbers like those shown in *Figure 14-2* are not necessarily the most definitive. Further refinement of the detail layout can cause items like operator quantity to change.

The preliminary calculations determined that about 2.7 operators were required per shift. Using the mix and volume parameters as a base, it is always valuable to question shop supervisors about how the manning might be arranged under current layout limitations to manufacture the same product. It would be expected that the total number of full and partially utilized workers under the current scheme would be more than what the data implies. Typically the number can be up to as much as 30%. But it must not be assumed that reductions in the direct labor population will be significant with cell implementation. Recall that one of the principles of good cell design is to have control and the physical presence of as many necessary people resources as possible within the operating cell. There will be some activities now performed by indirect people that cell operators will be assuming. These might include tool preset, maintenance, and scheduling. Further, as noted in Chapter Three, there will need to be more time devoted to training, cross training, team building, continuous improvement, and problem solving, all of which impact on required cell manning allowances.

Equally important at this stage in detail planning is any assessment of excessive underutilization of equipment. Recall that up until now machine utilization has not been a primary focus of the cell planning. Care must now be taken that equipment is used productively, or return on assets will suffer. In the present case, preliminary calculations shown in *Figure 14-2* indicate two modest concerns, the centering machine and the broach. Note that the number of machines required here, respectively, is 0.6 and 0.5. This might indicate two key areas for potential part redesign or reprocessing to eliminate the need for the machines. This is where the macro planning advocated in Chapter Six will be beneficial.

From that planning there would be sufficient detail available to find out if there is another cell that requires centering and/or broaching equipment. Reviewing the macro layout would reveal which other cells planned for the future might be located next to the one being detailed here. If there is a requirement in an immediately adjacent cell for broaching or centering, then the fit of the equipment and detailed layout of this cell and the adjacent cell might share those assets. Care must then be taken with the load analysis of succeeding cells to determine if both can be successfully serviced by the shared asset.

The question that must be asked is, "What is the cutoff point of underutilization of assets?" Unfortunately, there is no single correct answer. A rule of thumb for determining underutilization is that usage of less than 30% should give cause to look carefully at alternatives. In this specific case, 0.5 and 0.6 machines is not considered a serious enough liability to sidetrack the planning effort. The utilization of the balance of all the other equipment appears sufficiently high so as not to be an issue.

SELECT AND ASSIGN ALL EQUIPMENT (PROCESS AND SUPPORT)

One of the next major tasks is to locate and document the equipment currently used in the production of cell parts (or perhaps slated for use but currently dedicated to some other production effort). Depending on the current organization of the plant, these machines are likely to be scattered throughout a number of departments. There is also a high probability that they are utilized for one or more other operations which will have to be noted and accommodated elsewhere when equipment begins to be dedicated to specific cell groupings. Condition, age, and current operating dependability are important pieces of information to document as each asset is identified.

In many cases, records may be available for maintenance information as well as machine hours, setup times, and scheduled part numbers. This will all serve as good base data for determining which equipment should be considered for cell use. Other questions should also be considered:

- Is this the best machine given the volume and quality desired?
- Are there alternative methods that could be considered?
- Are other machines available to pick up noncell part production?

If the mix and volume of noncell parts is not significant, then an alternative routing may suffice and capacity might not be a problem. Regardless, all parts involved with the intended cell asset must be addressed. If there are significant volumes, the reprocessing effort to find a home for noncell parts must begin early. Tooling lead times and even the acquisition of additional equipment might be a constraint. All of these significant efforts must be included in startup costs for the cell. For a more detailed discussion of new equipment needs and specifications, refer to Chapter 15.

Also important to consider is whether parts destined for the cell provide for an overall loading of greater than 50%. If this is not the case, then greater flexibility may be necessary to accommodate more part families and increase the overall load. There may also be an issue of technology. If the facility already boasts the latest technical advances, then that will certainly be the approach within cell groupings. On the other hand, if production is operating one or more generations behind the current technology plateau, manufacturers should be wary of performing a technology upgrade and implementing a cellular approach at the same time. Trying to incorporate too many new processes while implementing cellular concepts will place too much demand on the organization (and can sometimes even be incompatible). The first thrust should always be toward gaining the cell experience. Many installations achieve remarkable results using current conventional equipment. Most often the proper blend of old and new technologies achieves the highest results.

Much of today's thrust for purchasing CNC equipment is for the reduction of direct labor. That is an incorrect driver. Often missed in its justification is the resulting increase in (often more expensive) support staff, like programmers, skilled electricians, and tooling support. In today's competitive environment, direct labor reductions should no longer be the driver. What is important now is maintaining quality, adding value quickly, and providing flexibility for customer requirements (change).

During the detail planning stage, it is worthwhile to develop costs for good used equipment (or rebuilding of current equipment). Comparing processing time differences for conventional versus CNC applications is also of value. The minimal differences in throughput time and labor loading may be surprising. In terms of return on assets and payback, significant variation occurs when the cost of the latest technology is inserted in the calculations.

This is not to make a case for avoiding new approaches, technology, or equipment. From an implementation (and need for a "success story") perspective, it is usually much easier to start up a cell using technology the workforce and support staff know how to manage and are most familiar with. Later, if the economics still look good, replacement of cell assets with newer technology is more easily implemented. If executive management wants all the bells and whistles at the outset, the effort will be harder. If not, the team should include reputable used equipment and rebuilding costs in the detailed planning effort.

Again, the emphasis now is to support detailing of real equipment placement. In Chapters Five and Six, the emphasis was on theorizing certain relationships to arrive at some conceptual cell arrangements. Applying the same techniques for real equipment needs helps formulate detailed relationships. For example, using three color codes—parts with weights from 0-20 lbs., 21-40 lbs., and those greater than 40 lbs.—part weight can be highlighted independently. (The fewer the colors the easier the interpretation). Similar techniques can also be used if some of the parts are bulkier than others, or if special handling considerations are necessary due to finish or painted surfaces. Whatever the case, special handling needs as well as critical machine placement can be diagnosed and graphically represented to guide the detail planning of each cell.

Support Equipment And Functions

Up to this point, a great deal of the detail planning has made use of prior techniques employed with both the macro planning and original cell conceptualization. The bulk of the effort has been in translating all affected part numbers into the matrixes and spreadsheet information to document the "real" operating parameters (as opposed to the previous estimated, order-of-magnitude approaches). With the application of simulation runs, some pertinent operating characteristics of the cell have to be defined more succinctly, with the planning process contributing firm batch quantities and processing volumes.

Armed with all of this information, the final confirmation of cell equipment needs can take place. This includes the definition of any support peripherals and activities that should be completed. Much of this detail will not come from charts and graphs, but from team member experiences.

As outlined at the beginning of this chapter, a significant effort should have taken place auditing bills of material information, blueprint data, and routing information. From this exercise there should have developed a familiarity with processing individual part families, as well as identification of special part needs or unique manufacturing applications.

Of special importance is actual operator input. Walking through cell flow, operation by operation (and in some instances, part number by part number), the operator team members can talk through and visualize how manufacturing would really be accomplished. These discussions need to concentrate on special needs like tables, part lay-down space, in-process gaging operations, part conditions (ie. wet, dry, dirty, full of chips), and any special storage needs for tools, documentation, fixtures, and the like.

234

OPERATING CONSIDERATIONS

Additional operational considerations for the detailed cell plan revolve around the new operating concepts and technologies employed, which are beyond the actual part processing. These are considerations like:

- Cellular operating procedures;
- Implementation impacts or considerations;
- Setup requirements (and any identified reduction requirements);
- Cell scheduling—parameters/methodologies.

From the original conceptual plan, each of these must be discussed in turn, taking the concept and developing concrete approaches within the cell operating parameters. Documentation should highlight such things as personnel, special equipment needs, and associated space estimates for benches, storage, desks, and offices.

The entire exercise may take several days or even weeks. In addition to a complete equipment listing, the cell team should come away with a firm and consistent idea of how the actual cell will operate. As done previously with the process volume and flow charts, care should be taken to estimate percent utilization of these support needs, with decisions made for support sharing between cells (or among a group of cells) to arrive at final plant configuration alternatives.

Cellular Operating Procedures

Collecting and organizing the cell operating procedures must include these operating areas:

- Product, process, and cell documentation;
- Chip and coolant handling;
- Equipment spare parts requirements and storage;
- Part storage;
- Service part packaging;
- Machine maintenance needs;
- Skills, training, and cross training requirements;
- Measuring performance milestones;
- Continuous improvement.

As the team talked and documented its way through individual cell operations, and the overall plant implementation timing, there should begin to be a secure feeling about how the changes will progress. It is time to start documenting procedures and guidelines by which the new cell organization will operate.

A number of documents should have been established containing most of the information. These include detailed flow charts, the implementation time line, and any simulation runs and detailed master spreadsheets for unit part and equipment information.

Reviewed against the overall project time line, estimates for when these productivity improvements should become measurable are established as milestones in the project plan:

235

- Manpower requirements,
- Hourly output estimates,
- Scrap and rework targets,
- Setup improvement programs,
- Lot size goals,
- Inventory reduction levels, and
- Floor space requirements.

Each of the cell teams should understand these milestones, which in turn become their targets and goals. Achieving these targets and measuring performance becomes a regular routine for the cell teams.

The implementation schedule, along with performance targets, forms the framework for new operating procedures. Each cell group, along with the original project team, formulates those practices and procedures that will apply to their new activities. While this might follow the format of existing job descriptions or manufacturing process instructions, the recommendation is to start with a clean-sheet-of-paper approach and formulate descriptions that follow the approaches developed to this point.

Drawing on the wealth of part drawings, routing information, and other documented detail, a complete reference can be compiled for each cell group. The use of numerous illustrations should be encouraged, since there can never be enough detail.

Individual circumstances may dictate the final documentation conform to a specific format (employee handbook, process documents, or the like), but this should be arrived at from the cell team's work and be representative of the ideas and approaches. Final acceptance (and the degree of commitment) is often directly related to the sense of ownership associated with the final documentation.

Cell teams are the beginning of the demise of the individual machine operator. As a part of the planning exercise, cell groups need to establish just how individual skills need to interact within the cell environment. For instance:

- Will the parts flow from machine to machine with attendant operators remaining stationary?
- Will the operators circulate with the part process, requiring multidisciplinary skills for familiarity with each process operation?
- Will the cell require a full complement of workers to maintain output, or can reduced manning accommodate reduced output (such as split lunch breaks, which will reduce output but ensure nonstop production)?
- Is quick changeover planned on the fly, or will a downtime period be required to effect a complete changeover?

The approach to cell manning will dictate training needs. Additionally, issues like maintenance by operators, in-process inspection, or packaging and shipping operations within the cell may dictate special training needs. Implementation timing and targets will dictate the intensity and timing of the training.

Another issue is planning for continuous improvement. Continuous improvement is the philosophy that continues to break down bottlenecks, lessen rejects and rework, improve on-time delivery and the velocity of orders, and eliminate waste. The process takes those disciplines used in achieving cell manufacturing and iterates them again and again across finer lines of problem solving. In reality, the first successful benchmarks of the new cellular environment are merely setting a new baseline.

To support continuous improvement, cell teams must be encouraged to continue the interactive process. Time must be allowed for the discussion of issues, and suggested changes must be supported. Cell teams must be encouraged to experiment, make changes, and evaluate outcomes. All the tools have been identified, but the atmosphere for change must be present to encourage continued activity.

Further training will be necessary to learn more focused disciplines. These will vary according to targets set, from machine maintenance to NC programming, to various statistical techniques recognized for controlling processes and the attendant quality. The extent of the company's commitment will be directly proportional to the degree of continuous improvement achieved.

Implementation Impacts

One key output of the detail planning will be a time-phased plan for relocation of equipment, including training and startup of the new cellularized factory. There are many sophisticated techniques (and software products) for multistep project planning, but to get a good first sense of the overall reorganization, the "cut and paste" approach is again hard to beat. Chapter 17 will cover these and other techniques. However, implementation features do have a huge influence on the cell planning stage for potential issues like:

- *Buildup of buffer stocks*: Since the timing of the actual cell installation or formation often coincides with a delivery requirement for the products that those pieces of equipment must produce on a daily basis, the cell planning team must be responsible for the continuity of product supply. Preplanning of an adequate inventory buffer is the usual method for this, but it might be modified through the availability of alternative process or component sources on an interim basis. Volumes, timing, quality, and flexibility impacts for these alternatives must be factored into both the FMC design parameters and the implementation timing.
- *Use of temporary equipment and/or process*: During the periods of cell installation, it may be necessary to identify some alternative equipment or processes to substitute for the downtime usually associated with cell implementation. Careful planning for ensuring process capability is necessary. For example, using a backup multistation, multispindle drill press and a manual VTL (vertical turret lathe) may serve as a temporary alternative while a machining center is being rebuilt or moved into position and debugged.
- *Maintaining uninterrupted production/schedules*: As mentioned above, given a mandate that schedules *are not to be interrupted at any cost* will

237

drive the cell planning team to meticulous detail in its contingency planning; this could impact space and manning parameters for the cell design.

- *Opportunity for correcting errors/deficiencies*: Too often, planning teams or anxious manufacturing managers inflict themselves with taking on two or more major changes at the same time. A commonly heard phrase like, "As long as that machine is going to move, why don't we ..." is cause for implementers to shudder. The additional changes attempted most often are to rebuild equipment, update machine controls, or go ahead with material handling upgrades. Any or all of these examples can be disruptive to the cell implementation but may be absolutely necessary. The point is that the cell design team must recognize these as implementation impacts and compensate the cellular design parameters accordingly.

Setup Reduction

Cellular rearrangement often involves the use of Just In Time, Kanban, and quick changeover techniques. Setup reduction has quickly become a topic all its own. The danger in treating these disciplines as problems to be solved rather than elements of a new operating philosophy is that they are often treated separately. Organizations, stretched to make improvements but mindful of limited staffing, often choose to take piecemeal approaches to broad projects. Targeting problems one by one, they attempt to put one issue to rest before addressing the next. These disciplines, however, are intertwined. Success in one area cannot be achieved without supporting successes in the others. Achieving quick changeover serves no purpose and achieves no real benefits unless it is tied to improved cell scheduling or business growth.

Key to any setup reduction program is the participation of the people who will perform changeovers in the plant. Fundamental to their success is a clear definition of what exactly setup downtime is—the nonproductive time from when the last part of run A is finished until operators begin making the first good part of run B.

Nonproductive time is when the equipment is not physically producing products. A six-step approach to setup reduction begins first through a process of challenges and adjustments, and only then considering a technological or financial solution.

1. *Document what is happening.* The first effort of any setup reduction effort needs to identify just what the problems really are. These differ organization to organization, and process to process. Experience has shown that typically more than 50% of changeover downtime is due to "organizational" issues. These are delays reflected by statements like:

- "What's the next part to be run?"
- "Where's the tooling?"
- "Why can't I find a fork truck driver?"
- "The operator went to get a soft drink?"
- "First-piece inspection isn't finished!"

238

All of these delays are issues of logistics and can be "managed" out of the process. The cost of managing them is probably next to nothing, with the result of more than half of this delay being eliminated.

The most straightforward approach to eliminating waste is to document typical changeovers within the factory. The documentation should clearly show:

- The time that the last part of run A was finished (beginning of changeover);
- Definition of people involved and their roles;
- Each activity that takes place:
  - Approximate start time;
  - Duration period of the activity;
- General observations:
  - Locating tools;
  - Piece part characteristics;
  - Handling of parts;
  - Schedule availability;
  - First-piece inspection;
- Beginning run of part B.

During the documentation, it is important to note everything that takes place which accounts for time. The team may wish to use a video camera to tape the changeover, and then review it for analysis. Wasted time may come from lack of information or coordination, not having proper tools, mechanical problems with fit and function during the changeover, and so forth.

2. *Challenge activities and tooling.* Once several changeovers are documented, the recordings should be examined to decide on directions for improvements. A typical approach is to arrange each activity in sequence as it occurred and note beside it the time it took to accomplish. Simplification will often challenge the effective use of quick clamping, universal location pins, and common setup plates. Universal handtools and their sizes can be applied (sometimes custom-designed) to ease the complexity of selection, proximity, and application. Although some obvious fixes may be seen, one additional step should be done before rushing out and making changes.

3. *Shift internal time to external.* Beside each activity, team members should note whether it is now an internal operation, which is performed while the machine is stopped, or an external operation, which is performed while the machine is in the run cycle. The next step is to challenge whether or not the current internal activities could take place during the time operators are still running production, external to the actual setup time. Adding up the times associated with these internal activities that could be external will probably show these manageable functions typically account for at least 50% of setup downtime. The nature of these activities is such that they can be shifted to production time with minimal expense.

Operators should try performing the setup again and again with these activities shifted to note the improvements this brings and other opportunities for improvement. Benchmarking this new performance standard completes the first milestone of a setup improvement program. Further efforts in this area should concentrate on shortening internal setup activities before turning to external activities. It is essential to remember not to compromise safety in accomplishing setup reduction.

4. *Practice, practice, practice.* The next step of the setup reduction program is "practice, practice, practice." Believe it or not, by simply getting good at performing the tasks, another 10 to 20% of the wasted downtime can be eliminated.

This practice compares to the meticulous practice at the Indianapolis stock car races. Not only does everyone in the pit crew have an assigned function for which they are prepared (tooling and practiced technique), they also understand what everyone else's function is (cross training). This includes where those people are expected to be standing before, during, and after the changeover.

This thorough choreography of the pit stop can shave seconds of precious time, and it can be equally as valuable in machine setup reduction. The team should document several more changeovers, noting those activities which must be better organized, and then revise and practice them. Again, this new standard should be benchmarked. At this point, eliminating up to 50% of downtime may have been accomplished without spending anything substantial.

5. *Design away changeover.* Differences in tooling styles and situations seldom permit noting many of the types of engineering changes which can be made. Each application will dictate the level of design change that might impact changeover reduction. Redesign usually involves a minimal investment, but it simplifies operational process times or even eliminates operations, as well as reducing setup times.

6. *As a last resort, spend money.* At this point, tooling becomes the main source of improvement. In reference to the race car pit crew, note some of the clever tooling used to save time and assist in the effort:

- One-man car jacks that do not have to be pumped;
- Knock-off hubs for wheels;
- Fast-filling gas cans with built-in funnels (with safety design);
- Power wrenches for spark plug changes.

Analyze the tooling changeover for the application of these types of items. The idea is for one person to do one task, resulting in quick, efficient change with a minimum of effort.

The key is to challenge every aspect. Provide the team with support and the resources to try various solutions. Not every one will be a winner, but from many of the suggestions will emerge some pretty clever fixes for problems that have historically been dealt with daily, and may even in fact be "frozen" into the present tooling design approaches.

As is often the case with manufacturing improvements, the benefits of setup reduction tend to erode over time if the procedures are not reinforced and made standard practice. A rigorous method of documenting any new setup procedures must be developed, including step-by-step instruction sheets, illustrations, and tooling arrangements. A method must be put in place to train new operators in the process.

A new performance standard developed using this approach has consistently yielded simple yet massive downtime reductions, easily surpassing 90%.

Cell Scheduling

Reducing setup downtime is of no benefit unless that change supports some other need. The benefits of setup reduction and FMC definition can be captured in one of two ways:

- Reduced lot sizes—The ability to run the same part variety more frequently;
- Increased capacity—Enables scheduling part variety with a greater output volume.

If markets are not growing and capacity is not an issue, reducing setup downtime gains little besides some minor labor reduction. In this case, reducing lot sizes offers the opportunity to improve throughput (velocity) and reduce inventory levels. The real effect on changeover then will be to do more of them. With the reduction of setup downtime, even though each occurrence will take less time, the increased number of changeovers will have a net effect of holding overall downtime to the same level, or possibly even increasing it. Understanding cell scheduling is necessary to keep these somewhat contradictory situations in focus so that overall benefits can be captured.

The first thing to understand about cell scheduling is that cellular rearrangement significantly reduces, and can eliminate, the normal discreet kinds of scheduling activities present in most manufacturing environments today. A cell combines a number of traditional work centers, dictating part flow (and next operations) based on layout. Discreet work center scheduling is replaced by simpler cell scheduling. In other words, the scheduler tells the cell what to make next and all operations occur automatically.

In a general cellular reorganization, scheduling activities can usually be reduced by a factor of 10. For example, consider 50 machines reorganized into five cells. Rather than scheduling and routing 50 individual processes, only five cells must be scheduled in the rearranged plant.

All the traditional "crutches" of production control, such as expediting, hot lists, and inventory audits, are removed with cellular approaches as well. Line-of-sight management (looking down the floor and knowing instantly what is busy and what is not) becomes the norm. Control is simplified immediately. Cell planning allows operators to be responsible for entire orders, including packaging and shipping. Shop floor scheduling in the cell environment deals more with availability and distribution of raw materials, and packaging and shipping. Furthermore, these simplified controls fall under the new philosophies of JIT and Kanban.

Such approaches as point-of-use material storage and Kanban squares are simple techniques for fool-proofing the control of material flows to and from cells. Each cell is dedicated to certain groups of parts. This never changes. Scheduling then becomes a matter of saying when someone wants those parts produced. Only one factor impacts scheduling in the cellular environment—

capacity. Overcommitting the shop can still cause delays, even in a cellular environment. Cells cannot cure bad judgement!

DEVELOPING THE PHYSICAL LAYOUT

The next step in the detail cell design is to start putting lines on paper and coming up with detailed cell layout alternatives. Most companies today are striving for more and better usage of CAD to capture all drawing activity. It is interesting to note that not many have updated plant layouts in their CAD systems (since most often the first driver for CAD is product design). While CAD is certainly the tool of choice, initial layouts are better developed with cut-and-paste techniques.

This may seem like strange advice, but unless the CAD operator is experienced with developing specific layout details from spacial relationship diagrams, the effective placement of equipment groupings will require someone monitoring the operator's every action. Since most activities to this point have made use of team approaches, many participants would be left out of this important activity phase if left to an operator. One of the more enjoyable parts of the project is seeing all the data and documents finally taking shape into definitive layouts.

Instead, the team should have the CAD operator produce a blank plant layout to a 1/4-inch scale (no equipment or aisles) including placement of I-beam columns, as well as scale drawing footprints of present equipment to the same 1/4-inch standard. The blank layout and machine templates will permit group sessions around a table, arranging machine cutouts to match flow and relationship criteria. If new machines will be part of the cell grouping, the team should also create a library of templates for these from vendor information. This will allow the CAD operator to have all the necessary information for when the team begins documenting the layout alternatives. Eventually, all layouts will be done and stored on CAD.

There are a few software packages that develop random relationships according to input criteria like space requirements or from-to analysis data, and they will even "rate" the generated output. Most of them follow the established procedure for "systematic plant layout." While this might be a good tool, it also takes away from the team approach at this point. Such a package can be used to evaluate alternatives the group arrives at, and possibly to look at machine-generated alternatives after going through the manual exercise. There is no substitute for the ownership achieved by actually going through the layout process from start to finish.

Proximity analysis and the relationship diagram should be the key drivers for the layout alternatives, but keep in mind that there is no single correct answer. For this reason, the team should develop a number of alternatives and evaluate their features. Here again, team members must push for density and "squeeze, squeeze, squeeze," but with healthy reality checks during the process. Refer to Chapter Seven for helpful tools to properly evaluate these layout alternatives.

Flow, Density, And Velocity: Reality Measurements Of Cell Layout

The main design drivers for these layouts continue to be *flow*, *density*, and *velocity*; they make up a developmental mindset that must be undertaken when designing manufacturing cells, from beginning to end.

Can one measure fluidity? Reality in the manufacturing world of cell implementation can be most effectively expressed in three simple conceptual measurements: flow, density, and velocity. If a cell is evaluated from design through implementation in these terms, there will be confidence that it has been done right and the performance measurements will show it. In an attempt to develop manufacturing layouts that are focused and as unidirectional as possible, these concepts are offered as the layout performance measurement tool to most effectively drive a competitive business.

Flow, density, and velocity are terms used to describe cell design, measurement, results, physical layout, and flexibility. Sometimes jointly referred to as the property of *fluidity*, they are used to define a physical environment conducive to achieving the desired results from cells—permitting product to move through the production steps with a minimum of lost time or motion. Let's examine a methodology for attaining a manufacturing cell that demonstrates fluidity:

- How to measure it:
  - On-time delivery (*not* early or late);
  - Minutes or hours of throughput time—not days or weeks;
  - High quality (measured in parts per million);
  - Straight paths;
  - Net effective space utilization.
- How one might see a lack of it:
  - Inventory buildup;
  - Pallets and containers, empty or full;
  - Fork trucks, trailers;
  - Warehouses;
  - Excess distance between machines;
  - Walls, aisles, offices, desks, paper, expediting departments, process orientation, supervisors.
- How to change it:
  - Focused cells;
  - Compression;
  - Simplification;
  - Linking;
  - Reduction;
  - Product orientation;
  - Waste elimination.
- How one profits from it:
  - Less inventory (capital released);
  - Larger market share;

—Lower scrap/rework dollars;

—Higher revenue per person/area/assets;

—Less cost of making transactions;

—Greater satisfaction;

—Improved management control;

—Faster throughput.

Use these tools to evaluate just how effective cell design can be. Some of these tools are more quantifiable than others, but cell planners should intuitively apply flow, density, and velocity characteristics to optimize layouts. Every cell has different attributes, but many fit into some manufacturing standards. Let's elaborate on each for the characteristics and measurements.

Flow. Manufacturing flow is defined as the minimum path a part must travel in a manufacturing plant to accumulate its value-added content. A cell environment will obviously help in attaining an improved flow. In more global terms, however, the cell interacts with the remaining parts of the plant. The concept of internal customer-supplier relationships is very important within the plant infrastructure. Flow is attained when the supplier (cell, machine, or operator) is immediately next to the customer (another cell, machine, or operator) to be able to hand off the physical parts to one another. Avoidance of double handling, setting down, or delay enhances flow. Flow must demonstrate:

- Straight, unidirectional paths:
    —Simple, visible, constant movement;
    —Linked operations.
- No backtracking:
    —Routes that are as short as possible;
    —Parts that always flow toward the customer;
    —No confusion about part status;
    —Minimal movement, maximum value-added content.
- No alternative routings:
    —No choices to breed confusion, doubt, delay;
    —Dedicated operations if possible;
    —There are no surprises;
    —Capacity always known, consistent.

The usual measurements of flow are time and distance. An assessment is made by evaluation of:

- Proximity charting;
- Operations greater than six feet apart;
- Number of backtrack paths;
- Straight line or U-shaped flow path;
- Movement forward to customer.

Knowing how important these features are during the design stage, doesn't it make sense to validate and track the cell effectiveness against those same parameters?

Density. Density defines the closeness of the operations to each other and how close the customer is to the supplier. How dense a cell will become will be

244

a function of the size and type of machines and parts the cell will process, and how tightly it is laid out. For example, a vertical pump housing cell will contain large machines and require major material handling equipment and clearances. A shaft machining cell may require nearly as massive equipment but may require some specifically oriented, long racking resulting in a long and narrow configuration.

The general characteristics inherent in the analysis of density are:
- Minimum space between machines:
  —Smaller than the space required to put parts down;
  —Smaller than the space required for a fork truck to go between;
  —Less than six feet between machines;
  —Making it tough to have an alternative;
  —Space made available for other productive use.
- Opportunity for automation:
  —Closeness can make robots feasible for handling;
  —Parts conveying can be automated.
- Enhanced communication, team building, and closeness:
  —Cell teams become competitive;
  —Customers and suppliers know each other, resulting in quicker reaction;
  —When people feel good, productivity increases;
  —Visibility achieved drives quality improvements.
- No inventory:
  —Closeness prohibits containers and pallets;
  —Parts cannot go anywhere but onward;
  —Few racks, shorter conveyors for queue of parts;
  —Direct part handoff.
- Ease of managing the business:
  —Cell operators can manage themselves in close surroundings;
  —Quality improves with quicker response and communication;
  —Future changes can happen fast.

Good density design in cells will result in floor space being made available. However, that freed space must be available in *one* contiguous area or it will not be productive—as a matter of fact, it could become self-defeating if it becomes a gathering spot for inventory or junk.

It is possible to define the "before" baseline for flow and velocity against which to compare progress. Measurements are throughput time, plotted part paths, and actual distance traveled. Density improvement is a much tougher commodity to quantify since most manufacturers begin from a traditional departmental and functional layout. But even intuitive comparison with old layouts makes the improvement obvious.

Professional cell planners have accumulated a fair database of "good" cell density designs over the years to establish some generic density ratios as guidelines. The portrayal of the density measurement begins with an understanding of the layout itself. *Figure 14-6* shows a typical cell layout that has had a boundary placed around the perimeter and has each individual machine work

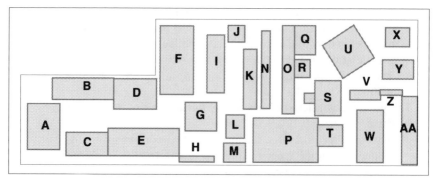

Figure 14-6. *Machine work area spatial boundaries for calculation of layout density.*

area (template area) highlighted; this will be used as an example for measuring and calculating a density ratio. The density ratio is defined by:

$$\frac{\text{Density}}{\text{ratio}} = \frac{\text{Sum of rectangular area bounding equipment templates}}{\text{Total area allocated for the cell layout boundaries}}$$

Interpretation of the resulting ratio must be tempered by the unique characteristics of the cell, such as the part size and equipment size. The mixture of equipment plays a large role, as do the characteristics inherent in cells that will combine machining, fabricating, and assembly. The table in *Figure 14-7* demonstrates ratios for some typical cell types.

A cell density ratio of less than 35% is generally too loose. As illustrated, ratios exceeding 50% are possible. Whatever space there is usually means that a company is paying people to walk past it in the course of their work every day. Geography (space) is a valuable commodity in most manufacturing plants and

Cell type	Parts	Cell area (feet)		No. of machines	Machine area (sq. ft.)	Ratio
		Length	Width			
Machining	Bar <1" dia.	90	52.5	8	2,381	50%
Machining	Bar 1"-2.5" dia.	101.25	47.5	8	2,089	43%
Machining	Fittings	110.5	45	16	2,402	53%
Stamping/ notching	Rotor/stator laminates	130	60	69+	4,180	54%
Formed plate	Pressure vessels	82.5	36	5	1,052	35%

Figure 14-7. *Examples of various successful cell density ratios.*

has seldom been challenged to full utilization. If expenditures were justified on occupied geography, most American manufacturers would be more competitive today. *Figure 14-8* contains two examples of actual cell density that challenge the concept of dense space.

Velocity. Velocity is the measure of *speed*. Speed is a popular term these days in America as a buzzword to link performance to customer service. That could be true, but in cell performance measurement, the interest is in speed or velocity as a measure of how quickly value is added to the product. The old term for expressing this same kind of measurement used to be "lead time" or "throughput." These terms have gotten so clouded with add-ons like storage, material delivery time, and engineering lead times, that seldom do two manufacturers ever measure them the same way. Most manufacturers, when asked for these measurements as they apply to their companies, can't answer with a concrete number; they will say, "It depends!"

Velocity is a simple measurement ratio, defined as:

$$\text{Velocity ratio} = \frac{\text{Value-adding time}}{\text{Calendar time from order entry to customer delivery}}$$

Good manufacturers today are in the 6 to 10% range for velocity. In the future, cells will require minimum velocity ratios of 20 to 35% to remain competitive. The basic advantages to this measurement are characterized by:

- Forces quick problem resolution:
  —Requires trained people close at hand;
  —Velocity allows little inventory buffer.
- Keeps parts flowing:
  —Kanban techniques become easy tools;
  —Value-adding time is usually a constant;
  —Nonvalue-adding time becomes equally important.
- Meets customer demands:
  —Quick, responsive delivery of finished product;
  —Part revisions can be quickly implemented;
  —Just In Time becomes reality.
- Improves cash flow:
  —Can the company sell its product before material invoices even arrive?
  —Work-in-process inventory is naturally eliminated or reduced, requiring less working capital.

Another objective of velocity is to keep the parts flowing to the next operation. This adds value as quickly as possible. This quickness or agility is one of the most important drivers in the marketplace today—get it to the customer quicker than competitors.

Recall the term "fluidity." Webster's Dictionary defines fluidity as being "able to move and change shape without separating when under pressure; that can change rapidly or easily; not settled or fixed; marked by or using graceful movements." These are characteristics that would benefit any organization, whether in manufacturing cells or office "cellular" cultures.

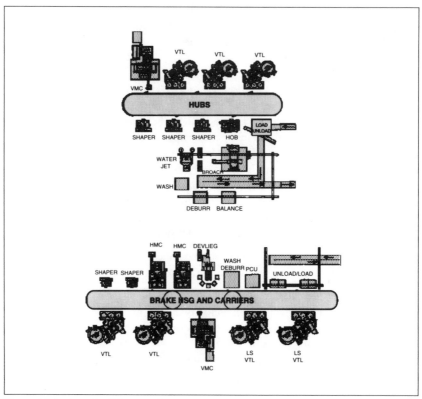

Figure 14-8. *Actual cell layouts illustrating adequate density.*

CONCLUSION

The team should seek inputs from operators on the alternatives that look reasonable (not just from team members). While it is expected that there will be some less-than-useful criticism here and there, it is also true that candid feedback and useful observations will point out things like, "This can't be done because..." and "What about..." Involving floor personnel early on also gets discussions going about the changes.

At some point, alternative groupings will be exhausted and the best two or three layouts will become apparent. Financial costing measurements as discussed in Chapter Seven will be a final measurement to help select the "best" grouping.

Detailed cell planning uses the same tools discussed in previous chapters to determine the finite details beyond the macro plan and conceptual cell. The difference is the level of detail that must be designed from the product base (all part numbers instead of representative parts) and the application of reality checks to the results. The careful "walking through" of the plan by operators will highlight improbable scenarios, and help document minor details that would have been overlooked in a purely mechanical and mathematical exercise.

||| 15 |||

EQUIPMENT SPECIFICATION AND
REQUESTS FOR PROPOSALS

Once FMCs are designed, and the necessary types of equipment for each have been identified, it will become apparent that some of the required equipment must be purchased or rebuilt. To make sure that the most appropriate equipment is secured, a carefully prepared set of specifications is mandatory, along with soliciting the best supplier to work with. Specifications are the company's way of saying, "This is what I want," and the terms and conditions of the request for proposal (RFP) which are wrapped around the specifications say, "This is the way I want it."

All too often, little time is spent on preparing specifications for properly detailing the equipment (or facility) requirements in advance. What is found — and a look around many facilities will confirm this — is a conglomeration of machines with different control packages, varying tool adaptation methods, no maintenance manuals, and fixturing of every conceivable variety. Appropriate specification is well worth the effort later on if an FMC is expected to quickly reap the benefits of the new layout and design. When new equipment is rapidly, haphazardly purchased and installed, the cell attendants are left with the laborious task of debugging it while struggling to get product out the door.

Figure 15-1 illustrates the steps required in the preparation of a meaningful specification and the accompanying request for quotation. (Note that the terms "request for proposal" and "request for quotation," or "RFQ," are used interchangeably.) This effort *must* begin simultaneously with the development of the engineering data. This is especially true if a company has not procured any significant new equipment in recent years and lacks familiarity with the process and the market. Some companies may lack the resources or knowledge to even identify the types of equipment needed and to construct an RFP; in this case, assistance should be acquired to first identify the required equipment before approaching too many vendors. Although vendors can provide a wealth of

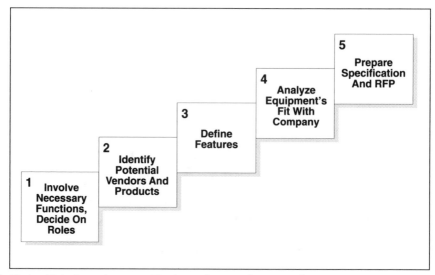

Figure 15-1. *Preparing a specification and request for proposal.*

information and technical assistance, manufacturers who let vendors tell them what equipment they require may pay dearly.

STEP ONE—INVOLVING ALL NECESSARY FUNCTIONS AND DECIDING ON ROLES

By now most engineers and managers are probably tired of hearing the cliches about employee involvement and the necessity of teaming. The best equipment specifications, however, will be developed by a team that is comprised of several of the major disciplines within an organization. Involving as many of the eventual users as is practical is beneficial, especially when their team responsibilities are clear. It's important also that all of the company's drivers be fully represented in this exercise. Striking the right balance between quality, cost, speed, convenience, flexibility, technology type, and future production considerations is essential to success. The amount of time required from each of the team members varies from near full-time for the coordinator to part-time technical input from machinists and other equipment users. Some of the participants in this task should be:

- *Purchaser:* The team leader makes sure that the specifications and request for proposal are properly prepared, becoming the liaison between the company and vendors. This role might be played by someone from the purchasing function, but it is increasingly being undertaken by the technical or advanced manufacturing engineering organization.
- *Engineers:* Manufacturing and industrial engineering are traditionally involved with the development of the overall rationalization of the need for new equipment. Their primary role has been defined in previous chapters, and continued participation through specification writing is necessary.

250

Involvement of design engineering is imperative to ensure that equipment capabilities will satisfy the required quality for manufactured components.

- *Finance:* The people within the company's finance function will typically want to contribute to the terms and conditions part of the RFP. Providing them with insights into the cost/benefit scenarios of the purchases will often result in their strong support, and better liaisons with the management group, which will eventually need convincing.
- *Shop representatives:* A representative (or more than one) from the shop is key. This end-user's input into selecting and defining the options necessary to produce quality parts is vital. If the specification involves machining, then a machinist should be recruited; an assembler should be included for specification of an assembly machine.
- *Maintenance/facilities staff members:* They will ensure that proper documentation has been requested in the quotations to assist them in keeping the machine(s) in good running condition. They should be most heavily involved in preparation of the spare parts and preventive maintenance programs. They also provide information on the necessary facilities requirements for installation, as well as integration with other devices such as cell controllers, material handling mechanisms, etc.
- *Tooling experts:* These are key players who provide input into cutting tool and fixture standardization programs. Any integration issues regarding new equipment can be resolved early in the vendor interface process.
- *Programming staff members:* These professionals should be involved with identification of the software compatibility/interface between the machine being purchased and existing programming methods.

In larger companies, there often is one additional group, *Contracts*, which may be called on to ensure the legality of the request for proposal document. In smaller companies, this task is often performed by the purchasing department with assistance from the company's legal counsel.

STEP TWO—IDENTIFYING POTENTIAL VENDORS AND PRODUCTS

The vendor base for standard and special machines changes rapidly. Acquisitions and mergers are occurring daily. Furthermore, most companies do not purchase machines on a regular basis; therefore, out-of-date supplier lists are common. Some of the sources for developing a comprehensive vendor list are:

- Trade journals, which are a source for finding out about the current technology suppliers have to offer.
- The *Thomas Register* and similar industrial directories, which provide company names by equipment category.
- Local equipment distributors, who supply data for the equipment they sell. Some may even be helpful in identifying other available equipment.
- Trade shows, which give some insight into emerging technologies and trends. Manufacturers should use caution, however, that they do not become "guinea pigs" for new high-tech features.

- Technical associations that promote the exchange of information and sharing of experience through newsletters and other publications, libraries, technical committees, etc.
- References from employees, corporate offices/other divisions of the company, other manufacturers, etc.

No matter what sources are used, putting together a vendor list is more than identifying the names of companies and their respective addresses. This is the time to begin building a matrix to be used for vendor evaluation (discussed in the next chapter). Key ingredients for the matrix that can be obtained now include:

- Information about the supplier: size, capabilities, financial situation, credit rating, service network, latest installations (and possible tours of these sites), ability to meet delivery schedules, and quality reputation. During later vendor selection stages, careful evaluation should be made of a vendor's financial stability, particularly if considering small, special equipment builders. By starting the process of verification with requests for Dun & Bradstreet reports, etc., *early*, a manufacturer can quickly eliminate some of the field and concentrate on the most likely providers.
- Vendor's responsiveness and level of customer service during a company's inquiry phase. If a vendor is somewhat disinterested or difficult to work with for any reason *before* he has a manufacturer's money, one must question the level of support that will be available afterwards. (Rating this level of interaction is often easier than people think it will be!)
- Features that are offered and the technology that's used. What do the available features do, and how would they apply to the business? What is offered as standard, and which features will result in add-on costs? (In today's environment, not too much comes standard.) The list of options and features and the way they are offered also varies with the brand and type of machine being purchased; the machine that looks like the best deal may end up at the high end by the time all the extras are totalled. For example, some of the many options/features to be decided on for a machining center are:
  —Chip conveyor,
  —Caution light tower,
  —Number of pallets in use,
  —Length of pallet change time,
  —Length of tool change time (cut to cut),
  —Part probe,
  —Brand of control,
  —Control model,
  —Flood coolant,
  —Through-spindle coolant,
  —Mist coolant,
  —Adaptive control,
  —Tool management, and
  —Automatic tool offsets.

After initially screening the suppliers and eliminating the ones that do not satisfy the criteria of the organization, the matrix can be used further to:

- Measure the individual companies' strengths and weaknesses *against* other supplier companies.
- Help the buyer develop a comprehensive list of the options and features available today so that he is knowledgeable and can ask the vendor about items that the vendor does not offer at first; many buyers work chosen elements of competitor's proposals into the proposal of the selected vendor.
- Add a structure to the selection process that allows the buyer to retain control of the process, instead of being dazzled by the vendor with the flashiest selling process.
- Ensure selection of the vendor based on the company's criteria, instead of politics or arbitrary preferences.

STEP THREE—DEFINING FEATURES

Once the team has gathered a listing of the available features and options and understands what they offer the business, the next step is to evaluate each item and categorize it as a need or a want, or an unnecessary frill:

1. *Mandatory.* What does the equipment *have* to do? If, for example, a central chip/coolant system is not in place nor is one envisioned, it becomes important to have a chip conveyor and accompanying coolant system. If several machines are being purchased and they all will be placed in one area, one can question/rationalize the need for a central system.
2. *"Nice to have."* What would be helpful for the equipment to do? It may be that, from a maintenance point of view, one brand of control system would make life easier than having several. When ordering standard machines, however, picking the brand of control is seldom an option. Based on the company's experience, there are probably some brands that maintenance cannot support very well and would rather not see. Features/options in this category do not eliminate or guarantee a certain piece of equipment, but they are given careful scrutiny in trying to meet less critical needs and major "wish list" preferences.
3. *Not necessary.* What would be a waste of money or an unnecessary complication? For instance, if a company has adequate fixturing and a good quality program, probing capability may not be a necessary option when weighed against the cost.

It's important that the decision method for categorizing features is understood and agreed upon at the outset; team members should not go away from this exercise feeling that their interests were not considered fairly and objectively. If there are many items within each category, a secondary ranking system may have to be used to put features in order within each group. And it's better to keep track of "not necessary" items than to ignore them as irrelevant. Sometimes a $250,000, top-of-the-line machine that "does it all" can contain more useless than useful features, whereas a much simpler machine ($70,000) can be modified to add the one feature ($20,000) that it was missing.

STEP FOUR: ANALYZING THE EQUIPMENT'S FIT WITH THE COMPANY

Before actually preparing the specification and request for proposal, there is still a little homework that needs to be completed. A *self-assessment* of the company is necessary to determine whether the equipment being considered is a viable alternative for the firm. Appropriate actions should be taken to meet any needs that are identified for making the equipment a success, whether through training current employees, hiring qualified people, or using the supplier or another outside service. Some of the issues that must be resolved include:

- *What are the company's NC programming capabilities?* Can it support the new machine? Should the company write the initial programs or have the supplier write them? Should it use a programming house? Is the company's programming language compatible with the new machine?
- *Does the company have the in-house maintenance skills to maintain the mechanical and electronic aspects of the machine?* Does it have a preventive maintenance program? Which spare parts does the company want to have in stock? How much does it have to (and want to have to) depend on the supplier for service?
- *Has the company standardized its fixturing methods?* What is its standard for cutting tool adaptation to the spindle? Should the company have the machine supplier equip the machine with fixtures or will it buy them from someone else? How many sets of tools does the company want provided with the machine?

When too little thought is given to these questions, the end result is a lack of standardization and the need for *extra* sets of fixtures and cutting tools dedicated to just one machine. Even if the knowledge to resolve these issues is resident in-house, making sure that the person has adequate time to get to these duties while still trying to make sure his own job gets done will require additional resources. Today's environment leads more companies to purchase turnkey machines (built, supplied, and/or installed complete and ready to operate) in order to quickly, cleanly resolve all of these issues.

STEP FIVE—PREPARING SPECIFICATION DETAILS
AND THE REQUEST FOR PROPOSAL

There are two forms of specification which may be necessary, functional and/or performance. The key differences are that functional specifications describe operationally how the equipment will work in terms like dimension, sequence, necessary interface, preparation, capacity, and loading. Performance specifications are primarily detailed features, such as accuracy, tolerance, efficiency, uptime/downtime, compatibility, and quality. The numerous forms that specifications can have will also depend on current internal policies, company size, and number of machines to be purchased. Resolution of these issues will help decide the format and the depth of detail that the specification and RFP should take. The depth of detail can be characterized from the two extremes:

- *Standard:* If the new equipment is standard (purchased as a catalog item with simple options/features to choose from), the level of detail in the specification is normally less.
- *Special or nonstandard:* If the new equipment is special (one of a kind, built by a special machine tool builder), the manufacturer needs to include more detail, especially about the overall construction of the machine and the control. He should consider purchasing a turnkey package which includes everything fully integrated, versus purchasing the base machine from one supplier, the fixturing from another, and the cutting tools from still another.

The following two case studies show the two extremes that specification documents can take.

Case Study 1

Company A embarked on a program to reorganize manufacturing from the traditional process flow to a cellular concept. This resulted in numerous manufacturing cells comprised of existing and new machines. Each cell was developed to accommodate a family of parts from raw material through completion. New machines required were a combination of CNC turning centers and CNC horizontal machining centers. The implementation sequence of the cells was established and purchasing of a new machine for the first cell undertaken. All the planning work outlined above was completed. The first cell contained one CNC turning center and Company A decided to request a turnkey proposal. Since the request would be for a standard machine, preparation of a lengthy specification was unnecessary.

A supplier list of builders of CNC turning centers was created. This list was shortened based on the matrix that was developed in step two; one of the main criterion selected was the proximity of the supplier's service organization to the plant. The accepted suppliers were notified of a prebid meeting to be held at the plant. (Using the one-time bidders' meeting approach eliminates the time necessary to contact each supplier individually, and it ensures that *all* suppliers attending hear the same information.) A presentation format was used; the type of data presented included:

- *History of the facility.* This includes an introductory statement detailing when the facility was built and acquired by Company A, the size of the facility, and the products manufactured.
- *Project background.* This consists of a description of the program—how it started, the direction being taken, and current status.
- *Project size.* For Company A, the program requires $16 million to $19 million in new equipment. They plan to purchase 35 new CNC machines over a period of three to five years.
- *Project goals.* Since this is a high-volume manufacturer, the goal is to produce high-quality parts in weekly lot sizes in a cellular format, ie., from raw material to finished components without WIP.

- *Cell description*. The several component cells are comprised of existing equipment plus new equipment such as:
  - —High-volume screw machines for some 20% of the high-volume part numbers within the part family;
  - —A new CNC turning center for the remaining 80% of the low-volume part numbers;
  - —Cleaning and deburring equipment.
- *Part features*. The parts fall in a diameter range of 0.375 inches (9.52 mm) to 0.937 inches (23.7 mm) and a length range of 0.574 inches (14.5 mm) to 1.727 inches (43.8 mm). They all are made of 430F stainless steel.
- *Desired machine functions*. Parts are to be machined complete in one handling, minimally manned, with minimal setup time and quick-change tooling, and have "crash" protection using a torque monitoring system.
- *Desired machine features/options*. The machine is to be equipped with:
  - —Automatic tool offsets;
  - —Quick-change tooling/collet/chuck jaws;
  - —Tool management system;
  - —Adaptive control;
  - —Light tower indicating messages for condition messages like "out of stock," "tool change," and "machine failure";
  - —3-1/2 inch floppy NC program input/output drive;
  - —In-process quality checks, and
  - —Standard options including chip conveyor, flood coolant, installation kit, etc.
- *Equipment's consistency with future cells*. Future equipment must incorporate identical tool adaptation, CNC controls, and overall machine design.
- *Turnkey vendor*. Supplier is to provide all parts, material handling equipment, and workholding devices. Supplier is instructed to quote separately on the initial cutting tool package, programming of the initial part, and programming of four additional parts.
- *Quotation instructions*. Vendor is instructed to:
  - —Price options and accessories separately;
  - —Price tooling and fixturing packages separately and itemize the lists;
  - —Provide statement of accuracy and repeatability for each axis of motion;
  - —Provide list of users of identical or similar machines;
  - —Express shipping date in weeks after receipt of order;
  - —Provide statement of position regarding OSHA requirements;
  - —State delivery FOB at the purchaser's plant;
  - —State that training is included in the price of the machine;
  - —Provide cost and duration of additional training, and
  - —State cost of runoff of production parts in buyer's facility.
- *Schedule*. This involves the presentation of milestone dates such as quotation receipt, purchase order release, and delivery of machine.

After the presentation, a brief shop tour was conducted and the vendors reconvened for a question-and-answer period. Packages of information were distributed, including:

- Facility layout;
- Part drawings;
- Volume and variety quantities;
- Copy of the presentation, and
- Contact names and phone numbers for questions.

For standard machine tools, this format works well. After the vendors' meeting, responsibility for the next step is in the hands of the vendor. The proposals will be similar enough to allow the buyer to make a comparative evaluation.

Case Study 2

Company B needed to purchase a multimillion dollar special manufacturing system to machine large aluminum weldments (8' by 8' by 20'). Both volume and variety are low. The large amount of machining, however, requires days to complete; machining is on six sides plus at various angles. For such a large investment, Company B decides to develop extremely detailed turnkey specifications. The specifications and RFP require two volumes of data which total over 250 pages.

Volume one: the quotation guidelines. The sections in the first volume consisted of:

- *Invitation to quote letter.* This is a standard letter of transmittal inviting the vendor to submit a quotation. The letter notes that the two volumes of data are enclosed and that, if awarded the contract, the vendor would be subject to compliance to the contents of both volumes of data. The letter may also highlight the contact's name and phone number and the proposal's due date, or other important information. It also tells the vendor who to notify and what to do with the specification if the vendor's firm declines to provide a quote; in most cases, it is best for the buyer to ask for the return of the information.
- *Instructions for quoting.* This section defines in detail how the quotation *must* be returned to Company B. This includes the number of copies, completion of any questionnaires/forms, concept drawings of the system, space and service requirements, control system details, critical timing, and estimates of all operations manning.
- *Conditions of quotation.* In this section, vendors are notified that it is the vendor's responsibility to understand and comply to what is contained in the documentation, through their own efforts or through further interaction with the buyer, and that any further significant information shared with a bidder will be made available to all bidders to ensure fairness. Proposals that do not conform to the RFQ's guidelines will not be considered. This section also specifies:

257

—The time period for which the proposal's prices must be considered firm and fixed;

—That quotations for the entire system *only* will be accepted, and

—That all information is to be treated as "private and confidential." Typically, if confidentiality statements are required, they are included in this section.

- *Quotation form.* The quotation form states that all quotations should be returned in a sealed envelope and the name and address of the person to whom they should be returned. The signature of someone who is authorized to sign quotes and make contractual agreements is requested; this party agrees to the terms and conditions mentioned in the documents.
- *Questionnaire.* This is a document to be completed by the vendor identifying the company's:

  —Finances and structure;

  —Liaison people—names, titles, phone numbers, and

  —Facility capabilities.
- *Pricing document.* This section declares that all prices will be fixed for the duration of the contract in U.S. dollars, and that all prices will be itemized separately in the following manner:

  —Prices related to machinery and equipment;

  —Prices related to computer systems;

  —Prices related to training, provision of manuals, and documentation;

  —Prices related to maintenance, spare parts, and parts lists;

  —Hourly rates for additional on-site work duly authorized by the manager of the program, and

  —Shipment terms and estimates.
- *Contract program—key dates.* Along with the quote, the vendor will submit a schedule of key contract program dates that will enable the company to achieve production quantities as planned. Adherence to the schedule will be a condition of the contract. Also stated is when the order will be placed. Any penalty clauses will be clearly indicated with terms, conditions, and results.
- *Notice of intention to use subcontractors.* In this document, vendors must state which, if any, part(s) of the work they intend to subcontract. The vendor is told that he is responsible for the subcontractor's compliance to all of the buyer's terms and conditions and is asked to show how he will manage and insure this relationship.
- *Acceptance criteria.* Included in this section are acceptance procedures/ criteria for:

  —All drawings and specifications;

  —Adherence to engineering and manufacturing requirements;

  —System operation/cycle time verification;

  —Prove-out requirements;

  —Verification and ultimate acceptance;

  —Quality specification—measurement and validation;

—Adherence to build/ship schedule;
—Written documentation, and
—Resolution of conflicts.

The vendor is requested to detail all information about test parts (quantities, when, and where) that the vendor might expect the purchaser to provide for development and runoff performance tests.

Volume two: the specifications. The second volume deals mostly with the details of the turnkey equipment specifications:

- *Overall concept of machining facilities.* This is a brief description of the program and the direction that the company is taking in terms of modernization and technology.
- *Requirements for manufacture.* This section contains the manufacturing parameters and requirements that characterize the company's plans for a manufacturing system. For example, advance planning within the framework of adopting CIM at a later date might be a consideration.
- *Functional specification.* These pages of documentation conceptually define how the manufacturing system could operate. Included are: the routing, and a general description of the machinery, the control system, quality requirements, fixturing, the material handling system, chip and coolant system, and tool preparation area. *Figure 15-2* is an excerpt of the quality portion of this section.
- *Performance specifications.* The manufacturing parameters and machining requirements itemized in this section summarize the features deemed necessary to machine the various models in random production by Just in Time scheduling. Computer control systems must be capable of standalone operation and might be linked to form part of a CIM system later.
- *Standard specifications.* This is a listing of the company's standard specifications, including electrical, computer, hydraulic, pneumatic pressures, paint color, etc.
- *Quality specifications.* This section identifies the quality procedures for both the quality of the manufacturing system itself and the expected quality of the machined parts. It deals with machine repeatability issues and conditions for a capability study. It describes the test conditions and the procedures that the capability study will be conducted under to establish process control.
- *Tool management specification.* This part contains a description of a system to control the inventory of tools and provide the data necessary for advance tool preparation. It describes how tools will be identified, tool standardization, tool setting systems, and tool change procedures.
- *Manufacturing control system specification.* This section delves into the system architecture, operations control system, and interfaces between the operation control system and the equipment. It describes in detail the actual control system of the equipment, including adaptive control, broken tool sensing, fault transmission, and spindle probing.

259

QUALITY SPECIFICATION EXCERPT

3.6.0   QUALITY REQUIREMENTS

The ability of the manufacturing system to achieve design specification quality consistently, and in the long term, is paramount in the development of the machinery and inspection equipment. The design aim is therefore focused on the capability to achieve the following:

3.6.1   The machine tools shall be capable of maintaining dimensional and geometric accuracy in accordance with the company's machine tool quality standards, Volume II, Section 6.

3.6.2   In order to ensure that the above specification is consistently achievable, it is expected that vendors utilize any or all available practical measuring and adjustment techniques, including, but not limited to, the following:

A.   Automatic tool breakage sensing (probes, electromagnetic, light sensing, etc.) on all features mutually agreed as critical

B.   Machine tool computer monitoring

C.   Special automatic quality stations with feedback to automatic tool compensation where applicable

D.   In-system weldment qualification with automatic data feedback, position compensation, and trend analysis

E.   Process capability analysis

F.   Automatic part/fixture identification for part program selection

G.   Use of statistical process control methods as applicable
Note: These features should be incorporated with the ability to manually override

Figure 15-2. Example of functional specification outline.

- *Maintenance specification*. This discussion contains a description of how the machine maintenance system will be designed to request planned preventive maintenance automatically at preset intervals for machine tools and other major equipment. Details on how the lubrication systems must conform to the company's standard specification are also included.
- *Installation and runoff*. This is a definition of the installation expectations from the vendor, and the buyer's people who will be required to be involved. It includes the fact that a runoff will be conducted both preacceptance at the vendor's facility and at the company's plant for final acceptance. This section also details in what state the equipment must be for a pilot run and final runoff.
- *Training*. This includes identification by the vendor of the training programs the vendor will conduct for *all* the disciplines involved in running the equipment, including maintenance, tooling, programming, etc. Submission of training manual samples, documentation, etc., is required prior to application of programs.
- *Documentation*. This discussion identifies the various drawings and other documentation needed for machine operation and installation—and when they are required and in what quantity. Foundation details should be required within a reasonably short period after final purchase approval.
- *Manufacturing system warranty and maintenance*. This section defines how long the vendor must warrant the system and who is responsible for the maintenance. It may require that the vendor supply emergency spares on hand for critical warranty parts, and will also specify conditions for warranty submittal for parts, labor, and travel-related costs.

Very rarely should a detailed specification like the one in Case Study 2 be the first thing a vendor learns of a buyer's project. Equipment appropriation at this level works best when a buyer "partners" with a potential vendor to simultaneously engineer the requirements together. Problems can be worked out before they become project-stoppers or major expense items. Schedules will become a common goal. In general, the buyer's staff will be much more informed and accepting than if the vendor went away with the specs and came back with a machine.

SUMMARY

There is, unfortunately, no cookbook for the preparation of specifications. The temptation to dismiss the process as a "paperwork effort" is all too real. It is, however, extremely important that these actions be taken to ensure that the appropriate machines are secured. Manufacturers who don't do this development work may end up structuring their thoughts and needs according to the vendor's agenda.

IIII 16 IIII

THE VENDOR SELECTION PROCESS

Evaluating and selecting suppliers for long-term, profitable relationships can be an enjoyable and educational experience. It requires some discipline and a few specific techniques to make the effort fruitful without being overly burdensome. Good vendors can supply a wealth of services and information if asked the right questions and given the necessary information.

The previous chapter discussed vendor relations to the point where the manufacturer explained what was wanted and the vendor went away to prepare the response. The manufacturer:

- Developed a potential vendor list and began to gather information for an equipment/vendor evaluation matrix;
- Defined machine features and functions and made sure that they were appropriate for the organization;
- Prepared the appropriate level of specification and business detail, and
- Sent out requests for quotations.

If equipment specifications were carefully prepared, and a broad cross-section of vendors was invited to the prebid meeting or given the RFQ, the team members will have had some initial exposure to the vendors themselves and listened to their questions. If the vendors were all placed in a room together, they'd know their competition, which should help sharpen their approach for the work ahead.

The remaining steps of the process will require maintaining the team effort to:

- Review proposals to complete the evaluation matrix;
- Identify the major candidates;
- Arrange vendor/installation visits to further develop the relationship;
- Select the vendor of choice;
- Finalize the purchase agreement, and
- Track the vendor's performance against the purchase criteria.

263

MAINTAINING THE TEAM

Traditionally, soliciting vendor proposals and undertaking vendor "bashing" (tough price negotiating) is a function of the purchasing department. These staff members may know little about the technical requirements involved, but they may hold the only keys to the drawer holding the all-important "purchase orders." Purchasing agents have traditionally relied on engineering staff members to write specifications and maybe even recommend one or two potential vendors. In this scenario, when a program demands new equipment purchases, the effort of contacting vendors, requesting quotations, answering vendors' questions (business *and* technical), and evaluating their proposals is most often done in addition to the purchasing person's everyday job. During new programs, that job also has the added burdens of processing prototype parts, going through costing exercises, holding review meetings, etc.

The team approach to evaluating vendor offerings transcends the functional pigeonholing of "engineer" and "purchasing agent." An example of just how effective this process can be with large capital purchases was depicted in the September 1990 issue of *Purchasing World*. Warren Consolidated Industries instituted the team buying concept during a $100 million project to design, purchase, and install a new continuous caster for the company's steelmaking and casting company. Project manager Dale Musolf indicates that the team concept permitted the company to evaluate information quickly and that "we didn't set out with preconceived notions of what we wanted....we developed a machine from what we saw" while visiting potential vendors. Purchasing manager Robert Stasko contends that due to the great success of the effort, actual savings are difficult to measure: "The pitfalls never materialized. Costs could have been significant and devastating....the team concept provides a fail-safe mechanism."[1]

PUTTING TOGETHER THE EVALUATION MATRIX

In the last chapter, the manufacturer began to collect vendor/equipment information, with items like options, delivery, prices, NC criteria, and so forth. Once the specifications and RFQs are given to vendors, this document needs to be formalized and converted into an evaluation matrix for comparing the quotations. There is usually some time to accomplish this, since most vendors require a few weeks to put together a proposal. The matrix should not be finalized too early in the process, since vendor questions will lead to the discovery of additional criteria, as may the receipt and evaluation of the first few quotes.

The main function of the matrix is to provide a means to compare vendor offerings in a structured, fair, and explainable fashion. The matrix is begun by listing machine *features*. Later, each vendor's offering is posted against these in a way that explains if and how the proposed equipment addresses the feature. The spreadsheet in *Figure 16-1* shows a typical layout and suggested headings which might be applied to evaluating the features of a machining center (disregard the first column for now). The top rows provide key data on each vendor; each column of information then pertains to that vendor's machine. This spreadsheet presents the features in a sequence perhaps typical of that found in a quotation package, oriented around machine features like dimensional data, tool changing, speeds, feeds, and control features. It is useful to begin organizing the information and, with simpler machines, may in fact be all that is needed to reach a decision.

Many times, however, a method is needed to reorganize what the vendor says into something meaningful to the company. There are various *functions* the company requires that these machines with their features fulfill. These functions have varying degrees of importance to the company. Typical functional considerations might be those listed in *Figure 16-2*. The company may rank their importance as shown in *Figure 16-3*. Going back to the spreadsheet then, each feature can be classified as falling within one of these functions; the first column of the spreadsheet serves as a "key" for sorting the features by function.

Figure 16-4 shows part of the features spreadsheet regrouped into a functional one. Later on, as the information blanks are completed through further interaction with vendors (eliminating as many of the "not specified" entries as possible), each feature within the functional group (or the functional group as a whole) may be assigned a score between 0 and 10 for each machine examined. The scale should be defined so that a score of, say, 5 means the machine meets requirements. A score greater than 5 would mean the machine exceeds requirements. A rating less than 5 means the machine falls short of requirements. The ratings can be averaged for the function and multiplied by the weight factor (the percentages from *Figure 16-2*). The sum of these weighted functional factors for each machine determines its overall, relative suitability.

The team's agreement is needed on the relative importance of each of these machine features and functions so that appropriate weight factors can be applied. Team members must not go into the process with preconceptions about the machines, their features, or the vendor. Nor should there be motives to rate highly the features needed for their departments to ensure their inclusion.

It is usual to use one of the popular spreadsheet software packages to analyze the criteria for evaluation and decision processing. If the spreadsheet is prepared properly, it can total points automatically (no math errors), and sort the criteria in different ways to provide a dynamic document that allows isolating key features for more detailed comparisons. Perhaps most important, the spreadsheet provides a simple tool for answering the inevitable "what if?" questions. Weighting factors for key features can then be varied, and a determination made as to the effect each might have on the choice of suppliers. (Another example of weighted factor analysis can be found in Chapter Seven.)

FUNCTION	Manufacturer: Vendor: Model/Configuration:	Manufacturer A Vendor X 1234-HMC	Manufacturer B Vendor Y 91011-VMC	Manufacturer C Vendor Z 1319BX-HMC
	FEATURES			
	DIMENSIONAL DATA			
A	Table/pallet length (inches)	15.7	3.5	15.75
A	Table/pallet width (inches)	15.7	19.5	15.75
A	Max. pallet load (pounds)	1,100	1,100	880
A	"Z" axis travel (inches)	19.7	17.7	18.11
A	"X" axis travel (inches)	27.6	27.6	22.04
A	"Y" axis travel (inches)	15.7	17.7	18.11
I	"Z" axis thrust (pounds)	3,300/5 minutes	3,300/5 minutes	Not specified
I	"X" axis thrust (pounds)	880	880	Not specified
I	"Y" axis thrust (pounds)	880	880	Not specified
A	"B" axis increments	1 deg.	0.001 deg.	1 deg.
Q	"B" axis accuracy	+/- 3 seconds	+/- 10 seconds	+/- 5 seconds
Q	"B" axis repeatability	+/- 1 second	+/- 10 seconds	+/- 3 seconds
I	Spin. hose to table surface/CL	16.6" max.	25.5" max.	5.9" min.
I	Spin. hose to table surface/CL	4.7" min.	7.9" min.	24.01" max.
I	Machine OAL	15'	9' 8"	8' 3/8"
I	Machine width	9'	8' 4"	11' 11"
I	Overall height	8' 9"	8' 4"	8' 9"
I	Weight (pounds)	15,400	18,000	17,000/20,062
I	Pallet height from floor	43.3"	Not specified	35.82"
	SPINDLE DATA			
I	Spindle diameter (inches)	2.95	2.95	2.76
V	Spindle taper	Cat. 40	ISO 7/24 No. 40	#40 Cat. "V" flange
I	Quill (Y/N), diameter	No	No	No
T	Minimum RPM, spindle	80	80	45
T	Maximum RPM, spindle	8,000	8,000	6,000
T	RPM steps, spin, (increments)	Not specified	Not specified	1 RPM
T	Horsepower—Continuous (.75 Kw = 1 HP)	7.33 HP	7.33 HP	10 HP
T	Max. horsepower—duty cycle	10 HP/30 minutes	10 HP/30 minutes	15 HP/30 minutes
	TOOL CHANGER			
A	Number of tools (std.)	32	20	31
A	Number of tools (option 1)	60	30	50
A	Number of tools (option 2)	120	40 or 60	100
T	Bidirectional (Y/N)	Yes	Yes	Yes
T	Tool change time (chip to chip)	10 seconds	10 seconds	10 seconds
I	Tool recognition (Y/N)	No	No	No
A	Maximum tool diameter (inches)	5.9	5.9	Not specified
A	Maximum tool length (inches)	11.8	11.0	11.8
A	Maximum tool weight (pounds)	22	22	22

A = Adequacy	R = Reliability/service/reputation/diagnostics
C = Cost	S = Setup time factors
D = Delivery	T = Cycle time factors
I = Information only	V = Versatility
Q = Quality/accuracy/repeatability	U = Unmanned operation

Figure 16-1. *Typical machining center evaluation by features.*

FUNCTION	FEATURES	Manufacturer: Vendor: Model/Configuration:	Manufacturer A Vendor X 1234-HMC	Manufacturer B Vendor Y 91011-VMC	Manufacturer C Vendor Z 1319BX-HMC
	POSITIONING SPEED/ ACCURACY				
T	"X" axis rapid traverse rate		590 IPM	590 IPM	629
T	"Y" axis rapid traverse rate		590 IPM	590 IPM	629
T	"Z" axis rapid traverse rate		590 IPM	590 IPM	472
Q	"X" axis accuracy		+/- 0.0002"	+/- 0.0002"	+/- 0.0002
Q	"Y" axis accuracy		+/- 0.0002"	+/- 0.0002"	+/- 0.0002
Q	"Z" axis accuracy		+/- 0.0002"	+/- 0.0002"	+/- 0.0002
Q	"X" axis repeatability		+/- 0.00008"	+/- 0.00008"	+/- 0.00008
Q	"Y" axis repeatability		+/- 0.00008"	+/- 0.00008"	+/- 0.00008
Q	"Z" axis repeatability		+/- 0.00008"	+/- 0.00008"	+/- 0.00008
	MACHINE OPTIONS				
T	Chip conveyor		Optional (included)	Yes, but unclear	Std. (screw type)
T	Flood coolant		Optional (included)	Optional (included)	Std.
I	Thru spindle coolant		Optional (not quoted)	Optional (not quoted)	Optional (not quoted)
I	Mist coolant		Optional (not quoted)	Optional (not quoted)	Optional (not quoted)
I	Work lamp		Optional (included)	Optional (included)	Std.
R	Adaptive control (Y/N)		Optional (not quoted)	Optional (not quoted)	No
Q	Tool management		Optional (not quoted)	Optional (not quoted)	Optional (not quoted)
S	Auto tool offsets—"Z" axis		Optional (included)	Optional (included)	Optional (not quoted)
I	Light tower		Optional (not quoted)	Optional (not quoted)	Yes, cycle finished
A	Number of pallets		2	Optional (not quoted)	2
T	Pallet change time		Not specified	N/A	20 seconds
T	Pallet index time		90 deg. in 5 seconds	N/A	90 deg. in 5 seconds
	QUALITY CONTROL				
Q	Part probe		Optional (not quoted)	Optional (not quoted)	No
	CONTROL FEATURES				
V	Control manufacturer		Mfg. X	Mfg. X	Mfg. X
V	Control model		0M8	11MF	15M model "A"
I	Number of simul. axes controlled		Not specified	Not specified	3
V	Linear interpolation (Y/N)		Yes	No	Yes
V	Circular interpolation (Y/N)		Yes	No	Yes
S	Memory capacity (10' = 1Kb)		260 feet	1,050 feet	264 feet
R	Battery backup (Y/N)		Yes—automatic	No	Yes—semi-automatic
R	Diagnostics		Yes	Yes	Yes
S	Cutter radius compensation		Optional (included)	Optional (included)	Yes
	DISPLAY				
I	— Actual position		Not specified	Not specified	Not specified
I	— Distance to go		Not specified	Not specified	Not specified
I	— Spindle speed		Not specified	Not specified	Not specified
I	— Feed rate		Not specified	Not specified	Not specified
I	— Tool offset		Not specified	Not specified	Not specified
I	— Tool number		Not specified	Not specified	Not specified
I	Smallest programmable increment		Not specified	0.0001"	0.0001"
S	Feed override (% to %)		Not specified	Not specified	0 to 200% (10% steps)
S	Rapid trav. override (% to %)		Not specified	Not specified	50%, 100%
S	Speed override (% to %)		Not specified	Not specified	50% to 120%
S	Number of tool offsets		99	99	99

Figure 16-1. *(continued) Typical machining center evaluation by features.*

267

F U N C T I O N	FEATURES	Manufacturer A Vendor X 1234-HMC	Manufacturer B Vendor Y 91011-VMC	Manufacturer C Vendor Z 1319BX-HMC
	OPERATION MODES			
U	— Manning	Fully automated	Manual	Semi-automated
S	— Execute (input memory)	Not specified	Not specified	Not specified
S	— Edit in background (Y/N)	Yes	Not specified	Yes
I	— Automatic (NC program)	Not specified	Not specified	Not specified
	OPERATION SUBMODES			
S	— Single block	Not specified	Not specified	Yes
S	— Dry run	Not specified	Not specified	Yes
I	— Block delete	Not specified	Not specified	Not specified
I	— Skip block	Not specified	Not specified	Yes
I	— Reference point	Not specified	Not specified	Yes
I	— Status	Not specified	Not specified	Not specified
I	— Interface	Not specified	Not specified	Not specified
	PROGRAM FORMAT			
I	— ETA RS2740	Not specified	Not specified	Yes, E/A RS-244
I	— ASCII (RS358)	Not specified	Not specified	Not specified
I	— ISO code	Not specified	Not specified	Yes, ISO 848
	DATA INPUT/OUTPUT			
I	— RS-232 C interface	Optional (included)	Optional (included)	Yes
A	— 3 1/2" floppy	Optional (not quoted)	Optional (not quoted)	Not specified
I	Cassette tape	Not specified	Not specified	Not specified
I	Punched tape reader	Yes	Not specified	Yes
	CONSTRUCTION			
I	Cast iron	Not specified	Yes	Not specified
I	Fabricated steel	Not specified	No	Not specified
I	Granite—epoxy matrix	Not specified	No	Not specified
I	Foundation required (Y/N)	Not specified	Not specified	Not specified
	SERVICES			
R	Nearest service center	20 miles	100 miles	350 miles
R	Nearest service technician	5 miles	100 miles	150 miles
D	**DELIVERY**	One week		Not specified
	COST			
C	Base machine price		$137,100.00	$235,000 est.
C	Price of options		$1,420.00	See quote
C	Total machine price	$235,590.00	$138,520.00	$235,000 est.
	SHIPPING			
I	FOB	Los Angeles, CA	New Jersey	
	UTILITIES			
I	Voltage	22OV/3Ph/60Hz	220V/3Ph/60Hz	Not specified
I	Total connected load	20 kVA	25 kVA	Not specified
I	Air (1Kg/sq. cm = 14.2 PSI)	57 PSI, 24.7 CPM	57 PSI, 24.7 CPM	Not specified

Figure 16-1. *(continued) Typical machining center evaluation by features.*

Adequacy	The machine is large enough to produce the product. The manufacturer offers at least the minimum features and options that are required to produce the parts. (Failure to satisfy this group of features precludes any further analysis.)
Cost	The cost of the machine plus options, foundations, and any special utility costs, itemized and totalled.
Delivery	The promised shipping date relative to company's target(s) as presented during the pre-bid meeting or in the RFQ.
Information Only	Machine specifications which are of interest but will not strongly influence selection.
Quality	The designed accuracy and repeatability of the machine; added options which allow for in-process inspection and correction of tool wear or other causes of dimensional drift.
Reliability	Location of the nearest service center and service-person. Availability of diagnostics. Overload protection. Reputation for prompt parts and service from current users of the equipment.
Setup Time	Features which shorten setup times, such as automatic tool offsets, quick-change tooling, tape debugging features in the control, and fixture mounting concepts.
Cycle Times	Features that affect the floor-to-floor production time cycle. Typically, these will include tool change times, horsepower, rapid traverse rates, number of spindles running at one time, and pallet shuttle times.
Unmanned Operation	This is specifically desirable with multiple pieces of equipment. It incorporates features which enhance unmanned or lightly manned operation of the machines. It would include tool management systems, light towers to flag problems, bar tubes or stock feeders, parts catchers, adaptive controls, etc.
Versatility	Features which enhance the probability of using identical equipment in other product cells. These will include the commonality of tooling and CNC controls. Other features are considered but vary with the specific machine type.

Figure 16-2. *Functional considerations for evaluation matrix.*

Function	Weight (Percent)
Adequacy	20
Cost	5
Delivery	5
Quality	20
Reliability	15
Setup time	15
Unmanned operation	5
Cycle time	10
Versatility	5
	100%

Figure 16-3. *Weight factors are assigned to the functions based on their importance.*

PROCESSING PROPOSALS AND IDENTIFYING MAJOR CANDIDATES

As the proposals begin to return, team members appointed to review them should carefully attempt to fill in the matrix with the pertinent data, clarifying any missing or unclear information with the vendor. The real world says that not everyone prepares their information in the same format, nor will they necessarily describe similar functions in the same terms. Evaluators should be prepared to spend a lot of time reading and rereading the packages to dig out the necessary information. Phone calls to vendors should be made as needed without hesitation. This step provides a more thorough understanding of each package and gives some insight on the ease of vendor interaction and the speed and adequacy of the vendors' responses. (Certainly a low-cost way to test the system!)

To ensure impartiality in the review process:

- Proposals should be handed out to all evaluators before any notes are made on the copies, and without any comments from the team leader like, "Here's a real bomb!"
- Some team leaders may try to ensure unbiased evaluations by blackening out the name of the vendor where it appears in the proposal to make it illegible, or copying the proposal onto plain white paper to diminish the impression made by the proposal's "packaging."
- Pricing information may also be deleted from the proposals that are circulated. This ensures that reviewers who do not usually make large spending decisions will not be intimidated by the dollar amounts and start dismissing options or automatically select the cheapest proposal. Proposals

F U N C T I O N	Manufacturer: Vendor: Model/Configuration:	Manufacturer A Vendor X 1234-HMC		Manufacturer B Vendor Y 91011-VMC		Manufacturer C Vendor Z 1319BX-HMC	
	FEATURES						
			Score		Score		Score
R	RELIABILITY (Weight: 15%) Adaptive control (Y/N)	Optional (not quoted)	5	Optional (not quoted)	5	No	1
R	Battery backup (Y/N)	Yes, automatic	8	No	1	Yes, semi- automatic	5
R	Diagnostics	Yes	5	Yes	5	Yes	5
R	Nearest service center	20 miles	8	100 miles	5	350 miles	1
R	Nearest service technician	5 miles	10	100 miles	3	150 miles	2
	SCORE (Average Score x Weight)		1.08		.57		.42
	VERSATILITY (Weight: 5%)						
V	Spindle taper	Cat. 40	3	150 7/24 No. 40	7	#40 Cat. "V" flange	4
V	Control manufacturer	Mfg. X	5	Mfg. X	5	Mfg. X	5
V	Control model	OM8	4	11MF	4	15M Model "A"	6
V	Linear interpolation (Y/N)	Yes	7	No	2	Yes	7
V	Circular interpolation (Y/N)	Yes	8	No	3	Yes	8
	SCORE (Average Score x Weight)		.27		.21		.30

A = Adequacy	R = Reliability/service/reputation/diagnostics
C = Cost	S = Setup time factors
D = Delivery	T = Cycle time factors
I = Information only	V = Versatility
Q = Quality/accuracy/repeatability	U = Unmanned operation

Figure 16-4. *Machining center evaluation by function.*

should not initially be excluded from consideration based on price; it may be that the best proposal does not have the most reasonable price tag, but that price can later be negotiated downward with the vendor.

- Evaluators should decide at what point they will begin to discuss the proposals with each other: after each one is received, a group at a time, or not until all are received and reviewed. There should also be a formal method to make sure that everyone records his opinion; it is difficult for someone to give his views after his boss has stated all of his own preferences and then asks, "By the way, what do you think?"

- Evaluators also need to remember to revisit the proposals they reviewed first to make sure their priorities and impressions haven't changed as they received more and more documents.

At the end of the analysis, the team would probably want to apply some subjective judgement as a sanity check on this somewhat mechanical method of making a decision. This is certainly a prerogative and a duty. Team members may feel more comfortable with the second highest point scorer than the first. This is the time to include what might be traditionally "intangible" factors—gut level, instinctive criteria, like vendor name or generally accepted "track record" in the industry, or base knowledge or experience of the company representative. These are all important criteria when making major decisions.

Those involved in proposal evaluation develop more knowledge about the equipment. They will demonstrate more credibility and confidence in any future discussions, as the package is presented and explained throughout the justification process. The disciplines described here, and the documentation which accompanies them, provide a means to "quantify" choices so others can understand the logic, but they allow flexibility to include those intangible elements so often key to sound decision-making.

MAKING VENDOR-ARRANGED VISITS

One or two on-site visits to the primary candidates or some users of the vendors' equipment is in order at this point. As part of the original package, vendors were requested to supply information about one or more users of identical or similar installations. As the evaluation matrix boils down the choices to several key leaders, there is nothing like a test drive to provide the final confirmation of choice. And if the vendor is a good choice, he will not have any trouble setting up a visit to show off another "satisfied customer." As with other activities, this one is best conducted by the critical members of the team.

Because of the team's mix of disciplines, team members can evaluate on-site installations from a variety of perspectives. They can talk to people actually using the equipment under consideration and find out their experiences with uptime, operator training, spare parts purchase, service availability, and the like. Buyers should take the opportunity to evaluate the vendor for compatibility, after-sale service, and flexibility. They should feel free to challenge vendors on all aspects of the design and application of the intended equipment. If necessary, team members should ask the vendor to leave them alone with the customer to make sure everyone can speak freely. New ideas about the equipment to be purchased will begin generating from these visits. The few days they will take is well worth the effort. (Note: If a vendor takes a potential buyer to a client site, the buyer should be as nonintrusive as possible, and thank the company promptly for allowing the visit. Buyers may also wish to consider allowing the selected vendor to bring visitors to their facilities when installation is complete, to return the favor and for promotional purposes.)

SELECTING A VENDOR

After all of the vendor proposals have been received and evaluated, and with all the information in hand, the team should determine that:

- The evaluation matrix and its associated factor weighting scale are still applicable, and
- Vendor responses were interpreted and scored correctly.

After this has been agreed upon, the best decision will many times be obvious; the team can then concentrate on preparing to present the matrix and the decision criteria to whoever (steering committee or managers) needs to approve spending the money that has been allocated. If it is a clear decision, the team should consider inviting the selected vendor to participate in the management presentation. To move the process along, any further negotiation for purchase terms to consolidate a contract should now be in the open and documented for both parties.

If the best decision is not that obvious, the team must identify its next step. Can it make a confident decision based on what it knows? Should management be asked to make the decision? (It may be that management has reserved this right.) Or is more information required? Is the team really considering the right equipment? Is compromise required—do the weights assigned to functions need to be changed, or do the required functions need to be relaxed to allow a choice? Should other vendors who were ruled out earlier now be considered? When faced with these options, the team may decide that it can make a decision after all!

To improve their performance in the future, the vendors who were *not* chosen by the manufacturer (the ones striving to improve, anyway) may ask for feedback on their performance; this is an excellent opportunity for the buyer to influence vendors and affect future interaction in terms of quality, price, delivery, customer service, technical offerings, proposal contents, selling methods, etc., and he should speak freely. Many buyers will include some of this reasoning in their letter or phone call to vendors thanking them for their participation but telling them they will not receive the contract; some invite the vendors to phone or ask for a meeting should they wish to learn more. It takes some effort on the buyer's part, but it can certainly be worth it when the next major purchase rolls around and more vendors are in tune with the company's needs and preferences. The manufacturer should also document these concerns for his own company files on the vendor, to assist others in the organization when they face similar interaction.

FINALIZING THE PURCHASE AGREEMENT

After the vendor selection meeting, the buyer's team and the selected vendor should formally meet at the buyer's site to detail and agree to the required terms and conditions. Those terms from the specifications that should be reviewed are:

- Equipment/cell/purchase goals;
- Part data (process, volumes, mix, features, materials);

273

- Machine features/functions;
- Performance requirements (setup, cycle, tolerances, uptime);
- Future compatibility;
- Delivery (progress dates, preapproval date, arrival date, setup/install dates, final acceptance deadline);
- Options/turnkey inclusions (tooling spares, handling devices, software);
- Costs (design, progress payments, machine/equipment options, installation, freight, training, warranty);
- Penalty and/or reward clauses (if applicable);
- Approval procedures (changes, progress, penalty);
- Runoff procedures, and
- Training (curriculum, documentation, timing).

For small equipment, this might only take a few hours. For some large, complex machine tools, it can take two to three days to iron out the final contract language and terms.

MONITORING THE VENDOR'S PERFORMANCE

This thorough selection process should help buyers choose vendors who routinely provide adequate service and products. Monitoring the vendor's performance to the agreed-upon terms should be a relatively painless process, if the buyer and vendor walk away from the final purchase session with common goals and the ability to profit from the partnership. Both parties should be working with the *same* time line schedule that documents all agreed-to dates. (Refer to Chapter 17 for some recommended time line formats.)

Setting up progress payments (where a portion of the fee is paid each time a milestone in the building process is met) will help ensure that the work is begun on time and is of high quality throughout the process. Thorough testing at the vendor's site and after installation at the buyer's facility will correct any last-minute problems. Adequate training will make the buyer's people as knowledgeable about the machine as the vendor, resulting in the greatest satisfaction.

SUMMARY

Vendors who want to stay in business will usually perform to buyers' expectations. But whether FMCs require a $5,000 washer or a $1.2 million machine tool, using some disciplined approach is necessary for organizing the assessment of needs, matching these needs to some potential vendors, documenting the agreed-upon purchase, and tracking the selected vendor's performance. The buyer will get what he wants if he is educated to the process, involves enough technical people, and can judge the viability of the vendor's match to his needs.

The purchasing process can always be improved upon. *Purchasing World* recently asked several experienced sales representatives what they would do differently if they were buyers.[2] Their top answers were:

1. Identify the *real* costs of the equipment to be purchased, including such issues as equipment life-cycle, maintenance, and operation costs.
2. Improve the needs assessment process, including clear input from the end user of the equipment.
3. Look more closely at the credentials and capabilities of vendors, seeking proof and validation.
4. Avoid making "safe buy" decisions; challenge the big vendors.
5. Use better motivational techniques on vendors, like recommendations or verbal praise.
6. Avoid over-specifying and over-buying for fancy new capabilities that are seldom, if ever, utilized.

ENDNOTES

1. "Team Buy Breaks Communication Wall," *Purchasing World*, September 1990, pp. 34-37.
2. Reich, Caroline G., "What Vendors Wish Buyers Knew," *Purchasing World*, December 1990, pp. 54-56.

INSTALLING FMCs

The best technical cell planning in the world will not result in a successful installation unless adequate basic preplanning takes place to make the installation happen as efficiently as possible and help monitor and manage implementation status. This important part of the team effort has the objectives of:

- Utilizing proven scheduling and project management tools to enhance team planning;
- Identifying all the installation steps and issues—including sorting them out, expanding on their detail, and extending them toward the true requirements;
- Using some traditional and nontraditional tracking mechanisms for verifying project schedule adherence.

The measure of a well-planned installation is the team's flexibility to adapt to unexpected changes that will happen along the way. Nowhere does Murphy's Law apply more than to actual installations. Effective teams must be so comfortable with the overall plans that they can anticipate, detail, integrate, and defuse as many unexpected situations as possible before they happen. This is why it's important that the planning team be kept intact for the implementation work.

CELL PREPLANNING

Planning cell implementation is no different from most other project management/installation approaches. This is illustrated in *Figure 17-1*. A cell installation plan takes on similar characteristics in logical sequences:

1. *Define project parameters*. In broad terms, the team should establish the boundaries of the cell project, which will determine the size, cost, time, and quality that go into the work. These help set the project priorities and

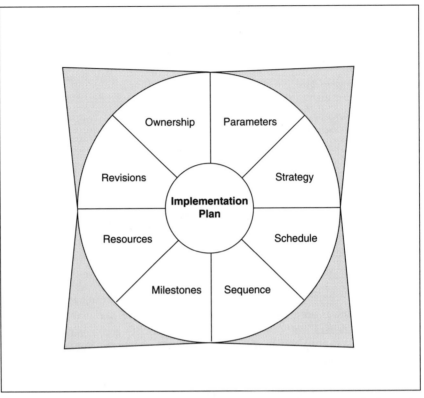

Figure 17-1. *The characteristics of a project management/installation approach also apply to cell implementation planning.*

measurable goals. A cell layout will assist in documenting many of the critical boundaries. Other parameters to consider are preliminary budgets, end goals, and available resources.

2. *Develop the project management strategy.* The project strategy determines the project management technique. Project strategy possibilities include one or more of the following:
   — Speed of installation at any cost;
   — Absolutely no production interruption;
   — Low-cost installation—no frills;
   — Use of in-house resources only; or
   — Quality supersedes quickness as a key installation criterion.

   The project strategy must be achievable and accepted company-wide for the project management technique to be effective.

3. *Create the project schedule.* Project details should be organized by task or specific activity level. The team should attempt to outline a schedule by breaking the project into logical subprojects. The work breakdown structure (WBS) format generally used to define key project elements will be discussed later in this chapter. From this can come more specific

278

detailed tasks. If the team is uncomfortable with structuring an outline first, it should start with a brainstorming session of all possible activities to get them on paper. The list can be organized into an outline later. (It's very likely that a logical outline will quickly become apparent after looking at the list of tasks.)

4. *Determine the move or task sequences.* Move sequences are usually influenced by constraints or primary drivers such as cost, time, etc. Many activities are dependent on another activity's completion or simultaneous start. Various techniques for logically determining appropriate sequencing are discussed later in this chapter.

5. *Define key milestone targets.* All plans must be tracked and reviewed on a regular basis. Key status review points should normally correspond to the achievement of a few significant results (milestones). Well-chosen milestones present clear targets to be pursued and natural review break points in the course of the project.

6. *Assign resources (people).* The team must next identify available resources and match them to appropriate activities, making sure that resources are defined to precisely reflect specific requirements. Things like skills, timing, and the number of people required are very important in making appropriate assignments.

7. *Review and revise the plan.* Fortunately, all good plans can be made better. Team members should not be afraid to continually review and revise plans. However, they must keep *everyone* informed and involved in changes.

8. *Solicit plan approval/ownership.* The best way to win plan approval is to directly involve the decision makers in the project throughout the plan's development. Subsequently, ownership by the entire organization must be solicited.

9. *Install the plan.* By now the actual implementation should be automatic, right? Don't believe it. Physical moves are the easiest to plan for and to adjust for revisions or problems. Changing people's attitudes and paradigms will be the toughest part, so team members shouldn't underestimate the impact of the changes. Communication, involvement, and training will be paramount.

BRINGING IT TOGETHER IN AN INSTALLATION SCHEDULE

There are three functions that project managers must emphasize:

- *Planning and replanning*: This is the actual development of information— manipulating the data and molding it into a workable plan, and not forgetting to change that plan as more becomes known. This goes on throughout the implementation.

- *Controlling or managing the plan*: This involves working to fulfill the plan, or managing the activities to complete the installation in the sequence and time allowed.

279

- *Communicating*: Communication means keeping everyone involved, motivated, and informed of developments.

A project manager's time should roughly be divided equally among the three functions, all of which can be performed better through good scheduling. There are three typical ways to organize and display an installation schedule—network planning techniques (Gantt charting), the critical path method (CPM), and the project evaluation and review technique (PERT). Neither the CPM nor PERT technique is very practical for cell planning projects, because they are usually too complex to manipulate and maintain. Both CPM and PERT rely strongly on:

- Dependency (precedence and successor) relationships:
  - Establishment of a critical path;
  - Known durations and associated costs;
- Ability to identify tasks that can be affected by added costs (known as crash costs):
  - Heavy computer calculation requirements.

CPM was developed by the Catalytic Construction Company in 1957. It was first used in 1959 on a joint plant construction project for Remington Rand and DuPont. Its main purpose was to develop a project schedule that minimized total project cost.

PERT was developed in 1958 by Booz, Allen & Hamilton, Inc., along with the U.S. Navy Special Projects Office. The Navy first used PERT to provide better and more intensive management of the Polaris Missile project. Its main purpose was optimization, with an emphasis on establishing a schedule having a known probability of successful completion within a given time constraint.

Even though CPM and PERT are cumbersome to use in cell planning exercises, there are several features of each that are valuable to use and display. The importance of the dependency relationships and the ability to recognize where and by how much the overall schedule can be influenced must be maintained even with the schedule planning of the Gantt chart.

Network Planning Techniques (Gantt Charting)

Gantt charting was developed by Henry L. Gantt, an associate of Frederick W. Taylor and a supporter of scientific management. It was introduced in 1911 to control repetitive manufacturing processes. The technique was also widely used in World War I. It allows activities to be depicted as a series of tasks to be accomplished, with each task associated to a length of bar on the chart, as shown in *Figure 17-2*. It has the advantage of displaying all tasks on one chart that is easily understood by almost anyone regardless of experience or hierarchical level.

The U.S. Navy expanded this technique in the 1940s and developed the "milestone method." This was done by attaching key time periods to the activities to allow graphics to highlight dependencies in a network plan. A typical plan is shown in *Figure 17-3*. Interrelationships between milestones demonstrate one task's precedence over another.

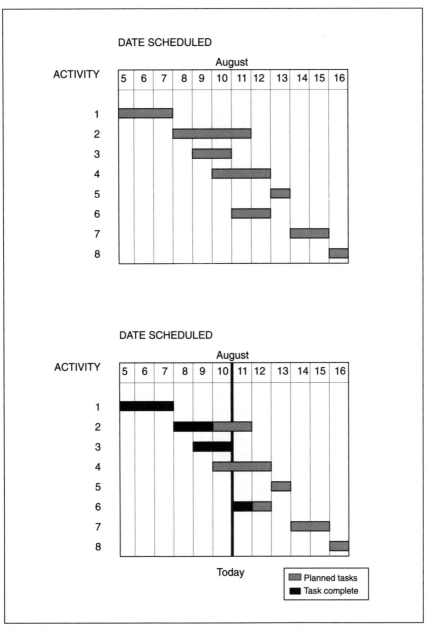

Figure 17-2. *The top Gantt chart shows the planned task schedule. Progress can then be tracked along the time line, as shown in the bottom chart. The team is falling behind on activity 2 and 4. It is ahead of schedule on activity 6.*

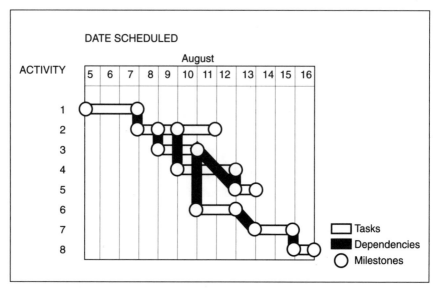

Figure 17-3. *Dependencies 2-3 and 2-4 are partial dependencies since only a certain amount of time must elapse after starting task 2 before beginning tasks 3 and 4. Task 4-5 is a full dependency. It requires the preceeding task to be completed before the succeeding task can start.*

In the installation process, all of the sequences and dependencies start to come together to form the game of scheduling. Natural sequences that fall in line can be carried out following the "domino" effect. Complex dependencies, however, will take on the characteristics of a strategic game of "checkers," where one event happens and several others (some not so obvious) can (and must) occur as a result. Moves in a dominoes sequence may also have a checkers effect in another sequence. To help ensure that tasks are undertaken as soon as they can be, team members should make notes for each task or activity about what other tasks must be accomplished or partially completed first. It is easiest to think of this structure in terms of the network pictorial shown in *Figure 17-4*. This format visualizes the activities that can happen simultaneously or in sequence in a natural progression toward schedule completion.

The first step in creating the schedule should be the construction of a simple, hand-drawn chart. An outline structure in Gantt chart style is recommended. This serves to rough-cut the size of the project. The cell layout should be used to keep the moves in perspective while charting the proposed schedule during the transition from existing to proposed locations. A handy technique is to use clear layout template overlays to view "before" and "after" pictures simultaneously. CAD system layouts, which are both simple and effective, will enhance graphic flexibility.

The project should be broken down into logical, workable subprojects and organized in the outline form referred to as a work breakdown structure (WBS).

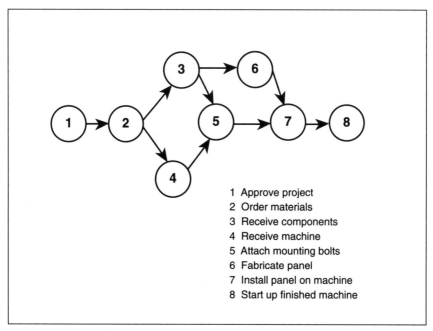

Figure 17-4. *A network diagram helps discern how tasks interrelate and how they can be sequenced.*

An example of a WBS appears in *Figure 17-5*. Statistics for each activity, identified resources, time/duration, costs, dependencies, and fixed dates should be gathered by the planning team and reviewed for accuracy and authenticity before the activity is scheduled. Supporting documentation for the activities should be organized. This is a good time for team members to perform a dry run of machine moves—to physically stand in front of the machine, envision preparing it for the move, walk along the path it will take, and imagine it being put in place and prepared for operation in conjunction with the rest of the cell. This will help to define important factors like foundation and utility requirements, overhead or aisle obstacles, material handling needs, and practical layouts. It may also help to avoid some simple but schedule-disrupting problems later.

The introduction of personal computers and simple, adaptive, user-friendly software packages to assist in documenting statistics and sequence planning has revolutionized planning methods. Before spending resources on software and computers, however, consider "keeping it simple." Presentation is essential, but understanding is even more important. Generating reams of schedules, graphs, and documents will accomplish little if neither the team nor management understand them.

Some of the prominent PC software packages are Symantec's Timeline, Harvard Total Project Manager from Software Publishing, Computer Associates International's SuperProject, Microsoft's Microsoft Project, Primavera Project Planner from Primavera Systems, Project Scheduler from Scitor, and Applied

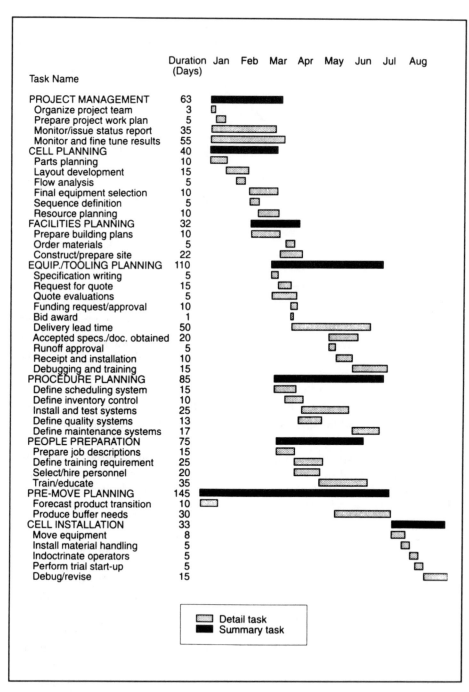

Task Name	Duration (Days)	Jan	Feb	Mar	Apr	May	Jun	Jul	Aug
PROJECT MANAGEMENT	63								
Organize project team	3								
Prepare project work plan	5								
Monitor/issue status report	35								
Monitor and fine tune results	55								
CELL PLANNING	40								
Parts planning	10								
Layout development	15								
Flow analysis	5								
Final equipment selection	10								
Sequence definition	5								
Resource planning	10								
FACILITIES PLANNING	32								
Prepare building plans	10								
Order materials	5								
Construct/prepare site	22								
EQUIP./TOOLING PLANNING	110								
Specification writing	5								
Request for quote	15								
Quote evaluations	5								
Funding request/approval	10								
Bid award	1								
Delivery lead time	50								
Accepted specs./doc. obtained	20								
Runoff approval	5								
Receipt and installation	10								
Debugging and training	15								
PROCEDURE PLANNING	85								
Define scheduling system	15								
Define inventory control	10								
Install and test systems	25								
Define quality systems	13								
Define maintenance systems	17								
PEOPLE PREPARATION	75								
Prepare job descriptions	15								
Define training requirement	25								
Select/hire personnel	20								
Train/educate	35								
PRE-MOVE PLANNING	145								
Forecast product transition	10								
Produce buffer needs	30								
CELL INSTALLATION	33								
Move equipment	8								
Install material handling	5								
Indoctrinate operators	5								
Perform trial start-up	5								
Debug/revise	15								

☐ Detail task
■ Summary task

Figure 17-5. *A work breakdown structure schedule.*

284

Business Technology's Project Workbench. These programs range in price from approximately $500 to $2,500, are compatible with most systems, and are relatively well documented. Some are more graphics oriented. Others emphasize size and capacity for task planning. The use of these packages contributes to an organized and appealing project presentation conveying not only schedule and costs, but professionalism resulting in clear communication and confident management.

MOVING TO THE NEW LAYOUT

All the cell planning and expended effort (plus the fate of future cell projects) may be ruined if the installation does not proceed according to carefully laid and communicated plans. The implementation must be skillfully orchestrated as a logical extension of the planning effort. A good plan serves not only as a road map for successful installation, but as a dependable reference and a confidence builder for the installation team. No amount of post-planning "installation scrambling" or "quick fixes" can replace the forethought of the well-developed plan, because there are too many variables that must be controlled. Most unsuccessful cell installations are spawned from an incomplete plan that overlooked or deliberately left to chance critical plan elements. It's too late to make critical changes when brick and mortar are cured, machines are on the receiving dock, or customer product deliveries are in arrears.

Every effort should be made to install the cell exactly as originally planned in the detail layouts. Occasionally, small plan deviations cannot be avoided during installation. When this occurs, the original planning team—which is now the installation team—can assess the nature of the deviation and its impact on the plan, and incorporate the change into the original plan logic. Each step in the installation should be reviewed before execution and audited closely after completion to catch any other changes that might have been made inadvertently. Changes should be recorded as they occur, so that the final project documentation is a true reflection of the work.

Before the installation begins, all data required to adequately describe the cell and the move should be completed:

- Detail layout drawings;
- Precise specifications of the production equipment;
- Precise specifications of any material handling equipment;
- Detail listings of all equipment involved in the installation;
- All specific utility requirements;
- Specific plans and timing schedule for equipment acquisition and moves to the new space.

The well-thought-out move plan ensures that the physical move will be accomplished with minimal stops and changes. *Figure 17-6* lists various procedures that should be considered in establishing a move plan. A detailed discussion of the specifics of the move plan elements is beyond the scope of this book, and they will vary significantly based on the nature of the business and

1. Select the cell installation team

2. Determine the information and data needed:
   — New physical cell area condition
   — Stores inventory, finished goods inventory, buffers, sequencers
   — Work-in-process inventory
   — Cell equipment:
     . Temporary replacement
     . Permanent
   — Quality plan
   — Performance measurements

3. Plan for accumulation of information and data

4. Establish overall timetable and progress schedule

5. Design necessary forms and procedures for recording information and data

6. Determine sequence of moving specific equipment

7. Check plans with each supervisor

8. Perform housekeeping

9. Determine moving methods

10. Place asset tags on each piece of equipment, tooling, etc.

11. Obtain layout with asset numbers on all equipment

12. Mark off floor area of new cell to designate location of each piece of equipment

13. Locate storage bins, buffers, sequencers, etc., in new cell

14. Establish equipment moving schedule

15. Tag all items to be moved

16. Inform vendors of firm arrival dates for all new equipment

17. Build new equipment arrival dates into equipment moving schedule

18. Adjust schedule to accommodate new equipment late shipment

Figure 17-6. *The move plan.*

19. Select an overall move coordinator

20. Design any needed forms and procedures for controlling move

21. Determine manpower requirements:
    — Maintenance
    — Cell team
    — Outside resources

22. Determine any container needs, and obtain and assign

23. Coordinate any vendor requirement changes with purchasing

24. Notify vendors and obtain agreement on changes

25. Determine any transportation handling requirements

26. Prepare specifications and directions for moves

27. Brief all move coordinators where applicable

28. Conduct employee move briefing and orientation—including visits to new location if appropriate

29. Check move list against each workstation equipment list

30. Make a dry run (if necessary)

31. Move

32. Install equipment, performance management system, quality plan

33. Resume production

34. Fine-tune operation

35. Bring processes into control

36. Stabilize production

37. Perform post-audit

38. Report project status:
    — Prior to move
    — After move
    — After production stabilizes

Figure 17-6. *(continued) The move plan.*

facility, types of equipment, amount of new machinery, and the like. However, the remainder of this section discusses issues associated with the move that must be coordinated.

Avoiding Production Interruption

If possible, the move should be conducted at a time that will allow the team to avoid undue pressure and provide considerable cost savings. Logically, the move should fall within one or more of the following time periods:

- During the annual shutdown;
- At a low production period;
- During a model changeover;
- After the new equipment arrival date is confirmed;
- During peak vacation period—using outside personnel;
- After any new utilities are installed;
- On weekends;
- On the second or third shift of a one- or two-shift operation;
- When known move resources are available.

In trying to avoid halting production, the team should first identify all normally scheduled downtime periods that might make equipment available for movement. Most manufacturers have at least one annual shutdown of one week or longer. Extending weekends to three or four days and, if necessary, scheduling longer work hours the other days can maintain minimum production during move times. In this case, employees may be happy to have a few extra days off—even without pay! Working resources around the clock for short, intense periods may frequently be another option, since people may be eager for the shift or overtime pay. When altering schedules like this, it's important to first ask for volunteers to work these odd hours. It may not be necessary to change anyone else's schedule. Cell installation work during these periods must be carefully planned to conclude in such a way that production can be resumed immediately.

Some utility work will probably have to be scheduled during off-peak periods. Move planners must also remember that contractors will charge substantially for overtime or off-hours scheduling. Gaining approval well in advance for move dates and blocking them out on everyone's calendars will help keep these periods from becoming controversial in later stages of the implementation. Production must be kept informed of any changes to the schedule and reminded about the shutdowns before each period begins.

If installing a cell disrupts the normal production of the items produced by the cell, it could result in:

- Late customer deliveries;
- Deliveries of products of inferior quality, due to the cell being put into production prior to adequate debugging.

Several alternatives should be considered to preclude inadequate customer product deliveries during and after the cell installation. If production must stop for the installation, the manufacturer may build inventory early in quantities adequate enough to cover the stoppage. Sometimes the financial impact of this inventory buildup can be lessened by convincing the customer to accept a portion of the buildup as an early shipment. Another alternative is to reroute the parts involved in the cell installation to other work centers or idle equipment. Because such deviations may result in overtime or production bottlenecks at the alternative work centers that will in turn stress normal operations, this temporary measure should be removed as soon as is practical.

The last alternative is to temporarily purchase the affected parts from an outside vendor. Such measures must be carefully considered because the vendor may introduce parts with unacceptable quality, causing more repercussions than temporary part shortages would have. Also, third-party vendors may prove unreliable if the temporary order ties up capacity required for their ongoing customer orders.

A customer delivery stoppage may not be avoidable in the cell installation process. If this is the case, it is extremely important that customers be kept informed of cell installation progress—even if there is the slightest possibility that their requested delivery schedule will be adversely affected. See Chapter Eight for a discussion on communicating cellular manufacturing plans and benefits to customers.

Assessing The Facilities

The cell team cannot underestimate the time, effort, schedule, or cost elements of the facility side of the installation equation. Production will certainly be interrupted if the team neglects to assess the physical facility in which the cell will be installed to determine its suitability. For example, the team should be asking: "Is the floor surface unimpeded, smooth enough, thick enough, and level? Is the ceiling height adequate for machine clearance or material handling cranes? Is ventilation required for special needs? Is there accessibility to outside entrances for point-of-use deliveries of materials? Is inside accessibility for material delivery by crane, fork truck, or conveyor feasible? How will removal of oils, coolants, material chips, and scrap be accomplished? And how are we going to move things from here to there?"

Today's EPA and OSHA restrictions have become serious obstacles in attaining timely cell installations due to last-minute discoveries about some environmental concern. A recent cellular installation was delayed six months due to an oversight in obtaining the necessary state air-discharge permits for an annealing oven. Involving plant facility or maintenance engineers early to identify permit requirements, asbestos regulations, disposal techniques, testing requirements, regulated cleaning procedures, and the like will help avoid these problems. Delays can be costly, particularly when key process equipment is involved and capital outlays have been made.

Another underestimated facility requirement is adequate utilities. Utilities include things like electricity, water, sewer, gas, coolant, scrap removal, and process gases or liquids. Frequently, these require major expenditures when long runs of mechanical supply lines are necessary.

Special foundations can also cause costly problems for cell designers. Initially, the need for special foundations should be challenged, since some plant floors are already thick enough to accommodate most machine tools. New concrete foundations usually require curing time prior to equipment placement, and, since most cell placement areas are currently occupied, staggered or multiple equipment moves may be necessary.

Facility constraints such as ceiling clearances are usually difficult to overcome. Overhead cranes are the most common restrictions. Manufacturers should consider under-hung rails if necessary, or localized jib cranes if practical. Actual roof penetration for cupolas should be a last resort. The clearances on access doorways should be measured. Cell planners should be certain to include additional time and cost if it becomes necessary to open special doorways or tear down equipment to pass through areas. Disruption may also be caused by the need to secure the access path through the plant used to move major equipment into place. External rigging contractors usually have special equipment that may save time and money in accomplishing move tasks. For example, special articulated hydraulic lifting cranes can be obtained to assist in tight plant interiors. Similarly, Bobcat-type endloaders may be used to replace manual operations, thus saving many days of manual labor.

Finally, finding available space to work with can make cell installation much easier. Is there extra "turnaround" space for making some temporary moves? Is there some underutilized equipment that is taking up space that can be temporarily or permanently removed? Are there storage areas that can be eliminated? Can any required foundations or trenches be installed early around existing operations? Clearing out clutter, making open space available, and doing all other possible preparatory work will facilitate the installation.

Guaranteeing That The Equipment Will Be On Time And Running

Being ready to move won't mean a thing if the equipment doesn't arrive as arranged. Cells are easiest to implement when they've made the best use of existing machines and require minimal new or modified equipment. Otherwise, the project may be dependent on new equipment that's subject to long lead times, negotiations, delays, and late delivery. Many vendors consistently demonstrate a 15 to 20% delivery time slippage in spite of adequate planning and lead time allowances. Cell planners can sometimes avoid this problem by giving an earlier delivery date than is required, providing they have room to store the equipment if it comes in on time!

Figure 17-7 illustrates a normal new equipment purchase timeline. Only slight changes need to be made to apply the same sequence to in-house equipment requiring rebuilding or new or modified tooling. The latter should not be minimized; too often new or modified tooling is sloughed off as the responsibility of the toolroom or tool engineer. The assumption that the tooling

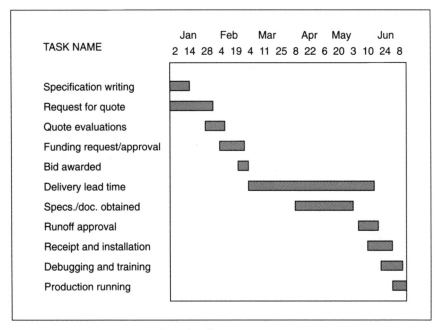

Figure 17-7. *An equipment purchase time line.*

will be ready and debugged whenever the equipment is ready is frequently overly optimistic. Tooling is often harder to obtain on time than machinery, and then it may need to be reworked several times.

Specification writing need not be an elaborate effort as long as essential requirements are sufficiently described. Specifications should not tie the vendor down to unnecessarily narrow options that are not recommended standard equipment, since standard (proven) equipment is the least risky and expensive. Submitting carefully prepared Requests for Quotations to a limited number of responsible vendors and allowing them adequate time to respond will help keep equipment purchases on schedule.

The cell team should require the vendor to supply all final equipment specifications—including operating manuals, foundation prints, and training documentation—prior to actual delivery. This allows the installation team the opportunity to proceed with foundation designs and gives operators the chance to become familiar with machine characteristics before vendor runoff. Insisting on vendor operating runoff (on the vendor's site) prior to accepting the machine makes it much easier to correct vendor defects or make modifications. Much more time is lost if the machine is shipped and installed before problems are found. Refer to Chapters 15 and 16 for more detail on developing equipment specifications and controlling and coordinating vendors.

Cell installers should never depend on production running smoothly immediately after the debugging stage. Complex new machine tools usually require as much as six months after installation to achieve full steady-state production, and the team shouldn't plan to be the exception. This is probably the primary reason not to depend on new machines in cell designs.

Keeping An Eye On Logistics

Throughout the steps of cell development, the concepts of material flow and equipment density, and their combined effect on the increased velocity of product throughput, are the primary drivers. Since production costs are reduced by smoothing out material flow, it's important to revisit these topics here as they apply to the actual installation. Cellular operations (smooth material flow with fast throughput) demand that raw materials of acceptable quality be present in the right quantities, in the right containers, and at the right time.

Vendors delivering material. Operation of a JIT cell places significantly different and tighter demands on vendors than exist under traditional operations:

1. More frequent shipments of smaller quantities, delivered according to a strict timetable;
2. Exact quantities packed in specific containers;
3. Explicit quality requirements—100% good;
4. Specific delivery points on the shop floor;
5. More complete and extensive record keeping and part traceability;
6. Much faster reaction time to unexpected interruptions or schedule changes.

Vendors cannot move from today's large batch deliveries to JIT concepts without some changeover time. It is imperative that the cell installation plan include proper time allowances for vendors to prepare for JIT deliveries. If only one or two cells are involved, with a small number of parts, getting JIT delivery from vendors may be simple. If an entire plant with numerous vendors is being cellularized, however, one to two years may be required to develop vendors into competent JIT parts suppliers.

Storing and moving material. As few materials as possible should be stored in the cell. Cell operators should constantly strive to eliminate cell materials in excess of those necessary to complete the one item currently in process. If the cell has been constructed properly, there will be little or no space to store completed materials inside its boundaries.

There should be no need to store large raw material parts in the cell. However, smaller parts such as nuts, bolts, fasteners, hooks, tapes, adhesives, and wires—usually purchased in lots or batches and used frequently—should be stored in bins or trays within the cell to provide the operator with easy access. These small parts will be replenished on weekly or biweekly time intervals. Many companies use Kanban replenishment systems for these types of materials.

Frequently, sequencers or regulators are built into cells or between cells to help regulate large parts that cannot be obtained or internally manufactured on a JIT schedule. Sequencers have been described as an interface between a

292

demand-driven, pull-type production operation and an MRP-driven material acquisition system. This interface might be physical—in the form of a rack, tray, or area of space—or it can be a control system that provides real-time judgements regarding the on-the-spot needs of the cell or the capabilities of the supplier. *Figure 17-8* shows that in reality, a regulator requires both. Sequencers can be large or small. Buffers may also be necessary to allow for the short-term buildup of materials between the manufacturing location of cell raw material and the cell itself. This may occur when one plant or cell is working one eight-hour shift, but a second works two eight-hour shifts.

Special attention should be given early on to part containerization. Transport and storage containers must be properly sized or parts will be lost or damaged. Additionally, an adequate number of containers must be acquired to allow for transport to and from vendors.

Building well-planned cell logistics into the cell installation process is a key element of effective cell operation. Because of the compound interrelationships between outside vendors, adjacent production functions, and material storage facilities, logistics planning should start early and continually evolve to support the cell as it changes to conform to changing product demand.

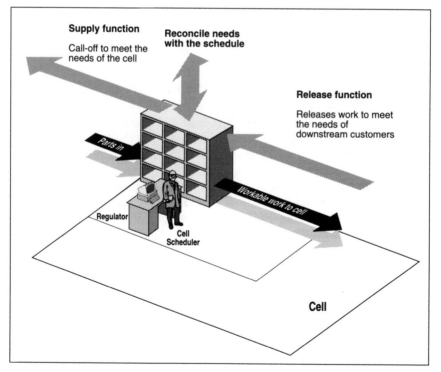

Figure 17-8. *A regulator enables the team to schedule material between multiple suppliers, customers, cells, or operations.*

Stressing The "Preventive" In Maintenance

A superior logistical arrangement will only be effective if machines are running correctly and consistently. Unfortunately, equipment maintenance is in many people's minds a necessary evil. In most companies it has been relegated to a centralized maintenance function to obtain cost savings, or savings believed to arise from specialization. These anticipated savings have not materialized, however. In most companies, maintenance costs are excessive and inordinate machine downtime is usually still prevalent.

Cellular manufacturing will not work if machine downtime is anything but minimal, and if it does not conform to planned norms. Therefore, it is imperative that cell operators perform the maintenance function differently than it is currently being done in many centralized, specialized, functionally oriented plants.

The concept of "total preventive maintenance" (TPM) is being adopted in many cellular manufacturing environments. TPM, which is cell operator-centered, brings many of the maintenance functions back into the cell. It's been found that cell operators have a greater interest than general maintenance people in ensuring that their cell's maintenance is done promptly, correctly, and thoroughly. This responsibility also helps develop pride of ownership in the cell. In addition, operators become familiar with a smaller number of machines and their maintenance histories, and they are much more accurate in identifying their mechanical problems and maintenance needs. Using this TPM concept, the maintenance department becomes the "backstop" and the cell is the front line in the effort to maintain a 95 to 98% machine availability. The idea is that machines are down only at planned periods for preventive maintenance and overhauls, and that breakdowns are unacceptable.

It is extremely important that the cell installation team have a plan to accomplish the transition from traditional maintenance to TPM during the cell installation process. This plan should at least consider the issues outlined in *Figure 17-9*. It is also imperative that operator-centered maintenance function well from the first day of completed cell installation. Breakdowns in JIT cell environments are so disruptive that they cannot be tolerated. Likewise, excellent maintenance is a key element in any good quality program, because machines cannot produce good parts if they are operating incorrectly or cannot hold tolerances. There is a strong requirement for appropriate feedback to the cells from the statistical quality program for gaining and maintaining process control of the equipment.

Planning For Quality

Chapter 12 describes the details of the quality planning necessary for optimum cell performance. Elements of the quality plan described there must be developed, built into the installation plan, and monitored as the installation progresses. It's important to remember that quality problems arise from the interrelationships between product design, materials, people, manufacturing machines, and the processes performed by machines. Quality performance in the

1. Identify cell machinery and equipment

2. Determine operating and preventive (PM) steps associated with each piece

3. Separate cell maintenance into the following for each machine:
   — Preventive maintenance
   — Overhauls
   — Breakdown response

4. List responsibilities for each:

   — Operators
   — Assistance from outside the cell (maintenance department)

5. Identify daily maintenance checklist:

   — Lubrication
   — Cleaning
   — Visual inspection, etc.

6. Develop PM schedule for each piece of equipment based on:

   — Time
   — Usage
   — Testing
   — Visual inspection

7. Develop operator checklist for the above

8. Develop records, display charts, and a recordkeeping and audit system

9. Obtain or start machine and equipment histories

10. Identify cell operators responsible for maintenance

11. Identify operator PM training needs

12. Develop training schedule to alleviate operator shortcomings

13. Build training schedule into cell installation schedule

Figure 17-9. *Steps for implementing a total preventive maintenance program within cells.*

parts-per-million (zero defect) range is no longer uncommon. Statistical process control methods of inspection solve some of the problem—as do elaborate quality control programs—but neither of these is the answer. They are not aimed at preventing defects as much as they are at measuring defects after they start occurring. The best solution is to design and install processes and cells that eliminate the possibility of producing or creating defective parts.

Incorporating Appropriate Management Systems

Similar to the questionable success of complex quality systems is that of complex computer systems, which have failed to improve management capability or reduce operating costs. This overriding observation was confirmed time and time again in the late 1980s. Usually the installation of computer-based management systems increases operations complexity, while increasing corresponding costs. One reason this occurs is that the newly introduced system has no relationship to the real problems and may even help to further obscure them. Likewise, obtaining more and more information in new formats complicates data gathering and interpretation for management, and nothing ever gets done to improve operations.

The important lesson for the cell installation team is that management systems must remain simple to operate and interpret. They must be inexpensive to create and, most of all, provide relevant and useful information. It is the responsibility of the cell installation project team to install the selected management systems, initiate their operation, and train users in all aspects of information gathering, presentation, interpretation, and reporting. These functions should be built into the installation plan and the system must be in place and operating the instant the cell is put into operation.

Chapter 18 discusses the importance of performance measurements in cell operation, control, and evaluation. The most important management concept the cell installation team must develop is the establishment of a proper balance between financial and nonfinancial performance measurement and evaluation. Financial measurements have more importance at higher management levels, and nonfinancial measurements are most useful at the cell management level. Both, however, are required to maintain healthy production. Nonfinancial performance measurements, which are often undervalued, are as available to cell managers as their own pulse rates. On the other hand, financial measurements tend to carry more import. They are like medical tests that must be processed at the laboratory (monthly closing dates) and returned to the user long after the need to react has passed.

The cell installation team must not only install and bring into operation the management systems, it must also train users to change elements of the cell's operation in response to the management system discovering undesirable results. The concept of a static system was abandoned in the late 1970s. As a result, all personnel must make operational changes as indicated by the management systems and change their management systems to reflect the adjustments.

Lining Up The People Resources

There are three groups of people to consider in cell installation planning: cell team members, in-house installers (doers), and outside installers.

The commitment of planning team members' time and effort to the installation project will be counted on in building the installation schedule. The team may rely on part-time, skilled in-house installers such as electricians, mechanical engineers, equipment riggers, carpenters, electronic specialists, plumbers, and general helpers. Because these craftsmen may have limited availability, installation schedulers must match their skills to the required calendar dates and guarantee resource availability. This may include reminding these people's managers of the cell project and ensuring their commitment to it.

Resources sought outside of the company to compensate for nonresident or unavailable skills also need to be identified, involved, and notified of upcoming activities. Early in the process, team members should begin making initial contacts with outside contractors and craftsmen to determine their availability, costs, quality, and—perhaps most importantly—flexibility. These arrangements may include blanket contracts on a time and material basis. Additionally, the team should consider relying on one or two at the most, since it is difficult to coordinate the efforts of multiple contractors. Also, most contractors will be willing to negotiate better rates if they know they are the primary source.

TRACKING SCHEDULE ADHERENCE

During cell installation, the project team must meet regularly with the project steering committee to communicate progress and exchange data. Aggressive goals or targets should be established, since seldom are schedules moved forward in this type of project without them. When doing steering committee updates, each team member should be given an opportunity to present material or field questions in front of the group on a regular basis. Project results are usually so positive that everyone wants to be involved in revealing them as they develop. This will help ensure project team members' commitment to the project during the critical implementation period. *Figure 17-10* shows a sample agenda. Status reporting should include a basic comparison of the baseline or original schedule to the actual schedule.

Traditional Schedule Tracking And Verification Tools

Cell installation project status is tracked as simply as possible. Three traditional tracking tools are project measurements, production measurements, and schedule adjustments.

Project measurements. Project measurements are usually based on those goals and targets established during the initial planning phase. Examples of these might be timing, equipment delivery status, overtime incurred against plan, committed versus spent funds, and so on. This level of detail is generally used by the steering committee for validation of the installation plan.

297

Section	Presenter	Time
Schedule status	Project leader	10 minutes
Process details	Team member 1	15 minutes
Layout status	Team member 2	15 minutes
Financial expectations	Team member 3	15 minutes
Wrap-up and questions	Project leader	5 minutes

Figure 17-10. *Steering committee meeting agenda.*

Production measurements. If production is impeded as a result of poor cell installation, the entire project might be condemned. It will not matter whether equipment was late, or machines would not run once moved, or that outside contractors did not respond. The project schedule should allow for contingencies. The effects of the cell installation project on operations can be observed by using the following measurements. The results of using these simple tools should be displayed on up-to-date, easy-to-understand charts at the cell location:

- *Disruption and/or downtime incurred.* The best way to avoid disruption is to plan for the desired outcome but allow for contingencies on an exception basis. For instance, if cell team members think that it will take five days to move a critical machine and get it going again, they should plan for five and allow an extra day or two as a contingency. Other plans should not hinge on the machine until after the sixth day. The team must also consider the downtime of dependent operations that rely on output from the machine being moved. Overtime hours can often be related to the effectiveness of installation planning if used as an installation overrun solution.
- *Inventory levels.* Tracking inventory levels in work-in-process storage areas will allow the team to monitor trends or cell impacts. (Premove buffers could distort this measurement unless excluded.) One interesting tracking solution is to try to physically collect all WIP inventory in one or two places and watch it go down. The need to monitor inventory level disappears after operations within the cell stabilize. *Figure 17-11* shows an illustration of inventory tracking.
- *Throughput or on-time delivery performance.* Getting products made quickly and to clients on time—important reasons for installing FMCs—will be impacted quickly after the start of the project. Tracking past to present performance on a simple chart can be very effective, as shown in *Figure 17-12*. Being able to show that the definition of "on time" has contracted is also impressive. An example might be cutting 30 days for customer orders down to two.

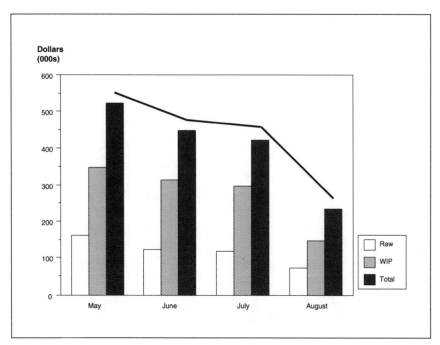

Figure 17-11. *Tracking decreasing inventory levels during cell installation is one way of verifying results.*

- *Employee head count.* Several easy measurements exist for finding out about the decline of the number of employees before and after cell installation, such as total labor hours or simply attendance. This is depicted in *Figure 17-13*. Evaluators should be sure to include indirect involvement such as material handlers and setup or maintenance people. Again, sensitivity must be exercised in relating these numbers; the team can't lose light of the fact that each of these jobs represents someone's livelihood. If the number of positions vacated through natural attrition is high, it might be a good selling point to illustrate.

Operating budget variances, or common traditional production measurements, are probably the worst indicators during cell installation. *Many* things that are happening at this stage cause negative variances. Extra consumables—including rags, oil, coolant, small tooling, and other expense items—will be used up in cell installation activities. Manufacturing management needs to be kept informed of the magnitude of the changes and to be aware of the potential for temporary budget increases.

Schedule adjustments. Schedule changes may be beneficial adjustments made as the project progresses and the team gains more knowledge, or they may be detrimental:

- Production schedule or product changes;
- Equipment lead time delays;
- Resource constraints;

299

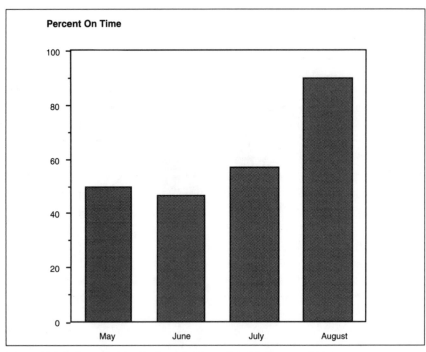

Figure 17-12. *Improvements in throughput speed and delivery performance present strong justification for cells.*

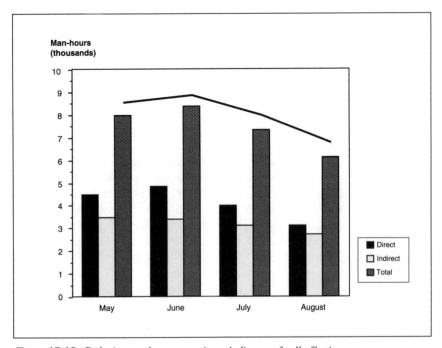

Figure 17-13. *Reducing employee count is an indicator of cell effectiveness.*

- Late funding;
- Late discovery of facility monuments—old foundations, latent asbestos, inadequate electrical power.

Installation schedule changes must be communicated to the entire organization as they happen. This includes purchasing, R & D, sales, and marketing. Any of these functions may be able to assist in dealing with these schedule adjustments without incurring added costs or constraints. If the plan is right, however, and it details the design requirements, none of these issues should be cause for undue frustration.

SUMMARY

This chapter has attempted to emphasize discipline in *thinking through* a cell installation, and to highlight a number of techniques for use in dealing with the myriad of people and things that can help make cell installation successful. Scrupulously following such disciplines is essential during a first-time cell installation. Discipline may be relaxed in subsequent installations, because the value of cells will have been proven within the company and the team's level of understanding will be better.

Cell installation is really no different from any other planning effort that involves people, time, and spending money. Nonetheless, it must be emphasized to anyone beginning his or her first installation that the physical movement of machines is easy—the management of the change process is the real key to success. Machines can be taken apart, put back together, and restarted, and parts can be picked up, moved, and stored, but people don't deal with change the same way. They are much harder to plan for and predict—especially people in someone else's department!

Team members must undertake a vigorous effort to communicate the plans, the changes, and the progress of the cell installation to everyone possible, not just those responsible for moving the machines. This should include managers, clerks, fork-truck drivers, store handlers, laborers, and those who will make cells a success—future cell employees.

IIIII 18 IIIII

AUDITING CELL PERFORMANCE

When the cell installation project described in Chapter 17 is complete, everyone will want to know the results. While it is normal to be eager to determine the immediate, short-term outcomes of a rigorous project, it is more important to have planned the project to allow ongoing day-to-day performance measurements. Finally, cell personnel need to know what adjustments must be made to keep the ongoing efforts parallel with ever-changing strategic company and plant goals.

The ability to quickly and accurately determine results is critical to the economic well-being of any manufacturing company. Determining cell performance is implicit in any equation attempting to further define plant or company success, because the cell is the smallest manageable unit of both. If managers successfully control the workings of the production cell, then by combining all the cells, they have made a significant contribution to the success of the business.

Managers can effectively control cell operation by converting business strategy into successful cell actions. The success of cell actions can be determined by establishing what successful actions are and comparing actual results to the predetermined norms, as depicted in *Figure 18-1*. The measurements in this model are called "performance measurements" (PMs). This chapter discusses the importance of performance measurements with the following objectives:

- Develop a general understanding of performance management and performance management systems.
- Apply this understanding to the *short-term* question, regarding the cell installation: "How well did we do?"
- Also apply this understanding to *ongoing* cell performance measurement to control ongoing cell operations.

303

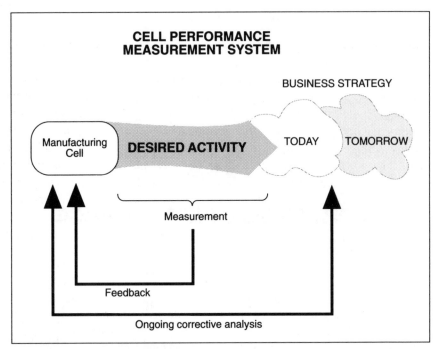

Figure 18-1. *Cells should be judged successes or failures based on their contributions to achieving the business strategy.*

This examination of performance measurement issues is very important to those contemplating cell design and installation. The performance measurement process tells managers how to concentrate on the activities that are important to cell control. This philosophy supports the well-known axiom, "You get what you measure." While motivational theory is not a topic of this book, managers must remember that measurements showing poor cell performance may be caused by:

- Poor cell design,
- Poor measurements, or
- Employees who aren't working toward the goals that are being measured (see *Figure 18-2*[1]).

PERFORMANCE MEASUREMENTS AND MANAGEMENT SYSTEMS

In an ongoing manufacturing operation, performance management is part of the control feedback loop. Performance measurements are the devices that detect the outcomes of cell operational activities, the same as an automobile speedometer detects speed or a thermostat indicates engine temperature. The detection of nondesirable results activates corrective action to adjust the activity so that desired outcomes are obtainable.

This book does not attempt to present a detailed study of performance management. It is appropriate, however, to examine PMs as they relate to cell operation, beginning with the development of a general understanding of them.

"THINGS THAT GET REWARDED GET DONE"

In his book *The Greatest Management Principle In The World*, Dr. Michael LeBoeuf described how he arrived at some answers for managers wondering why, despite all of their efforts, they weren't getting the results they desired from employees' work. "It all came together when I heard the parable of the man, the snake, and the frog.

> A weekend fisherman looked over the side of his boat and saw a snake with a frog in its mouth. Feeling sorry for the frog, he reached down, gently removed the frog from the snake's mouth and let the frog go free. But now he felt sorry for the hungry snake. Having no food, he took out a flask of bourbon and poured a few drops into the snake's mouth. The snake swam away happy, the frog was happy and the man was happy for having performed such good deeds. He thought all was well until a few minutes passed and he heard something knock against the side of his boat and looked down. With stunned disbelief, the fisherman saw the snake was back—with two frogs!

The fable carries two important lessons:

1. You get more of the behavior you reward. You *don't* get what you hope for, ask for, wish for or beg for. *You get what you reward.* Come what may, you can count on people and creatures to do the things that they believe will benefit them most.

2. In trying to do the right things it's oh so easy to fall into the trap of rewarding the wrong activities and ignoring or punishing the right ones. The result is that we hope for A, unwittingly reward B and wonder why we get B."

A manager who understands this key principle must naturally ask LaBoeuf's corollary, "What's being rewarded?"

Figure 18-2. *Successful performance measures can't occur when employees aren't being motivated to work toward achieving them.*

Basic Assumptions

Performance measurements should be derived from the business strategy. The business is guided by insights into customer needs. These needs are converted into the high-level business strategy for servicing the customers' needs as well as the business's own growth and stability. Therefore, the lower level strategies of day-to-day operations should directly support overall business strategies selected by top management. Cell PMs must provide linkage between cell operations and the local facility needed to support the overall business strategy.

Performance measurements must be hierarchial as well as horizontal. For performance measurements to work effectively, they must be interrelated throughout the organization. As the PMs extend from the top toward the bottom of an organization, they must consistently make the transition from broad to more specific, relating to narrower planning horizons. At the cell level, many performance measurements are monitored and reported daily. Cell-level PMs must support cell operation but not interfere with the operation of other cells; it would not be wise, for example, for a machining cell PM to emphasize quantity of output if the cell's customer found only half of the machining cell's products to be of an acceptable quality.

PMs must support the business's multidimensional environment. At the cell level (as well as at the business level), performance measurements are biased toward internal and external cell effects and cost and noncost effects, and they can be characterized as such:

- Internal effect: percent on-time delivery,
- External effect: number of customer complaints,
- Cost: relative labor cost, and
- Noncost: education level.

Specific Characteristics

Performance measurements should help measure progress toward controlling activity costs. PMs can be developed so that they directly relate to problems that cause significant operational costs. It is true that many cost drivers are nonfinancial because they may relate to issues not directly controlled in the cell, like product design or specified processes. However, other cost drivers have a direct impact on cell-level costs:

- Overtime hours,
- Machine downtime,
- Amount of rework,
- Scrap dollars,
- Setup reductions, and
- Throughput time.

Performance measurements must not be considered permanent. The competitive businesses in the 1990s are flexible and rapidly changing. Any change in items such as product, delivery, machines, or tooling may well cause changes in the need for certain types of performance measurements. Companies move rapidly into and out of markets and otherwise change business strategy to adjust to ever-changing competitive pressures, and existing performance measurements must be continually reviewed in response to these changes. New performance measurements may be needed to replace PMs that have become out of touch with important business drivers.

PMs must be easy to understand. Performance measurements are used to influence human behavior toward desired positive results. Therefore, there must be no mistake in the meaning that performance measurements convey. The relationship between the activity measured, the PM's objective, and the performance measurements themselves must be clearly understood by all; this understanding must have an unmistakable link to corrective actions that result in the desired measurable change.

Performance measurements must be few in number. People are able to effectively focus on relatively few objectives at one time. When too many performance measurements are established, the attention given to any of them lessens, and the impact on all objectives is weakened. Cell PMs should number around three to six.

PMs should be reported on a timely basis. Effective corrective action will not occur unless the performance measurement is reported near to the time the measured activity occurs. In a cell, this converts to hourly, daily, and weekly. The appropriate time is that required to take effective corrective action in relation to the significance of the activity. For example, unplanned machine downtime must be immediately corrected. Overtime hours, however, must accumulate to form an identifiable pattern that can be analyzed to formulate corrective actions.

Performance measurement results must be clearly visible. Changes will occur only when people see and understand the results of performance measurements. These results should be displayed prominently in the cell work area in a manner that clearly communicates the information to cell employees. Again, the results must be easily interpreted into corrective actions.

Management must be committed to supporting the PMs. Clear commitment to supporting the performance measurements must be demonstrated by all management levels. Managers must communicate their expectations about the performance measurement results and require the necessary changes indicated by them. The performance measurements must be the primary vehicle running the cell day to day.

Performance measurements must be clearly linked to the employee evaluation process. Performance measurements should support company goals. Employees should be evaluated in relation to the success they obtain in helping the company attain its goals. Therefore, the success of the performance measurement system depends on how well the performance measurements are incorporated in the evaluation process and impact employee behavior.

Attributes Of A Good Measurement System

Wide scope multicell installations will require the design, development, coordination, and installation of a comprehensive performance measurement management system. Successful PM systems should:

- Be mutually supportive and consistent with the business's operating goals, objectives, critical success factors, and programs.
- Convey information through as few and as simple a set of measures as possible.
- Reveal how effectively customers' needs and expectations are satisfied and focus on measures that customers can see.
- Provide a set of measurements for each organizational component that allows all members of the organization to understand how their decisions and activities affect the entire business, and
- Support organizational learning and continuous improvement.

Many different techniques can be used to determine what measures are important. Climate surveys can be used to determine how an organization perceives the existing measurements relative to their importance. If measures are out of alignment with drivers, survey data can be used to determine the priority of the measurements that need to be changed. External supplier and customer surveys can be conducted to determine the alignment of internal activities with external drivers. As with internal surveys, the data are very valuable in setting priorities and establishing action plans.

Traditional Performance Measurements: Obstacles To Improvement

Manufacturing companies have used performance measurements to indicate results (and the extent of them) for years. Recently, the need for improved manufacturing capabilities has caused manufacturers to examine exactly how performance measurements have been used and to what effect. Generally, experts tend to agree that traditional performance measurement systems fail to support the needs of modern-day companies. They fail to be accurate indicators of day-to-day activities and, more importantly, continuous improvement.

Computer-Aided Manufacturing, Inc.'s (CAM-I) *Cost Management Conceptual Design Project*[2] indicated that traditional cost accounting and performance management systems fail in today's business environment because they:

- Ignore inventory levels in assessing production process effectiveness.
- Focus on machine utilization and report direct labor by operation.
- Frequently include rework/replacement efforts as part of earned value when measuring productivity (i.e., exclude quality as a factor in performance measurement).
- Measure and reward individual work center productivity, rather than tying measurement and reward to the productivity of the shop, assembly line, or plant of which they are a part.

Used inappropriately, performance measurements can be counterproductive to achieving management goals such as continuous improvement. Managers historically have focused on only one element of multifaceted performance measurements, not realizing that achieving high success in it may well mean an overall negative business impact. *Figure 18-3* lists some traditional PMs and explains how they inhibit business success and improvement.

Traditional Measurement	Action	Results
Machine utilization	Machines run in excess of daily unit requirements.	Excess inventory buildup, improper mix produced, eventual obsolescence, quality problems.
Purchase price	Purchasing increases order quantity to get lower price.	Excess inventory, increased carrying costs, supplier with best quality and delivery may be overlooked.
Variances	Focus management attention only on unfavorable variances.	Focuses attention on symptoms, not real issues.
Standard cost overhead absorption	Supervisor overproduces WIP to get overhead absorption in excess of his expenses.	Excess inventory. Improper inventory mix. Obsolescense.
Direct labor efficiency, utilization, and productivity	Produce more standard direct labor hours.	Excess inventory. Improper inventory mix. Obsolescense.
Direct labor reporting	Focuses management attention on direct labor.	Does not identify largest cost reduction opportunity, overhead.
Responsibility accounting	Managers focus on cost center or responsibility center, not business as a whole.	Opportunities for reduction are masked because the reasons for high costs are not identified.
Earned labor dollars or hours	Management keeps workers busy to maximize earned hours.	Inventory increases. Output is important. Meeting delivery schedule is lower priority.
Direct/indirect labor ratio	Ratio becomes management's focus.	Does not control total cost. Does not identify reasons for high cost.
Quality cost	Focuses emphasis on quality department budget.	Does not focus on real problem causing quality problems, e.g., product design, tooling, training, preventive measures, failures themselves.

Figure 18-3. *Traditional performance measurements that inhibit improvements.*

This chapter will discuss more appropriate performance measurements that enable forward-looking manufacturers to better control their businesses; the emphasis on effective control of manufacturing must move away from the measurement of worker performance to that of establishing and maintaining a smooth flow of small quantities of first-time quality work through the manufacturing shop in a predictably reliable manner. More importantly, the focus must change from one that stops judging the short-term production out the back door to one emphasizing the long-term measures associated with concepts of continuous improvement and low-cost, defect-free production.

Financial And Nonfinancial Performance Measures

It is becoming increasingly clear that managers must incorporate more nonfinancial performance measurements into control systems than were traditionally utilized. To establish desired control, systems must be interactive and responsive to sometimes immediate change. Accounting data generated by most traditional cost accounting systems is more appropriate for higher level decisions. The closer to the shop floor the decision point, the more likely it is that the best data is nonfinancial. This occurs because cost accounting information is:

- Irrelevant to most shop floor decisions, because indirect cost allocation bases are inappropriate;
- Too vague, because unit-measured shop activity has been converted into dollars in the system;
- Too late for appropriate action to occur;
- Too obscure, due to commingling with data associated with other activities.

Financial measurements are very important at the strategic and tactical levels of business. It is at these levels that actions at the shop floor or operational level become converted into the results of operating the business—profit and loss. Lower level performance measurements should lead to immediate actions, while higher level performance measurements should lead to less immediate action and more analysis of planning and underlying causes.

SHORT-TERM: HOW WELL DID WE DO?

Most successful cell installation projects are directed by a well-thought-out development and installation plan. Components of the installation project usually include business strategy, a manufacturing strategy, a cell layout, and plans for quality, logistics, training, control systems, and the necessary manpower/skills, among others. The elements are combined into a financial and operational analysis and justification package commonly composed of various combinations of measurable control elements, as shown in *Figure 18-4*. While some companies deviate from these measurements, most still limit project justification to a selection of the items on this list.

• Head count reduction	• Greater conformance to specification
• Throughput time reduction	• Improved ratio of machine downtime to scheduled time
• Lead time improvement	
• Physical space reduction	• Budgeted installation cost achieved
• Setup reductions	• Safety improvement
• Quality improvement	• Yield improvement
• Inventory reductions	• Overtime reduction
• On-time delivery improvement	• Reduction of part distance traveled
• Installation schedule adherence	• Return on investment
• Reduced number of operations	• Payback
• Faster response time to customer feedback (customer service)	• Internal rate of return
	• Discounted cash flow
• Lower bill of material levels	

Figure 18-4. *Measurable control elements.*

The importance of tracking and verifying the FMC's progress against planned benchmarks and objectives was discussed and illustrated in Chapter 17. If an adequate job of monitoring occurred during the installation, little work will be necessary in the post-audit process. It is academic to argue the attributes and benefits of project justification after the fact. Once the project is justified and approved, those approval criteria are "frozen" forever; project success evaluations must consist of a comparison of results to the justification attributes as they were developed in the justification package.

It is important that post-audits be initiated and completed within three to six months after completing the cell installation. If the initial results are undesirable, a comparison of the cell installation plan elements to actual results will give rise to the nature of the problems. These problems can then be rapidly addressed, adjustments can be made, and minimal plan benefits lost before corrections are completed. If a project is planned from the onset in accordance with company strategic objectives and appropriate reporting guidelines, the post-audit should be anticlimactic.

311

The post-audit team should be a multidisciplinary group of three to five employees, including one or two members of the original project team. The cell installation plan should detail the specifics of the post-audit to at least include team composition, audit timing, reporting structure/level, and leadership. The post-audit team should document the audit results in the form of a report to the managers who approved the cell installation plan. This report should describe the project results, plan element by element, and specifically address each item within the elements. Where appropriate, the report should suggest next steps to further incorporate objectives or backfill shortcomings in the project planning and execution.

ONGOING PERFORMANCE MEASUREMENT

It is mandatory that companies that change their physical operations structure also change their performance management systems to fully support the physical change. For certain, the introduction of cell manufacturing techniques and philosophies into a traditional manufacturing company—and even the redesign of current cell manufacturing facilities—disrupts in-place performance management systems. Performance management system changes should be documented in the original cell installation plan. The system changes must help the cell attendants to answer basic performance questions such as, "What should I do?" and "How well am I performing?" With timely answers to these questions, the cell operators can then analyze feedback, develop corrective action plans, implement the plans, and monitor the results.

Once in place, the system should be re-evaluated and modified where necessary each time a new product is introduced, processes are changed, significant changes in product mix occur, or the company's strategic drivers change. The cell installation plan should define circumstances requiring system re-evaluation. There are three elements of performance measurement:

- System structure,
- Measurement elements, and
- Timeliness of measurements.

System Structure

Successful cell direction from a higher layer of management depends on how well the company or plant performance measurement system mirrors the physical shop floor. This means that the structure of the physical operation must be the same as the performance measurement system. In addition, responsibility for the control of a physical entity like the cell must be placed on one individual, such as the cell foreman, cell lead person, or cell supervisor.

When a physical operation with a congruent management system exists, results are produced by the operations system that are easily measured and reported by the performance measurement system. Deviations, positive or negative, are easily identifiable and locatable. The one person in charge of a

312

specific cell can be solicited to make the necessary corrections. In most instances, problem identification and resolution will reside in the cell; the flow of reporting information will be cell-outward.

Measurement Elements

It is imperative that the cell members be responsible for selecting performance measurements as part of the original cell design. Ownership of all aspects of cell activity is enhanced by including cell performance measurement as an integral responsibility of the cell team. By making cell members responsible for both financial and nonfinancial cell performance, management ensures that operational and profit objectives are met at the company's lowest basic business unit level. Then the cell becomes the cornerstone of product cost and performance control.

Focusing PMs on cell goals gives rise to cell success and can be tailored and changed as demand requires. Below are examples of focusing to meet the needs of customer satisfaction, business objectives, and investment in the future:
- Customer satisfaction:
  —Schedule adherence,
  —Product quality,
  —Lead time,
  —Selling price, and
  —Service level.
- Business objectives:
  —Inventory level,
  —Least unit cost, and
  —Planned resource utilization.
- Investment in the future:
  —Training/skill level,
  —Number of new ideas generated,
  —Number of new ideas implemented, and
  —Employee turnover.

A typical manufacturing machining cell might select the performance measurements shown in *Figure 18-5*. (Initially after cell installation, inventory level may be an important measurement. However, as the cell stabilizes operations and inventory is forced out of the cell, the cell is capable of Just in Time operations and monitoring of inventory levels will probably cease to be needed.) These cell performance measurements adequately support the three main cell requirements. In addition, they are easy to measure, understand, chart, and relate to cell success and a company's strategic drivers.

Timeliness Of Measurements

Timeliness in summarizing and analyzing cell performance measurements enables rapid identification of problem issues. Some issues such as product quality defects must be immediately identified and corrected. Analysis of

313

Performance Measurement	Goals
Achievement of schedule	Customer satisfaction
High product quality	Customer satisfaction
Lowest transfer price	Business objective
Short lead time	Customer satisfaction
Complete machine information	Business objective
Increased training/skill level	Investment in future

Figure 18-5. *Performance measurements for a typical manufacturing machining cell.*

long-term quality defects gives rise to pattern identification that points to such problems as lack of training, the need for process simplification or engineering design changes, tooling and fixture obsolescence, and other hard-to-identify problems. Finally, what's timely at each level of company management varies. Most cell-level performance measurement is subdaily, daily, or weekly. Reporting to operations management will probably be weekly, monthly, or by exception. Business management reports will usually be monthly or by key exception.

MAKING IT WORK IN REAL LIFE

A major British manufacturer of aircraft wing components installed a well-designed performance measurement system. First, the assembly of aircraft wings was organized into four cells, and cell responsibility was placed with a cell foreman. The operation was demand-driven, requiring delivery of roughly 40 wing sets per month. The cells were constructed to accommodate peak demand of 100 wing sets monthly. Four levels of business responsibility were considered appropriate to control the business; at each level, reporting requirements were defined in terms of subject and frequency, as shown in *Figure 18-6*. Appropriate cell performance measurements were then devised as follows:

- *Due date performance*: This graphic analyzes and portrays the programmed date, actual date, and lead times for each set of wings. This report is circulated to the business and operations managers.
- *Attainment of schedule*: This is a comparison of expected time (not standard time) to complete, against actual time. This report is circulated to the cell foreman, shift manager, and operations manager, both after each shift and consolidated weekly.

Monitor	Frequency	Business responsibility
Due date performance	Set completion	
Business costs	Monthly	Business
Quality report	Monthly	
Due date performance	Set completion	
Business costs	Weekly/monthly	
Percent attainment to schedule	Weekly	
(high level)		
(shift level)	Exception	Operations
Quality item and cost report	Exception	
Inventory turns	Exception/monthly	
Machine information	Exception	
Work-in-process	Exception/weekly	
Percent attainment to schedule	Daily	
(shift level)		
Quality item and cost report	Weekly	
Inventory turns	Weekly	Shift
Machine information	Weekly	
Work-in-process	Daily snapshot	
Business costs	Weekly	
Percent attainment to schedule	Daily	
(shift level)		
Quality item and cost report	Weekly	Cell
Inventory turns	Weekly	
Business costs (own cell)	Weekly	

Figure 18-6. *Reporting requirements at varying levels of business responsibility.*

- *Quality:* These reports depict the incidence of rework, scrap, and concessions made. Comparison of these figures with the number of items processed through the cell gives management an analysis of the numbers and costs of quality deviations, which are the individual cell's responsibility.

315

- *Inventory turns*: Inventory turn compares the output of a cell with the inventory held in it, thus identifying the effective use of material held in the cell. Material components of aircraft wings are large and unwieldy. Batch size varies from cell to cell in the course of the manufacturing process, so regulators (sequencers) are used to control the flow. Therefore, target inventory turns become necessary to minimize cell costs and help ensure material flow synchronization from cell to cell.
- *Machine information*: Two of the four cells contain large NC machines; this necessitates the availability of machine information to indicate how much stoppages are costing and whether the level of preventive maintenance is adequate.

This properly planned and installed cell allowed cell members to identify all costs and performance measures influenced by the operation of the cell and charged to it. Cell teams developed ownership of the tasks and the resources available to achieve these tasks. Ownership carries with it responsibility; cell team members understand and are accountable for the effects of their efforts, the controllable costs, and the overall financial and operational cell results. Team ownership and understanding facilitated achievement of company target goals.

SUMMARY

It is highly important that the cell installation team understand the performance measurement process and how to properly include the appropriate performance measurements in the cell planning process. Traditional performance measurements are no longer viewed appropriate; they tend to be counterproductive to achieving management goals such as continuously improving operations.

The inevitable question of "How well did we do?" with the cell installation project is answered by a timely post-audit. Post-audit procedures should be described in the cell installation plan and address major plan success elements.

Long-term cell success requires an effective performance management system that mirrors the factory and company physical structure. The best system is simple, with performance measurements that are easy to measure, understand, and chart, and that support the cell's success and the company's strategic drivers. The system should provide timely data relevant to the appropriate level of management, and be flexible and responsive to change.

ENDNOTES

1. LeBoeuf, Michael, *The Greatest Management Principle In The World*, New York: The Berkley Publishing Group, 1986, pp. 22-23.
2. *CAM-I Cost Management Systems (CMS) Phase I Conceptual Design Document*, Arlington, TX: Computer-Aided Manufacturing, 1987, p. 19.

‖‖ 19 ‖‖

UPGRADING FMCs INTO
FMSs AND CIM SYSTEMS

It has been said, somewhat humorously, that an FMC becomes a flexible manufacturing *system* (FMS) when the cost of the software exceeds the cost of the hardware. This may be true, but upgrading an FMC into an FMS, or further, to a computer-integrated manufacturing (CIM) environment, is neither a trivial nor a short-term task. Starting from a traditionally organized functional manufacturing environment only makes the task more complex.

By definition, an FMC is task-oriented; it is geared toward producing a variety of clearly defined products. In the same way, an FMS implies that a higher level of automation and a comprehensive, coordinated means of control define the means of production. When all or most of the functions associated with operating an FMC have been automated, the FMC becomes an FMS; this level of automation usually includes scheduling, material storage and handling, machine loading and unloading, chip removal, dimensional or quality control, and tool control, storage, and delivery.

By including the manufacturing support tasks and adding a common database with a coordinated system of communication in a real-time network, the FMS becomes a CIM system. All three—FMCs, FMSs, and CIM systems—are designed to produce a defined range of products in minimum time, at an effective cost, with an appropriate level of quality. However, as shown in *Figure 19-1*, their impact on the business operation increases with each iteration.

An FMC may affect the functions that feed it material and receive its products, schedule it, provide it with work instructions, and receive data from it. Initially, an FMC, with its need for new skills and flexibility, might also influence training and recruiting activities. Forward-thinking organizations may even adjust the way they account for direct and indirect labor to reflect the new manufacturing methods. The operational impact from a single FMC on the whole business, however, is still fairly limited.

317

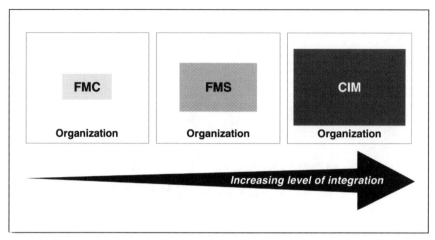

Figure 19-1. *Degree of impact on the entire business.*

Like an FMC, an FMS demands that scheduling, material input and output, and work instructions conform to its requirements; however, it also affects material handling, maintenance, disaster recovery, product design, schedule and priority modifications, and other support functions. Most of all, it requires data interfaces compatible with machine-level, not human-level, flexibility. CIM systems, on the other hand, affect the entire business—from sales to the shipping dock, from receiving and accounts payable to invoicing and accounts receivable. Upgrading to this level demands in-house expertise and competence, as well as the dedication of significant resources. While the impacts and potential benefits are enormous, the integration effort is equally staggering.

PERSPECTIVES: FMC, FMS, AND CIM

The Foundation: FMCs

The typical FMC impacts a narrow portion of the business:
- Direct labor,
- Indirect labor (due to improvements in material handling as a result of better product/material flow),
- Factory supervision and scheduling, and
- Inventory carrying costs (due to shorter lead times and less on-hand inventory).

While a single FMC may affect a portion of labor (direct and indirect), pieces of overhead, and portions of material costs, this usually represents no more than a 20% impact on the total cost of the product—especially if it is only a physical installation, with no organizational realignment. While a 20% impact is nothing to ignore, a broader application of FMCs will affect a bigger piece of the business pie.

The Next Step: FMSs

An FMS will affect direct and indirect labor, but its impact, especially on indirect labor, will be greater. Automating some quality functions, material handling, scheduling, supervision, maintenance, and manufacturing/industrial engineering will obviously impact a larger portion of total product cost. It may even have peripheral effects on purchasing and product design. While impact on product cost is somewhat larger and FMSs do have some effect on much of the organization, an FMS primarily concentrates on the manufacturing or production and support functions.

Total Integration: CIM

A CIM system truly impacts the entire business. In a CIM environment, design engineering typically uses a computer-aided design (CAD) system to develop product drawings and specifications. These become the foundation of the database to be used by manufacturing and purchasing. Orders for production scheduling are entered directly by salespeople in a computer-usable format for translation into material and other resource requirements. In some cases, the orders (or demand) are transmitted directly from the customer via electronic data interchange (EDI) connections. Production and machine instructions are produced using a computer-aided process planning (CAPP) system, and they are downloaded directly to the production machines and measurement equipment.

Material is received (and may be inspected) and processing is initiated by the automated manufacturing, scheduling, and control system. Process adjustments are made, tools are changed, and maintenance, when needed, is summoned automatically—all to keep production on time and within specification.

Direct labor is now a much smaller part of the cost picture; this demands that cost accounting change the entire basis of product costs from the traditional labor base to reflect the true sources of product cost within the business. Financial planning, in turn, should be tied to capital investment to support the changes needed to justify specific product development, technology, or forecasted volume changes. Even suppliers will have material specifications, container specifications, labeling requirements, certifications, and delivery windows that are accurately controlled. The orders will be issued and connected directly to production needs. In summary, CIM affects the entire organization, and all major cost components—direct labor, overhead, and raw material.

Since the implications of FMS and CIM involve the entire manufacturing operation, upgrading to this level is a complex task. Determining the scope of the project, selling it to the organization, establishing a multifunctional team, writing specifications, selecting suppliers, training, installing, starting up, and moving to full production will be measured in years, not months. These steps will consume a significant amount of corporate resources and require access to other experienced outsiders.

GETTING STARTED

An FMS or CIM project is not likely to be difficult—initially. Planning and implementing an FMS parallels the FMC planning process. Nurturing the project through a period of tough questioning about its justification *will* be difficult. The total expenditure will probably be much larger and spread over more time than most other project expenditures. For this reason, traditional financial justifications designed for a quick, easily measured payback or ROI may have to be scrapped in favor of justification that takes a broader view. Because most FMS or CIM efforts follow closely on the heels of FMC projects, traditional justification methods are strained further, since the FMC justification probably claimed the significant cost savings.

This is not to imply that an FMS or CIM project cannot be financially justified. These projects *can* prove extraordinarily profitable. However, the justification and ROI may be spread over five or more years. Typical short-term project philosophies don't evaluate the implications and product costs beyond the initial payback period. In a project of this nature, a longer-term vision must be adopted.

For example, one objective outside the direct financial payback criteria is improved customer service; reduced product development or production lead times can help maintain parity with or exceed the capabilities of worldwide competitors. Preserving or improving market share may also mandate improvement. Of course, with improved customer service, as with some of the other benefits, there may be an indirect financial reward as well. Another objective closely related to improved customer service is increased delivery reliability. While shorter lead times are important, the company's performing to those lead times *consistently* allows the customer to better estimate needs (and to supply better information about demands). Ensuring that the production process employed in the FMS or CIM environment is consistent and comprehensive minimizes *total* production time.

Other benefits of FMS and CIM applications not always fully considered are product quality improvements. With CIM, the process begins with more consistent designs. These designs result from standards developed to permit efficient operation of the CAD-oriented design function. Established features can be re-used effectively; fits, tolerances, and interferences can be tested before patterns and tools are made, materials are ordered, and the work hits the shop floor. Translation of CAD-produced drawings through CAPP will detect basic manufacturing problems like tool interference, and will provide a consistent manufacturing method not normally found in a manual production environment. Repeatability becomes a significant part of the automated process due to:

- The ability to automatically compensate for material inconsistencies and tool wear using adaptive controls;
- The ability to dimensionally probe and make process corrections for a part still on the machine;
- Handling material in a controlled environment designed to minimize part damage or distortion, and

- Using automated coordinate measuring machines (CMM) to detect trends which could produce out-of-tolerance parts.

Do The Right Thing For The Right Reasons

Replacing existing equipment with an FMS must be done for valid reasons. One reason that is not valid is to achieve planned or specified quality or volume. It is unlikely that replacing the current equipment with more automated, more complex equipment—which also demands more sophisticated skills to operate and maintain—will change the environment or system performance significantly.

Increasing utilization of direct labor or equipment is also an invalid reason for justifying the installation of an FMS. Direct labor, probably the smallest of the major portions of product cost, is not the first place to look for cost reductions. With a total impact of 10% of product cost (maybe), eliminating direct labor costs will not radically change product cost. A reduction in *total* staff level—direct labor and support (overhead)—is potentially a valid objective, providing that other increased costs do not offset these savings.

Another frequently cited objective when installing automated equipment in factory and office environments is improved records accuracy. While record accuracy is convenient, it does not necessarily contribute to the bottom line. However, if record accuracy helps reduce inventory and makes deliveries to customers more reliable, it may indeed be value-adding. Nonetheless, the contribution of recordkeeping accuracy to total system payback is likely to be very small.

Performance Measurements

Any proposed measurement indices for the system should be tied directly to its justification; these determine whether the system is performing adequately or if adjustments are required. Valid performance measurements must be visible outside of the organization. If the justification does not result in improvements that are visible to the outside world, it should be re-evaluated. These include things like:
- Product quality,
- Faster and more predictable deliveries,
- Competitiveness in desired markets, and
- Responsiveness to customer demands.

A series of measurements which indicate how well a manufacturing cell is performing was outlined in Chapter 18. These measurements are equally valid in measuring an FMS and, along with some others, can help to measure performance in a CIM environment. Since CIM involves traditionally nonmanufacturing sectors of the business, it requires some additional performance measurements:
- The calendar time required to introduce new products, from the beginning of the process to availability to the customer;

321

- The calendar time required to process customer orders, from receipt to delivery;
- The size of the support staff required for a given level of business (the overhead level);
- The amount of inventory required to support the business, and the obsolescence write-off needed on an annual basis;
- The cost of processing the scrap and rework that results from the manufacturing process, and
- The traditional measurements of the total business: profitability, return on assets, and others.

PLANNING FOR IMPLEMENTATION

Shop Floor

Planning ahead for the transition to an FMS or CIM when working to install a cellular manufacturing environment requires an incredible amount of foresight. Space and arrangement planning for future shop floor upgrades is one area where long-term vision is important. In the cell arrangement, human flexibility can overcome the effects of bad planning. However, upgrading to an FMS and a CIM environment may be more of an exercise in starting over than upgrading if cell planners fail to consider areas such as space and flow paths for automated transportation of parts, tools, and chips. More specifically, there are several options when upgrading a metalcutting FMC:

- Operators with manual machines can be replaced with DNC machines with pick-and-place robots;
- Forklift trucks and pallets can be replaced by an automated conveyor system;
- Inspectors with gages and surface plates can be replaced by on-machine inspection probes and an integrated CMM;
- Setup operators with tool delivery carts can be replaced by a tool delivery gantry robot and an automated tool store, and
- Forklift trucks and chip hoppers can be replaced by AGVs or an in-floor chip conveyor system.

Without appropriate planning at the FMC stage, each one of these replacement steps could require a new layout and a new facilities and services plan, instead of having the potential for being carried out on a *modular* basis.

The details of the planning steps for flexible cells described in this book are also applicable to the implementation of FMS and CIM. Some planners may be tempted to overlook some of the details (and the details *can* be excruciating) because "automation" or "software" will take care of the problems. In fact, the scope of the change, particularly when going directly from a traditional environment to an FMS or to CIM, is much larger and demands more thoroughness and logic than the change from traditional manufacturing to cellular manufacturing.

322

Computer Systems

In addition, a new manufacturing system may require more computer power. At the FMS level, the computer system will require specialized, application-driven hardware linked together to make the FMS work. Some FMS designs will have direct data interfaces with the existing "business" computer. If, however, the entire FMS is installed over a relatively short period of time and is self-contained with all internal functions, all that might be required outside of the system itself may be communications interfaces for downloading machine programs, part and equipment status reports, schedules, and quality data. These requirements should be considered as early as possible—during the planning stage for the FMC if feasible. Many FMS installations have been put in by vendors as turnkey programs that include, internally, all necessary hardware, software, and interfaces. Besides new manufacturing applications, the hardware should support entirely new business-level applications. The computer may be required to communicate with a CAD-applications computer or support the CAD application itself.

Other applications such as CAPP or quality and maintenance management may require similar support. There will probably be a new database for the organization, or at least additional applications tied to the database, which will require hardware support. In addition, the size or character of the business could be changing at the same time. While it is possible to plan for most of the variables in a straightforward manner, the interaction of the variables, in addition to unforeseen changes in the business, will almost certainly require adjustments throughout the CIM planning phases.

The computer software and communications must be the subject of intense early planning. Each of the functions in an FMS and each application to be included in a CIM environment must be thoroughly examined. Functionality and communications must be analyzed, performance specifications must be written, and the best matching applications and communications software selected. While each of the packages will have clearly defined and documented formats, startup and use will never be executed as cleanly as planned. Good planning will reduce the problems, but business changes, personnel changes, software version changes, undiscovered bugs, and unpredictable interactions will cause enough problems without letting poor or incomplete applications software and communications planning contribute to the system problems.

STRUCTURE: PROS AND CONS

The change to an FMS or to CIM, if justified, should have more advantages than disadvantages for the business that implements the changes. An FMS will add:

- More structure to the production process;
- A more clearly defined and, hopefully, better logic to the method in which parts are manufactured;
- A discipline to the manner in which work must be done.

CIM will do essentially the same for virtually the entire manufacturing organization, since a common database is shared by the entire organization. This will quickly improve communication, the quality of production, throughput time, and direct labor and overhead rates.

Unfortunately, people who find change difficult will point out the downside to the implementation of an FMS or CIM system. Structure *can* have disadvantages. People must adapt to the system and its requirements for timeliness, methods, and standards of performance; if the current staff cannot adapt, they cannot coexist with CIM. New personnel or personnel in new positions, who have not been part of the design and development effort, may also find the environment baffling and difficult to work in.

Unexpected product changes may negate the entire effort. For example, a highly efficient metalcutting FMS may be of little use if competitors have found a way to make the parts from injection-molded plastic at a fraction of the cost. Changes in volume, product mix, or location of the market also may negate the need for a large capital investment. Supplier dependability and standardization, planned as a requirement to cost-effective and good functional use of the new system, may not materialize because suppliers find that they cannot justify the changes.

The amount of preliminary planning required for an FMS or CIM implementation could also be considered a disadvantage. For the duration of the planning (and implementation) exercise, there is a project leader, a planning and implementation team, and the involvement time from the rest of the organization to be underwritten. The elapsed time can stretch into years, and it means time and other resources taken away from the execution of tasks necessary to the core business—selling and delivering the product.

In addition to the time and resources, planners must design in "hooks," which are means of accommodating change in the future, as seen in *Figure 19-2*. This ability to change must be planned into the system, whether the change is in the product itself, the way the business operates, or the need to digest new software or upgrades, new communications requirements, new processes, or the process to replace temporary system resources (those that were not able to be permanently implemented when the FMS or CIM system was initially installed).

A concrete example of the need for such hooks is the lack of firm standards and well-defined protocols for network communications.[1] One of the objectives with applications software is to remove one module and simply replace it with an upgraded one. With everything done correctly and the cooperation of the software suppliers, this might be possible. More than likely, however, other changes to the system will be required. Some of the same planning will have to be repeated when a new piece of software becomes available that is apparently effective enough to warrant installation.

The issue of up-front cost for an FMS or a CIM installation is another potential negative that needs to be addressed. The justifications discussed earlier may pale as the negative cash flows begin to mount and the benefits are not yet in sight. The temptation to stop the effort will be great, but the cost of stopping,

Process Hooks	• Space for additional equipment
	• Equipment adaptable to process:
	  — Parts outside planned size envelope
	  — Other than specified materials
	  — Tighter tolerances
	• Common tool controls
	• Standardized fixtures and tools

Software Hooks	• Neutral interfaces • Universally recognized communications protocols • Variable length record formats • Accessible data, data base

Computer Hooks	• Open operating systems
	• Computers:
	  — With upgradable processing capability
	  — With upgradable on-line storage
	  — Linkable to other CPUs
	  — With upgradable communications

Figure 19-2. *FMS/CIM hooks.*

in terms of organizational morale and the unlikely event of recovering any costs from a partially completed project, should be a sufficient deterrent to that course of action. Adequate up-front planning and taking changing business conditions into account ensures that the original reasons for starting the project remain valid.

While most of these and other anecdotal arguments should be addressed individually as a part of the business strategy preceding the implementation plan, many of the same, potentially devastating, occurrences can happen in a traditionally organized business. A tightly organized business, with a well-thought-out strategy and a manufacturing capability tailored to support that strategy, has a better chance of surviving than one that has not recently or not as rigorously planned for the "what ifs."

Verifying The Expectations

Expectations about the performance of an FMS or CIM system can range from the wildly optimistic "It'll do everything!" to "It'll never work!"

The most realistic expectations which should be part of the justification are obviously those grounded on something more than instinct. Where simple, straightforward, single-path systems are involved, equally straightforward time and cost estimates can serve as performance predictors. With a system as complex as an FMS having multiple components and subsystems and a variety of possible interactions, a computer simulation is probably the only way to achieve a reasonable estimate of system performance. With a CIM system involving much of the organization, segments of the system will probably have to be modelled individually.

The model should not be more complex than is required to create a realistic, representative image. For example, if a limited number of part families make up the overwhelming majority of the parts to be processed in a system, it may not be necessary to model every possible part combination or every possible staffing level and shift arrangement. *Figure 19-3* shows some variables that could be included in a typical manufacturing simulation. Additional variables are certainly possible; graphs, histograms, and reports can be generated to reflect the behavior of each variable as the simulation is run.

Creating a model of the existing process to validate the modeling techniques, particularly if the use of simulation models is not routine in the organization, may be a desirable preliminary step in the modeling process. While many users of models simulate manufacturing processes, few use them to simulate the "soft" side of the business, the nonmanufacturing aspects. By designing the model correctly, those aspects of the business can also be modeled. For example, by determining the types of communications to be originated and the reason for each type, the level of message traffic in a communications network can be modeled. Similarly, sales order processing, purchasing, production engineering, and other functions can be reasonably represented using simulation models.

Besides determining resource requirements and projecting throughput times, modeling can help with another aspect of verification; at various points throughout a model, one of a series of processes can be selected based on the conditions at the time. Thoroughly reviewing each decision point and what should happen at each point will help with the specifications for the same decision points in the live system. Accurate models require significant skill in the use of modelling software. If that expertise is lacking in-house, it can be purchased from systems suppliers, machine tool suppliers, or consultants.

Once a system configuration that meets the company's needs has been established, the ability of the system to handle both organic and inorganic growth must be determined. *Organic growth* is growth within the current organization and current product range. *Inorganic growth* is the addition of new organizations and products to the existing business. Potential combinations of change and the effects on the new FMS or CIM system can be evaluated by using the basic simulation model (or by testing the estimates used to evaluate the proposed configuration if a simulation was not used). If it is decided that the potential for growth is great enough to warrant making system changes to accommodate the growth, those changes will have to be made. The changes may be as simple as providing "hooks" to make implementation of new capabilities relatively easy. The changes also may be as complex as adding entirely new processing capabilities.

- Parts—variable types and quantities

- Processes—representing most types available

- Queues and buffers—most types and sizes

- Cycle times

- Setup times

- Teardown times

- Wait times

- Staff—variable classes and quantities

- Priorities—variable by process

- Shifts—single or multiple with variable working hours

- Routings—variable by part

- Transportation—conveyors, AGVs, fork truck

- Speed—loaded and unloaded with acceleration and deceleration

- Loading and unloading of transport

- Transport capacity—size and weight

- Tools and fixtures

- Attributes—part-specific elements which determine how the part will be processed

Figure 19-3. *Potential simulation variables.*

WHAT'S AVAILABLE OFF THE SHELF

Selection of a system or systems resources from available alternatives can be accomplished when the system functionality has been defined, the system justified, a team selected, and the specifications written. This is true not only of processing and material handling equipment, but of system control hardware, application software, and communication software. It is possible to buy:

- Machines and controllers,
- Terminals and data (bar code) readers,
- Cell controllers,
- Applications and general business computers,
- Communications hardware, and
- Material handling hardware.

These packages usually require little or no customizing to use in a wide variety of applications. The amount of hardware and software readily available for use in FMS or CIM environments is surprising. *Figure 19-4* shows the typical areas and the number of suppliers found.[2]

Building custom-designed *equipment* is an alternative way to fulfill system needs; generating customized or heavily modified *software* is another way. There are two overpowering reasons not to use these alternatives; first, untried logic and construction can be difficult to debug and correct. Secondly, *significant* time and people resources are required to develop and implement custom-designed software and hardware. Using custom-designed hardware and software does permit the user to be a pioneer, a reputation that some companies cherish. Since there is enough risk in systems implementation without adding more complexity by using untried systems, it is good to remember that a pioneer has often been defined as the person found face down in the mud with arrows in his back.

Supplier Area	Number of Suppliers
AGVs	30
Automated storage and retrieval systems	39
Automated test equipment	7
Cell controllers	32
Computer numerical controllers	23
Conveyors	40
Factory floor computers	40
Machine vision	73
Programmable logic controllers	25
Robots and robotics	48
Sensors	247

Figure 19-4. *Suppliers of FMS and CIM hardware and software.*

INSTALLATION OF FMS AND CIM

In Chapter 17, the traditional steps of project definition and strategy were detailed—the creation of a plan with a schedule and key milestones, resource assignment, handling in-progress changes, plan approval and ownership solicitation, and actual implementation. The FMS or CIM installation process steps, in general terms, are the same as those for an FMC. Obviously, the broader scope of an FMS or CIM system requires that the details within the outline steps be different.

The additional material handling, quality assurance, and other automation of an FMS requires definition, specification, and planning. In a CIM installation, there are not only additional design elements to be considered, but different members of the organization must also become active participants. A CIM installation, due to its scope, is also likely to be a much lengthier installation containing many more diverse system elements; it will have the facility, equipment, and arrangement considerations of an FMS plus many more considerations away from the manufacturing floor.

Among the practical considerations of CIM and FMS installations are the differences in lead time for obtaining various components and the resource considerations of trying to get it all done at once. Both can cause serious problems in either type of installation. There are, however, tools available to get around these problems. In the case of people and skill resources, assistance is available in the form of contract employee services or consultants. Most firms find it neither desirable nor affordable to maintain the skills and staffing levels required to complete a major project internally, especially when, as the project progresses, the skills requirements shift from conceptualizing to detailed planning, to education and training, to implementation, startup, full-scale operation, and progress audits.

Use Of An Emulator

Nonsynchronized timing in component or system availability or readiness is another problem. Since the level of integration is high for both FMS and CIM, it is most desirable to have all system components available for testing simultaneously. Because of the timing or resource problems in integrated testing, a compromise is usually required. One of these potential compromises involves the use of a software emulator—a software module that takes the place of an unavailable component of an FMS or a CIM system. An emulator provides a limited number of programmed interactions with other system components until the system or active module it is replacing can be installed.

Testing a machine tool, for example, may require the tool to interrogate a scheduling system through the FMS controller and the factory communications network, then perform some action based on the specific response received. If the actual scheduling system is not yet available, an emulator can recognize the interrogation routed through the FMS controller and easily provide one of a series of simple, pre-programmed responses that will test both the communications compatibilities and protocols, and the actions of the machine tool. *Figure 19-5* shows how the emulator might be hooked to an FMS system.

Turnkey Systems

Purchasing a turnkey system is another possible solution to the complexities of implementation. With this option, the system integrator (contracted to put all of the components together and ensure they work as a unit), if carefully selected, will have seen the situation before and repeatedly worked out solutions to problems. For the same reasons, the system may be assembled and work more logically than if done by in-house resources.

There are also disadvantages to a turnkey approach. While the systems integrator may be familiar with the environment in which the application is being used, he or she will probably not know the business—the product or the markets—as well as staff members do, and could miss a key requirement as a result. In addition, the in-house staff will not have an intimate understanding of the system, because they had little or no role in the implementation. It is also likely that staff members will feel little ownership in solving the inevitable startup problems and may not be as willing to make adjustments to keep the system working well. The solution to this disadvantage is to utilize the vendor "partnering" approach outlined in Chapters 15 and 16.

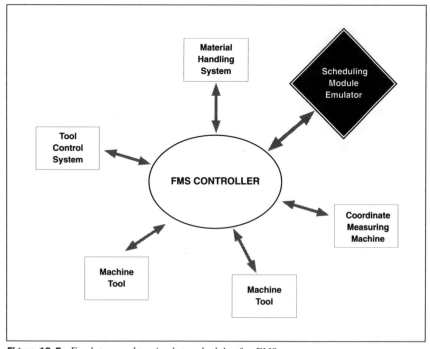

Figure 19-5. *Emulator used to simulate scheduler for FMS.*

330

MANAGING THE TRANSITION

Once the key system elements of an FMS or CIM system have been put in place and tested, there must be a transition from the old way of doing things to the new way. Organizations and reporting relationships will change. Some jobs will be added; other jobs will disappear. Furthermore, the content of the jobs that do remain will necessarily change. This means that new positions and skills must be considered as part of the implementation plan, and appropriate communication, outplacement, hiring, and education programs must be implemented.

Starting on the right foot with communication is especially important. Since the cultural impact will be significant, the communications plan must be completely prepared ahead of time, including answers to anticipated questions. Nothing will shake management's credibility as much as the inability to answer straightforward questions related to the implementation and its impact on the workforce. Thoroughly evaluating the skill requirements for new positions, agreeing to training requirements, setting training objectives and measurements, designing the curriculum, and securing qualified instructors run a close second to communicating well. Too few employees placed in new positions have a clear understanding of what is expected of them. All of these tasks can and should be identified as part of the overall implementation plan.

Another problem encountered in the transition to a new FMS or CIM system is "what to do with the old stuff." One example is the change from a manual drawing system to a CAD system. In such a transition, there will be hundreds or possibly thousands of old drawings that may have to be digitized or redrawn for the new system. If new customer orders consist mostly of new products, that may not be a problem—the new drawings can be generated as orders are entered. Most organizations are not so lucky; if time allows, new drawings for existing products can be generated as the need to use them arises or as part of a total update program—resources, cost, and time permitting. It doesn't sound difficult and it doesn't have to be—if the planning is done ahead of time (even for service parts orders for the *really* old stuff). There are now drawing services available that will convert old drawing files to CAD-compatible formats to accelerate the process. But this can be very expensive and still require some additional interpretation modifications after the conversion.

Planning should also include new drawing standards for use with the new system, and a way to incorporate the handwritten notes and corrections that are now tucked away on old prints or in machine operators' work benches—two conversion problems that digitizing probably will not solve. Similarly, old routings will have to change. Old equipment and methods will disappear and new processes will have to be planned and documented. New tools and fixtures will have to be purchased.

There will be open customer orders, purchase orders, and shop orders which may also have to make the transition from the existing system to a new one. Some will have partial shipments, invoices, payments, cancellations pending, and a whole variety of current information and transactions kept in them. This information is the lifeblood of the business, and someone at the company must always be able to talk intelligently when customers call to check the status of their orders. Any process attempting to move these orders between systems must be carefully controlled and audited; an order-by-order, item-by-item audit is not out of the question.

Managing the implementation schedule requires a unique blend of dedication to get the job done and an understanding of circumstances that require adjustment. It means understanding the difference between inconveniences or difficulties and a roadblock that requires additional time or resources to remove. It also means pushing hard while maintaining a cooperative attitude. That and a myriad of other details are major reasons why managing the project and the implementation schedule is a *full-time* job and requires a dynamic leader.

ONGOING SUPPORT FOR FMS AND CIM

Once an FMS or CIM system is installed, the job is not over; top management support is needed to see that the old, presumably less effective or more costly methods do not creep back into the organization. Besides management support in a general sense, much more specific support will be needed in other areas. An organized preventive maintenance program is key. In this highly integrated environment, a single equipment failure can significantly degrade system performance or cause the system to fail altogether. Equipment suppliers can be of considerable assistance in planning preventive maintenance programs and advising on what emergency spares to keep on hand.

With tightly controlled processes, tool and fixture support requirements are much more rigid. An essentially manual system with a low level of integration can, in some cases, use a variety of production methods to overcome problems with tool and fixture design or availability. Few such options are available in an FMS or CIM system. Since operations are highly standardized, the system expects to have a particular style of fixture and individual, preset tools available for processing. If the tool supply system is not performing to the same standards as the rest of the organization, production stops.

As is evident from the emphasis on training and education in this chapter and previous chapters, not only are the standards of performance changing, but the types of tasks to be performed are changing as well. With a higher level of computerization and automation comes less reliance on routine replication of information, calculations, and tasks in general. Employees are asked to make decisions in nonroutine circumstances, and to interact with other functions inside and outside the firm.

SUMMARY

The significant differences between FMCs and FMSs and between FMSs and CIM systems are mostly in the levels of control, automation, complexity, and sophistication. FMCs and FMSs both affect narrower areas of the business. A CIM system impacts the entire business—both the organization and its operation.

Justification for higher levels of integration, such as CIM, will generally have to come from more nontraditional areas like improved customer service, quality improvements, and flexibility to respond to changes. Some less-than-appropriate reasons for spending capital for this elevation of automation include increased utilization, increased records accuracy, or meeting current planned quality levels.

Converting a traditionally organized manufacturing environment, whether from noncellular production or previously installed FMCs, is an arduous and lengthy task requiring experienced and intensive resources. One must never underestimate the magnitude of the tasks ahead, setting the expectations in realistic terms. Verification through simulation methods is always recommended.

If possible, currently available hardware and software that has been previously proven and tested should be used. Turnkey conversions, if at all feasible, should also be considered, but an approach that will stimulate the *maximum* involvement and eventual ownership by the people who will interface with the system is essential in this implementation scenario. The key to successful automation or integration implementation will rest with the accuracy and the thoroughness of the planning.

ENDNOTES

1. McCarthy, J.J., "How to Wait for CIM Communications Standards," *CIM Review*, Spring 1990, p. 39.
2. *Directory of Factory Automation Systems Suppliers*, DataPro, May 1990, pp. 101-124.

‖‖‖ **20** ‖‖‖

FUTURE TRENDS FOR FMCs

Predicting a trend is not easy because current data from which to project is skimpy. The FMC concept *is* relatively new; it was conceived in the early 1980s. At that time, only a few of the more aggressive companies—or maybe it was companies under undue competitive pressure—became involved with the concept. The number of companies adopting the concept increased rapidly and most dramatically in the latter part of the 1980s. The bulk of the media reports about the FMC concept came from the metalworking sector of manufacturing. A few more came from the electronics sector. Noticeably absent were reports from industrial sectors such as plastics and furniture. Oddly enough, there were even a few articles about the concept being installed in *office* functions.

THE FMC CONCEPT IS HERE TO STAY

In spite of the slow initial acceptance of the FMC concept, FMC use accelerated. A reasonable estimate in the metalworking manufacturing sector is that as much as 20% of the total potential application opportunities are being exploited today. In other sectors of manufacturing, the figure is probably substantially less.

At the same time, both batch and high-volume producers in the Orient make extensive use of the concept, as shown in *Figure 20-1*. Similarly, FMCs are seeing some use in Western Europe. A recent survey of 300 metalworking manufacturers in England, for example, indicated that half of the respondents are now using FMCs.

The day will soon come when the intricacies of cell design become common knowledge and engineers receive training in them at educational institutions. FMCs have become an intrinsic part of the industrial landscape. The FMC is *so* value-adding and *so* sensible that 10 years from now it may be seen as *the* single

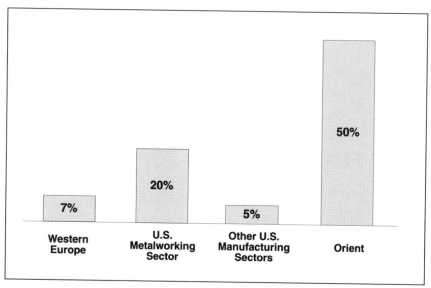

Figure 20-1. *Estimates of the present use of FMCs.*

most significant technique that led the Western manufacturing community out of the downhill slide that it faced from increased global competition in the 1970s and 1980s.

Because FMCs represent a "new" organizational "technology," they should be expected to go through considerable metamorphoses as they catch on through the 1990s, particularly as equipment vendors make them easier to install and use. Automation and transfer lines for high-volume, low-variety production have used the concept for decades. It is also the same way work has been accomplished on progressive assembly lines, at least since Henry Ford's time. An FMC merely takes the principle of continuous product flow and applies it where flexibility is required and low volumes are the norm.

The simplicity of this technology will make it increasingly easier for a wider variety of people to accept—engineers and operators as well as executives and managers. The only thing that makes it seem complex is that it's *different*. Paradigm busting *is* difficult. This chapter deals with the internal and external influences that dictate positive trends for the FMC concept as it evolves over the next decade.

Seven predictions can now be made with some confidence:

1. The huge economic value of FMCs will not be ignored any longer, despite the difficulty determining it.
2. The concept will become easier to use as machine and equipment suppliers respond to the new markets FMCs will create.

3. When new plants are built, they will include the FMC concept from the outset.
4. FMCs will finally open the door to increased use of robots, further enhancing their (the robots' and the cells') financial impact.
5. Enthusiastic operator acceptance of FMCs will help spread the use of the concept.
6. The up-to-now largely illusive dream of CIM may be realizable.
7. Perhaps most importantly, FMCs will impact the nonproduction facets of many companies, often leading the way to truly world class business operations.

The evidence available today that supports these seven trend predictions is detailed in the balance of this chapter.

1. ECONOMIC VALUE

The huge economic value of FMCs will not be ignored any longer, despite the difficulty determining it.

A sufficient number of FMCs have been installed and their financial benefits established so that the number of doubters is shrinking. At the beginning of the 1980s, when the FMC concept first emerged, few believed in their financial efficiency. Understandably, it *was* difficult for people to imagine how the costs of having their present shops completely redone (re-laid out and disrupted) could possibly be recovered. One result was that FMCs caught on very slowly. In fact, it took many manufacturers 10 years of experimentation before they became convinced that it was a fiscally sound idea. Even now, many manufacturers remain unconvinced; most shops still are not using cells for anything other than high-volume, low-variety production. Increasingly, as the financial track record of FMCs becomes more widely known and distributed, a growing number of companies will adopt them.

The clarity of the financial advantages of FMCs will continue to emerge from their historical and present state of confusion. The financial value of FMCs *is* confusing because:

- FMCs pay off in a fashion that is new and unusual for most companies. That is, the justification for the investment in them can be said to be "peculiar" compared to investments historically made in factories for new machines and equipment upgrades.
- Engineers (and many managers) have a poor understanding of what might be called the "real behavior of costs."

Anyone experienced in the FMC concept knows that benefits from them build as successive ones are installed. *Figure 20-2* clearly depicts this concept. If six FMCs were to accommodate the manufacture of all of the products made in a factory, the first cell installed would not produce one-sixth (16%) of the total benefits. Rather, the figure might be closer to 5%. Similarly, the sixth cell might be seen as capturing 40% to 50% of the benefits. In short, FMCs cannot be viewed incrementally for two reasons.

337

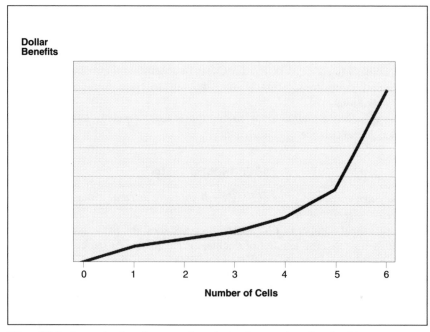

Figure 20-2. *Increasing benefits occur from the implementation of each additional cell.*

Mushy Justification

The first reason is that the justification of FMCs (the benefits) is mostly in "mushy" cost areas, such as:

- Quality,
- Inventory,
- Throughput time, and
- Floor space.

These are all *overhead* costs, the burial ground for all costs nobody knows what else to do with. For instance, no one seems to be able to calculate the cost of quality. The calculable parts of quality cost such as scrap, rework, and warranty payments *are* relatively easy to capture in many companies. However, estimates of all other related costs (hidden costs) range widely, reaching up to five times the calculable costs.

Similarly, inventory carrying cost estimates range from between 10% and 50% per year of the cost of the material and in-process value added, depending mostly on differences of opinion from one company to the next regarding what should be included—definitely not an exact science. Of course, no one knows the value of an acceleration of throughput time; it impinges on numerous costs and can have a large impact on market share. Finally, the real value of floor space

338

reductions depends on what's done with the floor space; empty space has no, and actually negative, value if left unutilized or allowed to accumulate inventory and junk. So to begin with, the primary areas of justification for FMCs are difficult to pin down. They are subject to opinion, interpretation, and circumstance from one company to the next.

Fractional Effects

The second reason for not being able to incrementally justify FMCs is the fractional effect they have on costs. Engineers, the primary architects of FMCs, as well as others, don't understand, nor were they ever trained in, what might be called the behavior of costs. Ignoring the accounting principles from the engineering side has caused serious misconceptions within the business world. *Figure 20-3* illustrates a few.

"Inventories are assets. It says so on the balance sheet."
The truth is, of course, that inventories are one of the worst liabilities! They certainly cost more than most balance sheet liabilities.

"The cost of carrying inventories is only 10% to 15%."
The truth is that the real cost is much closer to 30%-plus when all of the costs allocable to holding inventories are figured in. Often, these carrying costs can exceed 50% per year. Either figure is often much more expensive than long- or short-term debt!

"An hour of an operator's time is worth $75 on this machine because of overhead."
An hour of an operator's time (or anybody else's) is worth only what they are paid plus fringes, plus maybe 5% for "handling costs." The rest of the $75 is overhead.

"Because we can buy this part from a vendor for 10% less than our cost, we ought to buy it."
The truth is that the "our cost" figures are outstanding if they are accurate within 25%. Taking out parts increases the effective overhead for the parts that remain. The final truth is that the our-cost figure is usually based on the old non-FMC method of manufacture.

Figure 20-3. *Misconceptions about the behavior of costs.*

In the example of the six cells, the first FMC installed will perhaps eliminate the need for half of a fork truck and all of its associated costs. This means, of course, that nothing is actually saved by the installation of the first cell in the way of fork truck costs. Maybe by the time the second or third FMC is installed, however, a *whole* fork truck can be eliminated. But the third cell didn't make the elimination happen. There are literally hundreds of individual areas of cost that are reduced in this fractional fashion:

- All of the floor space savings potential may not occur until the fifth or sixth cell is installed. The freed-up space can then be used for other value-adding activities.
- If there are two linked cells handling the same parts, very little will be saved until *both* are operational.
- If all six cells are involved with making parts for a single product, no throughput-related savings will occur until all six are installed. Perhaps an entire storage area or warehouse can then also be eliminated.
- The same logic applies to other areas of potential benefit, such as reducing layers of management or the numbers of people in affected functions such as production control, quality control, or expediting.

So the peculiar, nonincremental benefits of FMCs require understanding of the timing of their achievement. In spite of such confusion, an increasing number of engineers *are* learning the financial value of the FMC concept as more are installed and proving their worth. In fact, more companies are making the conversion to FMCs on faith; that is, some companies are installing FMCs *without* the laborious justification exercises that traditionally accompanied FMCs. Increasingly, the statement can be heard, ''We must install the concept simply because we *know* it is the right thing to do.''

Experience And Understanding

Some corollaries have evolved from this experience level with FMC installations: As more people become experienced with FMCs, fewer consultants are looked to as the only sources of this knowledge, reducing overall costs. Throughout the 1980s, sensing a void in know-how, consultants rushed in to lead manufacturers through the intricacies of cell development, justification, and implementation. Such activity accelerated the use of FMCs, but also added the financial burden of high consulting fees to the justification equation.

As with all such voids after the passage of time, however, there are now a large number of nonconsulting people in the manufacturing community knowledgeable in the intricacies of FMCs. Although many of these people will still make use of consultants on FMC projects to add skilled resources, make use of greater experience, and audit planning, they will take a much more dominant role in FMC installations. This should make FMCs even more financially attractive to their potential users.

People changing jobs may prove to be the most common cause of the proliferation of the FMC concept. The mobile nature of many people in the manufacturing sector will spread understanding of FMCs and their design and implementation. As people move between companies, they will be bringing previous successful experiences with them into their new environments. This is especially true for people changing jobs between market sectors of manufacturing.

Top managers have been obstacles to FMCs, many times because of the investment they require, but they appear to be more routinely accepting the value of FMCs. As increasing numbers of FMCs are installed, the number of managers pushing for even more installations *must* increase, because the favorable financial impact will prove too tempting for them to do otherwise.

The exception to this will be those managers who really aren't interested in making a manufacturing business more competitive. These top managers make their living trading companies, redeploying assets, and leveraging money in general. Unfortunately for those people interested in improving operations, the late 1980s saw an increase in the number of such top managers. Perhaps the demise of the easy credit, junk bond market of 1989 and 1990 and a lessening of leveraged money-making within the manufacturing sector will promote an increasing number of top managers whose objective it is to *fix* companies rather than trade them.

2. SUPPLIERS RESPONDING

The concept will become easier to use as machine and equipment suppliers respond to the new markets FMCs will create.

There is already ample evidence of this trend. Most people designing FMCs know that batch-type operations such as painting, washing, heat treating, and plating required in the middle of a sequence of work steps can detract substantially from the potential savings of an FMC installation, as depicted in *Figure 20-4*. Suppliers of such batch equipment (plating to only a limited extent) have already responded to the need for small, inexpensive units to perform the process operation *within* an FMC.

In the old process-focused method of arranging a factory, it is common to trundle all parts that require heat treating to a sometimes huge central area where parts are batched and processed through large heat-treat furnaces. This practice is dictated by equipment; these large furnaces are the only equipment *available* to perform the operation. This creates material handling chaos, work-in-process inventories, poor quality, and a host of other nonvalue-adding costs.

With the FMC concept aimed at *eliminating* such nonvalue-adding costs, increasing demands were placed on heat-treat equipment suppliers for small units that could be placed in line with the other operations that had to be done on parts. Another sector of equipment suppliers—machine tool builders—are also responding to similar demands. The number of single-purpose machine tools can be expected to dwindle as companies seek multipurpose machine tools for their FMCs. The market demand for machining centers, which are capable of performing a number of operations, will increase. Furthermore, creative

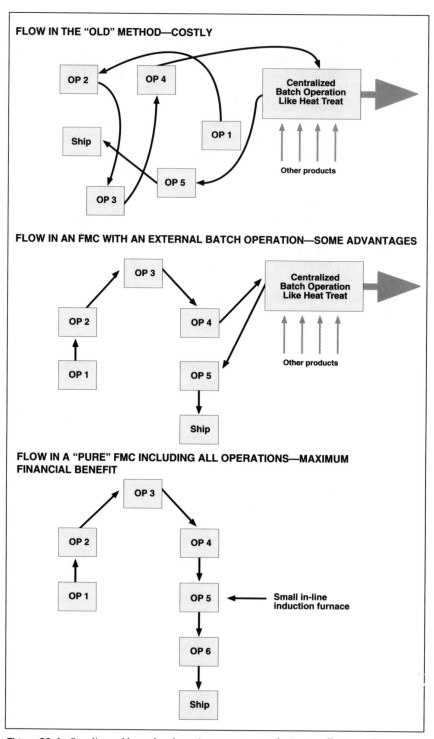

FLOW IN THE "OLD" METHOD—COSTLY

OP 2

OP 4

Centralized
Batch Operation
Like Heat Treat

Ship

OP 1

Other products

OP 5

OP 3

FLOW IN AN FMC WITH AN EXTERNAL BATCH OPERATION—SOME ADVANTAGES

OP 3

Centralized
Batch Operation
Like Heat Treat

OP 2

OP 4

OP 1

OP 5

Other products

Ship

FLOW IN A "PURE" FMC INCLUDING ALL OPERATIONS—MAXIMUM FINANCIAL BENEFIT

OP 3

OP 2

OP 4

OP 1

OP 5 ← Small in-line induction furnace

OP 6

Ship

Figure 20-4. *Suppliers of large batch equipment are contributing to effective cell development with the introduction of small, inexpensive units that can be included in the cell.*

engineering minds will probably create more multipurpose machines that will add to the forces making the FMC concept easier and more economical to install.

Additionally, machine tool suppliers have already responded to the requirement for no setup time between part runs that is intrinsic to the FMC concept. It seems reasonable to predict that these suppliers will continue to create ways to eliminate the need for setup time. These kinds of demands on batch-processing equipment manufacturers can only continue to intensify, and the suppliers *will* respond accordingly, because market demand still controls the world of economics.

3. FROM THE OUTSET

When new plants are built, they will include the FMC concept from the outset.

There can be little doubt that the FMC concept *will* be incorporated in new plants. What this means, of course, is that FMCs will prove to be even more financially attractive in these new facilities, since full advantage can be taken of them. Existing plants require costly efforts to move machines and equipment in order to incorporate FMCs, but the return on these costs is still invariably attractive. In a new plant, the payoff from FMCs will be even more outstanding since all equipment will be put in the right place the first time.

4. ROBOTS REVISITED

FMCs will finally open the door to increased use of robots, further enhancing their (the robots' and the cells') financial impact.

The news media often publishes articles showing the U.S. as being far behind in the use of industrial robots. Frequently, this disparity is used to explain the competitive edge the Japanese enjoy. Nothing could be further from the truth! Robots are merely a symptom, not the cause; it makes little or no sense to park an expensive robot in front of a drill press, loading and unloading it. The Japanese wouldn't use a robot this way either (unless the operation produced a hostile environment, anyway). On the other hand, the use of robots begins to make increasing amounts of economic sense if they are loading and unloading a *bank* of machines in an FMC, as seen in *Figure 20-5.*

FMCs will provide an increased number of opportunities for the use of robots. Once FMCs become widely used, the U.S. will catch up and exceed those who use robots in profusion now.

5. ENTHUSIASTIC OPERATORS

Enthusiastic operator acceptance of FMCs will help spread the use of the concept.

Where FMCs have been installed, their acceptance by the people who must run them has been exceptional. There is good reason for this; "they are fun to

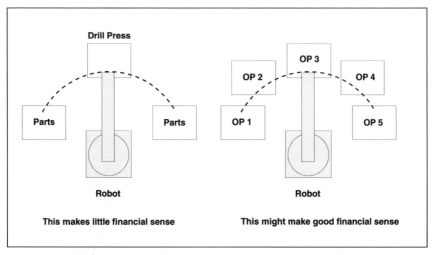

Figure 20-5. *Robots cannot be effective in the wrong environment; cells will help accelerate the demand for robotics.*

work in" when compared to the work environment they replace. *Figure 20-6* describes this traditional factory environment; this picture of a typical day in the factory is not in the least an exaggeration, as anyone knows who has performed such menial, repetitive tasks.

The FMC concept has a profound effect on this drab, if not hostile, environment. The FMC environment usually permits shop floor operators to perform more than one task. Experienced operators find that they have a more diversified job to do, and that they must think and plan. In addition to being responsible for executing the processes in their FMCs, operators running many FMCs today handle tooling, maintenance, fixtures, quality, and scheduling matters, all items that were once the responsibility of specialists in other shop departments. Many FMC installations have been able to place so much responsibility on operators, in fact, that supervision is no longer necessary.

Most operators enthusiastically embrace such added responsibility because it represents an escape from tedium. It is not unusual in a positive FMC environment to see operators coming in on their own time to fine-tune their FMCs! It is reasonable to conclude, after witnessing such happenings on numerous occasions, that machine operators, often in a union environment, haven't lacked the desire to have fun and be challenged at work, but rather lacked only the opportunity.

In the old batch production shops organized by process, an operator's workday consists of, for example, turning a run of parts on a lathe. When the operator finishes these parts, he or she sets up the lathe to turn the next batch of parts, and so on, year in and year out. When the operator isn't doing one of these two jobs, he either has nothing to do or spends his time trying to find the next batch of parts to run.

In many shops, operators are not even allowed to set up their own machines. Setup "specialists" do this work. This narrow specialization of factory tasks gives rise to work standards, which indicate that the operator is supposed to turn "x" number of parts per shift.

The operator in this environment is, in effect, himself a machine. He is doing a boring, repetitive job requiring the least use of his intellect or enthusiasm. Coming in to work in the morning to face eight hours of such dull work is hardly inspiring. Exactly at shift's end, the operator lines up in front of the time clock with his coworkers, who spend their days performing equally depressing jobs, so anxious is he to escape the monotony of his work.

Figure 20-6. *A typical day in the traditional factory.*

6. CIM REALIZABLE

The up-to-now largely illusive dream of CIM may be realizable.

The computer hardware and software has been available for several years now to accomplish computer-integrated manufacturing. At best, only a handful of manufacturing companies have been able to apply CIM economically. It should be no surprise that so few find it practical; if it cannot be done sensibly manually, it certainly should not be computerized.

Mastery of FMCs should permit more use of CIM. The work flowing through FMC batch factories will be organized so that shop floor transactions and manual interventions will be substantially reduced. Computers then ought to be able to track the flow of work. If CIM does prove economically feasible, it will not be in its present hugely expensive, complex form. This is because an FMC environment is simple, and today's CIM software (and MRP software) is designed for the complex flows of today's nonFMC environment.

A principal problem with both robots and CIM today is that they were both designed before batch shops knew how to become orderly and organized, with genuine product flow—that is, before FMC use became increasingly widespread. Further enabling the eventual use of CIM is the impact that FMCs will have on the rest of the organization, which will allow the factory to be integrated with other parts of the business.

7. APPROACHING WORLD CLASS

Perhaps most importantly, FMCs will impact the nonproduction facets of many companies, often leading the way to truly world class business operations.

FMCs are *not* the whole answer to enabling Western countries to compete successfully with offshore competitors. A shop that is producing most of its products using the FMC concept usually *will* see an excellent return on the investment in FMCs. But many offshore manufacturing concerns have taken the FMC philosophy and applied it to the rest of their manufacturing businesses, and they will continue to enjoy an edge until their Western counterparts do likewise. Many manufacturers who master FMCs quickly discern that they can begin to link other factory and nonfactory work steps with the linked work steps of an FMC, as seen in *Figure 20-7*:

- *Downstream*, some are finding that parts can drop off the end of an FMC into a shipping container and immediately be shipped to a customer or distributor. This permits instant invoice preparation and the elimination of finished goods inventory.

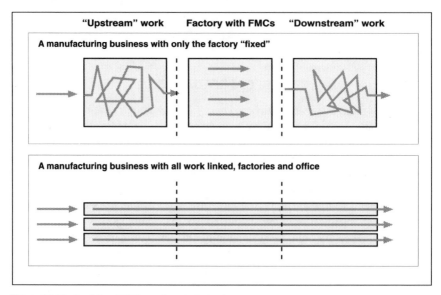

Figure 20-7. *Applying FMCs to the whole business.*

- *Within the factory*, some are finding that parts can be fed *directly* to an assembly line or another FMC. This practice can eliminate an entire assembly parts storage area and interim component buffers.
- *Upstream*, many are finding that if order entry, credit checking, and engineering steps are linked with an FMC, large amounts of calendar time can be extracted from office functions and the order fulfillment process.

Linking other functions to FMCs can sometimes permit carving up a manufacturing business into minibusinesses with their own P&L statements. Installing other techniques such as employee empowerment and its philosophies can also help a company become lean and quick, truly approaching world class status. But without FMCs leading the way, many companies will never get there.

SUMMARY

The seven trends predicted in this chapter clearly support a positive future for the FMC concept. By the end of the 1990s, few manufacturers will not have taken advantage of FMCs. The idea of physically linking the accomplishment of serial work steps in a batch environment is already a permanent part of industrial practices.

Whether an equally large number of FMC installations will be technologically upgraded into FMSs or CIM installations is not as clear. Some will. Those enterprises that, in the 1980s, jumped to FMS or CIM *before* mastering the intricacies of FMCs and failed will certainly slow down the rate of technological upgrading. Yet it should be expected that competitive pressures will eventually push the majority of manufacturers to adopt *anything* that provides a genuine edge over the competition. Effective competitive weaponry (like FMCs) is not likely to *ever* become obsolete.

CASE STUDIES IN CELLULAR MANUFACTURING

The following case studies include FMCs planned and installed by Ingersoll Engineers' clients. These projects were mostly three- to five-year rearrangements begun in the mid- to late 1980s. As part of our normal follow-up on such long-term projects, these clients are sharing some of their successes and the lessons they learned. These companies are forerunners in manufacturing, management, and those committed to investing in their futures.

Case Study #1:

IMPLEMENTING MULTIPLE MANUFACTURING CELLS TO ACHIEVE COMPETITIVENESS

Ingersoll-Rand Company
Engineered Pump Division
Phillipsburg, New Jersey

Introduction

Ingersoll-Rand Company's Engineered Pump Division manufactures centrifugal and hydraulic pumps for the mining, military, flood control, and municipal water supply industries. In 1984, a management evaluation concluded that the operation was antiquated and too large to maintain its competitive position. For example:

- Machining, assembly, and test departments were scattered throughout four buildings up to half a mile apart.
- Machines were grouped together by type, resulting in inefficient material flow.
- Existing equipment included a low quantity of NC/CNC machines.

Specifically, the evaluation indicated that lead times were high due to manufacturing methods, planning, and part flow, that inventory levels were high, and systems were outdated. In short, the facility was a low-tech shop trying to produce a highly engineered product with consistent quality and tight tolerances. *Figure 21-1* illustrates the path and distance that two component parts travelled during manufacturing in the existing layout of functionally grouped machine tools.

This evaluation led to a corporate commitment to improve the operations by creating a "factory of the future." Initially, a thorough analysis of the existing business was conducted, including future markets, old and new products, and sales levels. Next came an in-depth analysis of parts and their manufacturing strategies, and existing equipment, and the development of part families to support planning for cellular manufacturing. A team-oriented structure was used throughout the planning and implementation process. Employee groups, including direct labor workers, supervisors, and engineering support, worked to define the needs, secure capital equipment, and implement the physical and philosophical changes.

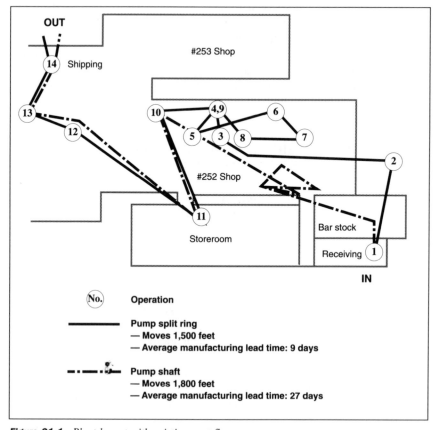

Figure 21-1. *Plant layout with existing part flows.*

Approach

The result of these preliminary planning steps was a divisionwide objective—to consolidate and modernize the Phillipsburg operation to reduce:

- Product cost,
- Manufacturing lead times,
- Space requirements and overhead costs, and
- Inventory levels (while improving inventory turns).

While the primary emphasis was on the implementation of cell technology and conversion to Just in Time manufacturing, several other subprojects were also completed:

- Facility upgrades (new floors, shipping and receiving areas, and manufacturing offices),
- Acquisition and installation of new manufacturing equipment,
- Implementation of DNC (direct numerical control) technology,
- Relocation and modernization of:
  - Support functions (such as hydrotest, shipping, receiving, and inspection),
  - Storerooms, and
  - Test stands/loops.

The overall program was planned as a five-year effort. An extensive portion of this time (four years) was to be devoted to planning and detail engineering. Final implementation was to take place during the fifth year. (Part of the "planning" phase was actual implementation of a pilot cell to debug the approach, identify critical roadblocks—either people or technology—to ultimate success, and give everyone firm successes before undertaking the full-scale implementation.)

The team members going through this process encountered a number of psychological barriers, including a "resistance to change" attitude within the division. Some of the diverse disciplines now interacting together had years of isolation to overcome with regard to participating in any division improvement activities. In some cases, no one had ever asked these people what they thought before!

A part family analysis indicated that 12 manufacturing cells would be required in the manufacturing operation. Parts were first grouped by geometry (complexity), size, and annual production quantity. The resulting cells were first analyzed on paper, and then detailed in a scaled 3-D model. This model gave the team the ability to rearrange the individual machines within the cell and all the cells to optimize both the internal cell flow and overall factory flow. The model was also a display tool for the various vendors and contractors, and a teaching tool for internal personnel. Since it was decided that production would not be slowed during the rearrangement, a detailed proximity and sequence of machine tool/area moves was developed. The 12 cells that were developed and implemented were:

- Four ring cells,
- Grind cell,

- Shaft cell,
- Impeller cell,
- One small and one large VTL (vertical turret lathe) cell,
- Prismatic cell,
- Material prep cell, and
- General machining cell.

In excess of $10 million was invested to purchase CNC tools, build shipping and receiving facilities and new test loops, and rebuild selected equipment. Additional expenditures were for small equipment and tools, and DNC implementation.

Results

Nearly five years after the project began, the results of the cells are evident: reduced lead times, part manufacturing cycle times, and work-in-process and finished inventory levels. The four ring cells, the first to be installed, resulted in specific productivity improvements:

- The introduction of new equipment with "live" tooling and right-angle milling and drilling capabilities eliminated secondary operation setups.
- Utilization of group technology techniques minimized the number of setups, while standard turret tool packages reduced setup costs.
- The use of check cuts reduced the constant adjustment of tool offsets, in turn reducing cycle time and improving quality levels.
- The improved equipment reduced cycle times because of the increased speeds and feeds.

Overall results were:

- A 40% reduction in setup time;
- A 30% reduction in cycle time;
- A 20% reduction in the number of operations;
- A 50% reduction in work-in-process and finished inventory.

Figure 21-2 offers additional detail on the results with two specific parts.

The facility space reduction project, also part of the factory-of-the-future effort, resulted in a revised facility with wider buildings, more height under crane hooks, increased crane capacity, and improved access. The new layout of the five shops (Shops 7 through 11) enabled work to move from shop to shop without exposure to the elements. Initially, the production process required 381,000 square feet; after consolidation, only 250,000 square feet were required—a 34% reduction. *Figure 21-3* shows the consolidated shop layout, where overall part flow is less than 10% of the original. Machine tools are grouped according to the family of parts being manufactured in each cell.

Conclusions

According to managers at Ingersoll-Rand Company's Engineered Pump Division, this effort provided many lessons, among them:

- The need for a completely defined commitment from management— constant changes in direction and plans are not recommended;

Part: Lock Nuts Annual Quantity: 200 Machine: Okuma LR15M CNC Lathe				
Area	**Before**	**After**	**Difference**	**Percent**
Operations	6.0	2.0	4.0	67
Cycle time (hours)	1.0	0.4	0.6	60
Setup time (hours)	4.1	2.5	1.6	39

Part: Gland Annual Quantity: 100 Machine: Okuma LC40M CNC Lathe				
Area	**Before**	**After**	**Difference**	**Percent**
Operations	4.0	2.0	2.0	50
Cycle time (hours)	0.7	0.6	0.1	14
Setup time (hours)	4.1	1.9	2.2	54

Figure 21-2. *Specific results of cellularization of two products.*

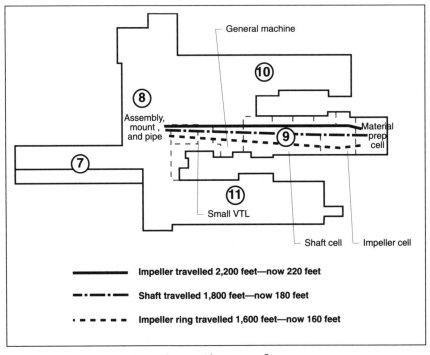

Figure 21-3. *Consolidated shop layout with new part flows.*

- The need for a more diverse project team, composed of departments affected most by the decisions.

Another important conclusion was that consultants should be part of the process to introduce new concepts, establish goals, and prepare and execute plans. At a certain point however, they should be phased out so the internal employee team can finish developing and implementing the plans, fostering ownership. Finally, because each operation is unique in its culture, goals, and personalities, reliance on textbook solutions should be avoided. A blend of textbook concepts and real-life experience will produce a plan that is realistic, achievable, and liveable.

The implementation of this factory of the future was a major undertaking for the division compared to the size and scope of previous projects. New manufacturing strategies were introduced—cells, JIT receipt of materials, production per customer order (not made for stock), DNC—and involved a divisionwide team effort with full corporate support. The successes at the Engineered Pump Division prompted other Ingersoll-Rand sites to plan to implement similar strategies.

Case Study #2:

MANUFACTURING CELLS IN A FOCUSED
BUSINESS ENVIRONMENT

Union Special Corporation
Huntley, Illinois

Introduction

Union Special Corporation manufactures industrial-grade sewing machines. The company once held a large market share, but over a period of five years, highly aggressive offshore competition decreased this share. In 1988, action was taken to halt the decline. Since the market was becoming a buyer's market, price, quality, and delivery became the sales drivers.

One product family representing more than 50% of the units sold was chosen for the first manufacturing improvement program. Within this family, over 300 different model variations were possible, and customers frequently required machine delivery within one week after receipt of order. At first, the normal lead time to produce one unit ranged from six to nine months, depending on the options selected. To satisfy all market demands, a substantial finished and semi-finished machine inventory was required.

Approach

An initial factory analysis established that the following actions were needed to improve Union Special's competitive edge:

354

- Decrease total manufacturing costs.
- Increase product quality:
  - Mechanical operation, and
  - Paint quality and consistency (aesthetics).
- Decrease manufacturing lead times.
- Eliminate finished machine inventory.
- Establish and maintain linear machine production capability.
- Improve control of the manufacturing business.

A focused factory approach with component cells was deemed the best method of achieving the program goals for the selected product family. This approach was chosen because the product family components were unique in comparison to other factory products, and the associated manufacturing requirements were also segregated from overall production operations. Several other characteristics of the product family indicated a large-scale focused factory approach would result in the most benefits:

- The product family's position in its product life cycle was mature.
- Sales volumes for the previous five years had been consistent but flat.
- Engineering documentation was current and accurate.
- Substantial in-house product knowledge and expertise existed.

The focused factory approach integrated a number of component cells, producing complete assemblies in a segregated area of the plant. Individual cells provided the necessary flexibility and response; their integration as a focused operation allowed for quick response to the various models demanded by the schedule. Long lead times for both components and finished units were reduced considerably, and better control reduced the problems of missing parts and large finished goods inventories.

A joint team of Union Special employees and Ingersoll Engineers consultants were assigned to this effort. Together, they conducted data analysis, performed detail engineering, created the program master plan, and directed implementation activities. During this three-year program, all of the required manufacturing activities were identified, integrated, and organized into a single area within the existing manufacturing facility, consolidating from long material flows shown in *Figure 21-4*. In addition to the physical changes within the facility, some changes in the fundamentals of how business was conducted also occurred within the focused factory:

- Decision-making authority was placed at the lowest possible level.
- Production was scheduled on an order basis only.
- The manufacturing operations were staffed as a business within a business.

The organizational changes instituted as part of the program were as important, if not more so, than the new equipment and factory arrangement. The concentrated nature of the focused factory operations required the concentrated attention of support staff. In a more typical organizational structure, the actions of decision makers could have actually lagged behind manufacturing's ability to react in a focused factory setting, and they might have lived up to their traditional classification as "burden!"

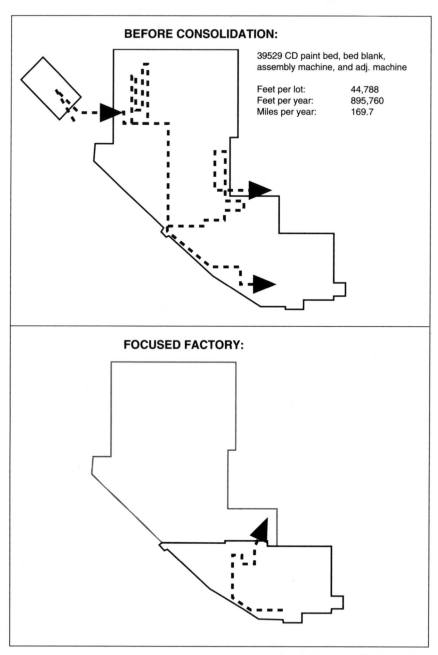

Figure 21-4. *Union Special's plant layout and component flow before consolidation and as a focused factory.*

Results

Program benefits began to accrue quickly after implementation. The production capabilities of the focused factory rapidly increased to planned levels, allowing excess inventory to be eliminated through sale. Within the first year of integrating the component cells:

- Overall product operating costs decreased 15%.
- Product inventories dropped 25%.
- Total quality costs decreased 90%.
- Cash flow increased $2 million.
- Manufacturing response time decreased 90%.
- Inventory turns improved 300%.
- Customer orders doubled.

Additionally, there was improvement in control of operations and related activities, and a large increase in labor utilization and overall manufacturing productivity.

From the planning stages through implementation, several key project considerations required attention. These issues can be categorized as "hard" (equipment-related) or "soft" (people-related):

- Hard issues included:
  —Determining equipment availability.
  —Analyzing production interruptions.
  —Designing new tooling and fixturing.
  —Outlining new production measurements.
- Soft issues included:
  —Conducting training and education.
  —Changing the functional organization into a focused one.
  —Cross-training the workforce.
  —Preparing new job descriptions and wage adjustments.

The project team also encountered a number of surprises:

- The factory workforce quickly accepted the focused business concept and operating philosophy.
- Excess inventory stockpiles decreased immediately and continued to decline without impacting delivery performance.
- All 300 model variations improved equally in their ability to be manufactured and shipped.

The improved factory flow, with focused factory cells placed in close proximity, dramatically reduced part travel distances. Factory layout was concentrated in a tight, effective area, as seen in *Figure 21-4*.

Conclusions

Reflecting on the experience, team members noted that the single biggest challenge was middle management's slow acceptance of the large-scale changes required to support the focused business program. Middle management believed that the changes required to accomplish the program goals would be mostly technical. Increased communication and education about the nature of the change was necessary to overcome this resistance.

Union Special's focused business has outperformed the remaining traditional manufacturing operations in nearly every category. Additionally, the focused business workforce has become an exceptionally motivated team and continues to improve its productivity, furthering the newly attained competitive advantage.

Case Study #3:

USING CELLS TO IMPROVE RESPONSIVENESS TO CUSTOMER DEMAND

Ingersoll-Rand Company
Centrifugal Compressor Division
Mayfield, Kentucky

Introduction

Ingersoll-Rand Centrifugal Compressor Division manufactures air compressors for industrial and commercial applications. The company supports a significant spare parts and service operation for the equipment it supplies as well.

In the late 1980s, Ingersoll-Rand's Centrifugal Compressor Division was in the midst of a period of unprecedented sales growth for both complete units and replacement parts. Five-year sales forecasts projected a more than 200% increase from the base year. To meet the challenge of increased volumes, improvements in shop floor operations and productivity were required.

Ingersoll Engineers was selected to work with Ingersoll-Rand to evaluate the existing manufacturing capability at Mayfield and define cost-effective improvements. Success and growth in the marketplace hinged on the components of customer responsiveness: high-quality products, on-time delivery, immediate parts service, and the fastest response time in the marketplace. These critical factors were translated into the specific objectives shown in *Figure 21-5*. Translating this mission of responsiveness into tangible goals led to an aggressive program of change.

Area	Program Baseline	Program Goals
Annual inventory turns	4-5	24+
Plant lead time	—	-50%
Service response	85% in less than 5 days	1 day
Cost reduction	—	20%-40%

Figure 21-5. *Project goals must be set early.*

358

Approach

The Centrifugal Compressor Division established a manufacturing strategy based on improving customer responsiveness by organizing to minimize flow time. At Mayfield, this required increased flexibility toward product line expansion or addition, and increased sensitivity to customer demands for finished product and aftermarket service. These critical customer responsiveness areas were linked together, and their integrated improvement directly impacted the ability to respond to customer needs.

Ingersoll Engineers consultants and Centrifugal Compressor Division employees conducted a major feasibility study directed at the development of a plant layout and a master plan. Redefining the existing plant layout, shown in *Figure 21-6*, was a challenging task; interdepartmental travel for complete units

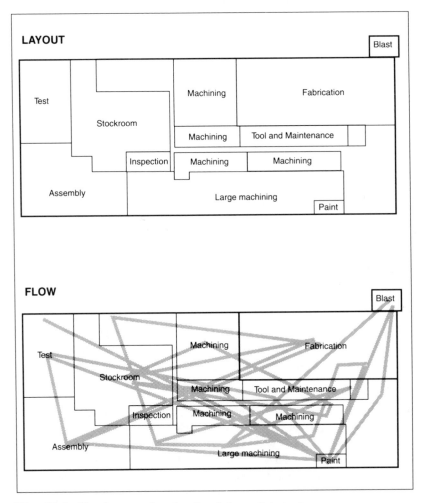

Figure 21-6. *Plant layout and product flow before cellularization.*

was 3,700 miles annually. While missed shipping schedules only became obvious in assembly and test, the focus became:

- Timely supply of parts:
  —Internal parts, and
  —Purchased parts.
- Shortages which caused substantial variances in standards;
- Supply first and assembly efficiency next.

Data used to evaluate part flow through the facility consisted primarily of part routings, standard process times, and labor reports. As is typical in many American factories, economic lot sizes of parts bounced back and forth from department to department, with flow times 20 to 50 times greater than the actual processing times. Lot sizes were usually larger than the number of parts needed at a given time, further increasing flow times and adding unnecessary inventory.

The strategy required a major reorganization of the physical plant, changes to existing buildings and grounds, and procurement of additional machines and equipment. The Mayfield plant already had one cell in operation and another was being installed; these cells provided the basis for further development. Initial analysis by the joint study team verified the six natural commodity families and their logical application for cellular manufacturing. These included:

- Bearings,
- Diffusers,
- Impellers,
- Coolers,
- Fabrications, and
- Major castings.

The layout strategy targeted the overall flow and integrated family cells into the manufacturing concept. Specific component cells would be integrated with improvements in fabrications and casting machining to support assembly and test. This, however, went beyond physical tools and equipment, to focus on the organization for rearrangement and improved work processing. "Unbolting and rebolting" the organization to minimize the flow time of decisions was as important to success as improving the physical side of the factory.

Moving to cell-based flow manufacturing, coupled with the business growth, required an investment of $30 million for new machines and equipment and factory rearrangement. While a number of major pieces of equipment were replaced, a previously expected building modification was not required.

Results

One year into implementation, benefits are being realized. While all program goals have not yet been met, progress is significant, as shown in *Figure 21-7*. In one specific instance, lot size was reduced from 300 to 1. In addition, plant lead time has been slashed from 30 days to 48 hours. Quality costs have been reduced by 75%. Significant potential savings, including an inventory reduction of 66%, are targeted over the next five years. The plant rearrangement cut the distance that components travel annually by more than half—from 3,715 miles to 1,693

Area	Program Baseline	Program Goals	Program Status
Annual inventory turns	4-5	24+	12+
Plant lead time	—	-50%	Less than 3 weeks (manufacturing)
Service response	85% in less than 5 days	1 day	1 day
Cost reduction	—	20%-40%	Not available yet

Figure 21-7. *Program progress is significant.*

miles—and proved less expensive and more implementable than other alternatives considered.

Conclusions

In retrospect, the plant controller notes that having a team of employees work closely with the outside consultants was a key to success. The primary benefit of having employees perform the majority of the research for the study was that they were in position to implement the results in a more efficient manner.

As a result of the Mayfield project, a similar effort has begun in Ingersoll-Rand's facility in Vignate, Italy.

Case Study #4:

TAKING A CELLULAR APPROACH TO
NEW AND EXISTING FACILITY LAYOUTS

Metalsa, S.A. de C.V.
Monterrey, N.L., Mexico

Introduction

Metalsa is a family-owned supplier of stamped components for the automotive industry. The company's specialty is pickup truck and light truck chassis, fuel tanks, structural parts, steering supports, and cross members. Metalsa mainly supplies the Mexican national market. This includes Big Three plants in Mexico, as well as German and Japanese operations.

During the mid- to late-1980s, Metalsa had the opportunity to enter the Class B stamping market. The company seized on this opportunity to reevaluate existing operations, and plan for expanded growth. Class B stampings would require a different level of quality compared to structural components. This would impact production operations, finishing, and even tooling itself.

Ingersoll Engineers was selected to assist with the project for new facility planning, tool room improvement, existing facility rearrangement, and even a pilot program for quick die changeover. New production plans would include more than 80 stamped parts. Reaching new quality levels would require retraining at every level of the company. New objectives aimed at becoming a customer-driven agile enterprise were set, and an aggressive timetable was put together to achieve desired results.

Approach

Since the overall program had a multifaceted agenda, separate teams were established to address the individual issues. Ingersoll Engineers, together with Metalsa's senior staff, served as the steering committee for the effort, with Ingersoll Engineers assuming the lead role for project planning and methodology. (See *Figure 21-8.*)

Not unlike many U.S. corporations, workers in staff positions were unfamiliar with problems facing operations managers and workers. Also, there was a general lack of appreciation for how decisions in one arena affected performance in another. Much of the beginning project work at Metalsa strove to overcome these kinds of obstacles; looking back, it is hard to imagine that the former environment could have ever existed.

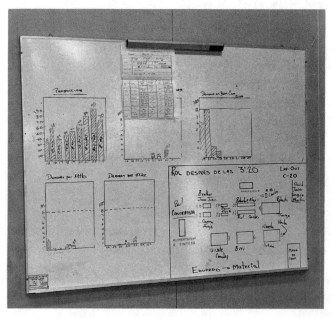

Figure 21-8. *Charted instructions and tracking play a key role in manufacturing operations.*

Of key interest was the rearrangement of the Monterrey facility into a cell-driven group of operations and the approach to setup improvement, which tied together existing part approaches as well as setting the direction for the new contract bid. Part travel through the existing layout, shown in *Figure 21-9*, was not unlike most manufacturers during the 1980s. Individual components traveled literally miles, and were held for indeterminate periods in queues, waiting for the availability of next-operation equipment.

Larger batch processing made scheduling difficult, and prediction of delivery and lead times was a guess in most instances. Quality problems were often multiplied in large batches because queuing frequently rendered obsolete or required extensive rework of entire run lots when problems were discovered. (See *Figure 21-10*.) In this environment, key drivers were:

- Shortened and predictable lead times;
- Reduced part handling and travel distances (See *Figure 21-11*);
- Minimizing of queues (elimination if possible, with continuous movement of components through the operations);
- Ever-decreasing batch sizes.

This last item lead to targeting faster changeovers, so more die changes could be incorporated in existing equipment, allowing more frequent but less voluminous batch runs.

Layout strategy. Data was gathered from a number of sources, including shop floor observations, to establish:

- Part routings,
- Machine utilization,
- Inventory levels,
- Average lead times,
- Normal batch sizes, and
- Special quality needs.

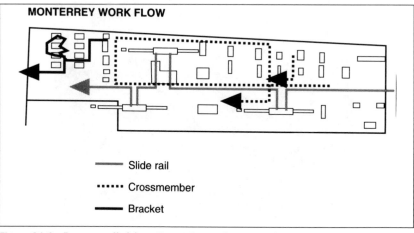

MONTERREY WORK FLOW

——— Slide rail

•••••• Crossmember

▬▬▬ Bracket

Figure 21-9. *Parts travelled for miles and spent long periods queued up for the next operation.*

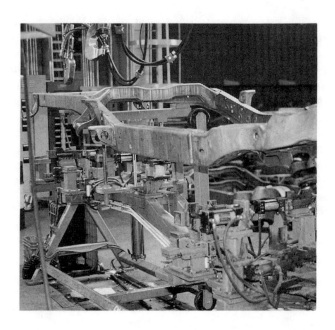

Figure 21-10. *Modular fixtures provide hole punching in finished frames, ensuring consistent location and tolerance.*

Figure 21-11. *Compact subassembly cells provide components as needed. Overhead welders clear floor space and discourage adjusting.*

A systematic approach was taken to chart special volume requirements, document representative part flows, and establish geographic and environmental concerns where special machines and operations might have to be isolated from each other.

An abstract representation was conceived, and then real department floor layouts developed to match the theoretical criteria. Final plans were then developed into a three-dimensional model so a decisive evaluation could be made.

A simple computer program was developed jointly by Ingersoll Engineers and the Monterrey Technical University to generate possible combinations of existing equipment and part routings. The basis was established through the floor operators, given a choice of what two or three presses they would select for each operation, in order of priority. These ratings considered a number of intangible elements such as ease of changeover, condition of equipment, or even existing geographic location. The importance of this is that most operations today have to deal with a history of existing tooling and processes. It is often impractical from an analysis point of view to even attempt to document all the peculiarities. From an implementation point of view, it is seldom justifiable to retool existing volumes because they don't measure up to the new criteria. Approaches are needed that can blend the fact of the old processes with the needs of new approaches.

New plant. Since the new plant layout necessitated a green-field sight, capturing "old" data was not required. What was a driver, however, was looking at the old methodologies and discussing which new approaches would be most applicable. This review asserted that progressive die technology would be most cost-effective over the former line die approaches. Tooling would be more costly, but capital equipment investment would be considerably less. Potential volume growth could be met without increased capital.

While progressive tooling would necessitate outside contracting (Metalsa designed and built the majority of its own tools), the exposure would bring the company to a new level of process understanding. The bulk of the effort could be focused on standardizing tooling design and equipment requirements.

The focus became establishing detailed equipment and tooling criteria:

- Shut heights,
- Feed line heights,
- Stroke length,
- Scrap removal,
- Location points, and
- Standardized knock-out and stripper designs.

These criteria set boundaries for future tool designs and supported the cellular approaches and quick-change needs.

Setup improvement. A pilot effort for quick die changeover was established to demonstrate to upper management what was possible with existing tooling and equipment. The effort also established with the workers the approach and methodology to be employed in bringing about the required change.

A key driver was to make use of in-house talent to establish a success story in which the reductions achieved met the criteria for improved changeover performance—reducing batch sizes as required by data analysis. In Metalsa's case, this meant reducing traditional four- to six-hour changeovers to a range of 20 to 25 minutes!

Ingersoll Engineers began the team sessions with a general statement that said most setup downtime could be improved 30 to 50% without spending a penny! This initial challenge set the tone. To establish a baseline understanding and overcome the language barrier, the first team session opened with a video of an Indianapolis car race to key in on the activities of the pit crews. The subject of car racing transcended most environmental and language barriers; most manufacturing employees could associate with the activity and understand the important criteria as they were pointed out:

- Pit crew members thoroughly understanding their individual assignments;
- Awareness of other team members' assignments and work positions (choreography);
- Special tooling design to approach certain tasks quickly and efficiently;
- Special engineering to make tasks more easily accomplished with the least effort and time;
- The requirement of frequent practice for continuous improvement.

From this basis, existing changeover approaches were documented and analyzed for opportunities to improve. *Figure 21-12* shows how shifting work that was internal to the setup (die exchange) to external work resulted in large time savings. Many of the interruptions to a quick changeover were "manageable" items that could be factored out without spending money. When this point was proven, the effort really took off!

Most important to Metalsa's demonstration was that *every* improvement didn't have to be addressed. Only those that contributed big chunks of time and paved the way toward the established goal were considered. For this reason, large capital expenditures for such things as rolling bolsters or automatic clamps were avoided; fork trucks and standard bolts worked just fine if the need for them was anticipated. (See *Figure 21-13*.)

Results

Metalsa won the new contract bid, and its Apodaca facility has grown to support many new contract opportunities as well. Monterrey's operations enjoy continuous improvement in quality approaches, and their recognition by the companies they supply serves as a record of that performance.

Revamped approaches have cut inventories 40 to 60% and lead times have kept pace with the automotive markets' continually demanding needs. Costs have improved such that Metalsa's continuing growth targets the Mexican market specifically, with little need for export to sustain improvement.

ELEMENT	MINUTES:	
	INTERNAL TO CHANGEOVER PROCESS	EXTERNAL TO CHANGEOVER PROCESS
Shut down	5	
Pull die	10	
Work area arrangement		10
Change material		10
Install new die	10	
Assemble gages		20
Run first cycle	5	
Adjust cycle	5	10
TOTAL	**35**	**50**

Figure 21-12. *Separating out those tasks that can be done external to the die change process yields a 60% improvement—without investment.*

Figure 21-13. *Chassis assembly cells incorporate similar features and density as subassembly cells.*

Probably most important has been the growth in personnel, where teams working together meet and exceed most challenges, making Metalsa a force to be reckoned with in the manufacturing community it services.

Conclusions

As Metalsa enters the 1990s, the opportunity for free trade with the United States and Canada offers new challenges. Mexico will no longer be a closed market. Unlike several Mexican companies, Metalsa is preparing to face these challenges. With dedicated people and the experience they have already gained with flexible manufacturing approaches, Metalsa should do well in the new free trade environment.

Case Study #5:

INTEGRATING TECHNOLOGY THROUGH
A CELLULAR STRATEGY

Automatic Switch Company
Aiken, South Carolina

Introduction

Automatic Switch Company manufactures, assembles, and ships literally tens of thousands of different electronically actuated valves for various industries. ASCO's products cover gas and liquid transfer, including some nuclear applications. Distribution is through numerous sales outlets, although some direct customer shipments take place. Production was split between two sites, in New Jersey and South Carolina. While there was some common product consolidation, there was considerable overlap, and unanswered questions on unique product placement.

The key operations were aluminum die casting, aluminum machining, screw machine turned parts production, and small parts assembly. Most of the equipment was 1950s to 1960s vintage, with NC machining being applied only marginally. In 1988, ASCO's management questioned the company's ability for quick customer (market) response with dependable quality at low cost.

Late in 1988, an initial assessment was made through a technology survey to address two key areas:

- A critical review of the manufacturing technology that ASCO was using, with an eye toward opportunities for quality improvement, cost reduction, and integration of operations;
- A review of components presently purchased outside, with an emphasis toward improvement in order to vertically integrate.

Bar stock parts distribution was skewed, as shown in *Figure 21-14*. Part numbers with any volume at all (which were supported by traditional eight-spindle screw machines) were only a fraction of the more than 1,500 active parts

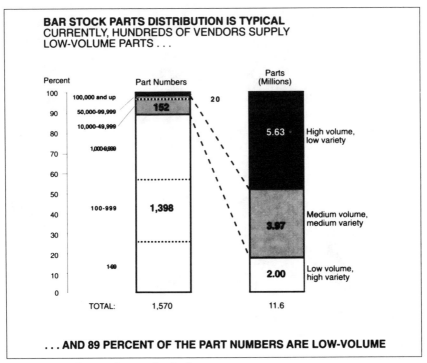

Figure 21-14. *Bar stock parts distribution.*

in use. Due to the lack of machine capability for short-run, low-volume, cost-effective production, remaining components were outsourced. In fact, 89% of the part numbers were in the low-volume category, and literally hundreds of suppliers were used to meet this demand! Also, because a number of the outside screw machine shops were no better equipped than ASCO's shop, they produced in yearly lot sizes, leaving inventory and obsolescence as additional cost factors.

Approach

The assessment would ultimately address the following objectives:

- Improving customer service by making production operations more flexible and responsive to customer needs. This would apply to new design introduction as well as producing existing designs. (See *Figure 21-15.*)
- Reducing the cost of manufacture for new and existing products by eliminating or combining operations, automating simple functions, and reducing the need for indirect labor in the production control, tooling, quality control, and maintenance areas. (See *Figure 21-16*).
- Reducing working capital requirements by permitting raw material and component inventories to be kept in their most versatile forms until customer orders were received, and then processing the material as quickly as possible.

369

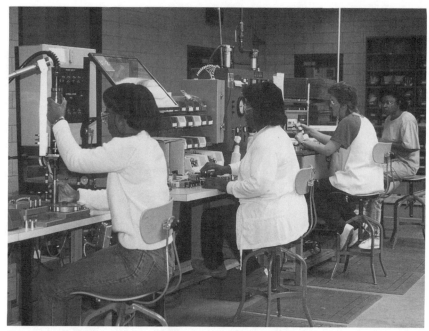

Figure 21-15. *This assembly operation was converted into a JIT cell for continuous flow.*

Figure 21-16. *The cell operator uses SPC to monitor quality within this conventional screw machine cell, made up of several machines that each perform an operation.*

The technology assessment quickly grew beyond just physical fixes, focusing on addressing other areas like:

- Product proliferation,
- Wide volume/variety patterns,
- Batch processing,
- Hundreds of suppliers,
- Lots of transactions, and
- Poor inventory accuracy.

Figure 21-17 illustrates what was in place, as opposed to some targets for how the business should operate.

The initial assessment was conducted by Ingersoll Engineers, coordinated through a few key ASCO people. It was the first phase of a three-phase approach. The remaining phases—developing the detail plan and carrying out the implementation—were carried out with "make it happen" teams, comprised of operators, managers, and key department personnel.

Kick-off meetings were held with each team to set up the process, review the current assessment, and outline goals. The individual groups then set out to address, in detail, their own work areas.

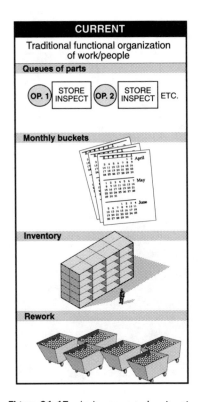
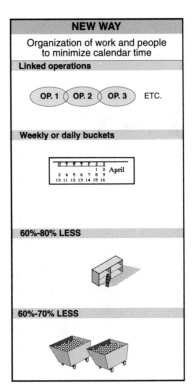

Figure 21-17. *As-is versus to-be situation.*

Together, the groups conducted detailed studies of every part manufactured or purchased by ASCO. These parts were eventually classified into part families based on similarity of part design, size, material, and operations. The groups presented their findings and recommendations to management through several working review sessions to rearrange the factory into manufacturing cells. This allowed a free interchange of ideas and questions, and provided continuing direction and attention to the project.

Existing (conventional) equipment was rearranged into cells to expand existing manufacturing capacity and begin experimenting with new concepts and approaches. The second step was to purchase CNC equipment to produce low-volume part families that were currently being purchased. To date $1.5 million has been invested in capital equipment. (See *Figure 21-18*.)

Overall, the program was based on improving flow—the flow of parts, ideas, and paper—using the following framework:

- Parts:
  —Inventory reduction—fewer queues;
  —Worker communication—quality;
  —Easier work—less motion;
  —Reduced rework;
  —Lot sizes of one versus batch.

Figure 21-18. *This CNC lathe is capable of performing all required operations, and thus is a cell by itself.*

- Ideas:
  —Improvement ideas welcome (and expected) from everyone;
  —Individual satisfaction.
- Paper:
  —Cell scheduling versus detailed machine workcenter scheduling;
  —Fewer work-in-process parts;
  —Inventory pull system—no production schedules.

Results

After the first year of a five-year program, results are already apparent:
- Standard bar machine production has been consolidated onto 30% of the original machine tools:
  —Part family grouping helped to increase available machine hours.
  —Recognition of product characteristics allowed improvement in traditional machine tool setups.
- New CNC technology has allowed for in-house production of nearly 200 of the original 1,500 bar stock components that were outsourced.
- Casting machining has been consolidated into just five cell groupings, including new CNC equipment (from the previous 30-plus work centers).
- Machining of part characteristics has eliminated some die cast core changeovers and associated inventories of stocked parts (unique requirements are literally provided on demand).
- Standardization has eliminated changeover of the back end of die cast presses, cutting setup time and changeover labor.

The project manager noted that the changes caused many elementary problems to surface that had to be corrected before moving forward, such as tooling shortages, incorrect method sheets, and incorrect routings. However, as each of these was addressed, the results met or surpassed expectations:
- Improved scheduling—there were less than 20 cell groupings (before the implementation there were more than 50 work centers).
- In some cases, setup improvements were dramatic. In a traditional screw machine product cell, changeover plummeted from eight hours to less than 15 minutes.

One of the most important changes wasn't directly related to a machine—emphasis was now placed on finding permanent solutions to nagging tooling and engineering problems, rather than installing temporary fixes.

Conclusions

Project implementers agree that an even greater effort is needed to explain the depth of such a project to upper management and other nonengineering areas. Before manufacturing can really be improved, every element that influences and impacts it must be detailed, and the interactions fully explained. Upper management needs to know, understand, and *commit* to the costs in time and manpower to set up manufacturing cells and follow through to making them work.

Case Study #6:

ADDRESSING CUSTOMER/SUPPLIER RELATIONSHIPS:
THE PRODUCT-BASED CELL

John Deere Harvester Works
A Factory of Deere and Company
East Moline, Illinois

Introduction

John Deere Harvester Works, East Moline, is the leading producer of grain harvesting machines, commonly known as combines. Its factory, which has been in existence almost 80 years, evolved and expanded with a traditional, functionally oriented layout for product flow. Accordingly, machining operations were clustered in one section of the factory and sheetmetal processing was at the opposite end in another building.

In late 1988, a new combine model was introduced, requiring significant building renovations in the main assembly building, known as V-Building. These changes resulted in about 160,000 square feet being made available for other uses. In early 1989, the company decided to move more basic, primary manufacturing operations into the available space and closer to the final assembly line.

JDHW managers identified a phased series of objectives. The first was to review which products could logically be relocated to V-Building from elsewhere in the facility. As illustrated in *Figure 21-19*, a combine can be broken down into a significant series of small subgroups. In-depth evaluation of the current manufacturing approach, and estimated improvements, highlighted

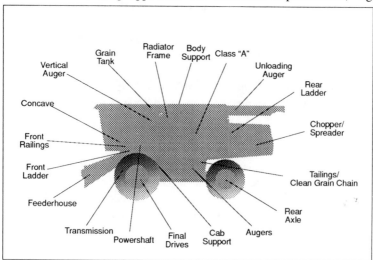

Figure 21-19. *Selected combine component subgroups.*

374

subassemblies with the greatest cost-reduction potential. Senior management wanted the review to consider products and subassemblies beyond those traditionally included in the building's assembly-only operations.

Another major objective was to create cells as close to the point of use as possible. This provided the setting for locating the "supplier" and "customer" next to each other; in the long run, it would encourage the evolution of work teams. This was all tied to a final objective of creating smaller business units within a larger factory to reduce complexity and increase manageability, while concentrating on continued quality improvement. (See *Figure 21-20.*)

Approach

Deere and Company has been a recognized leader in the application of group technology (GT) concepts in manufacturing. Most plants within the corporation had become well versed in creating GT cells. However, attempting to use a GT cellular approach inside the V-Building complex was eliminated because of space restrictions—there simply wasn't enough room. To more closely address customer/supplier relationships, a higher level of cellular manufacturing, product cells, was chosen. The guiding principal for the formation of product cells was to create as much vertical integration within the cells as possible (see the discussion comparing group technology and product cells in Chapter 5).

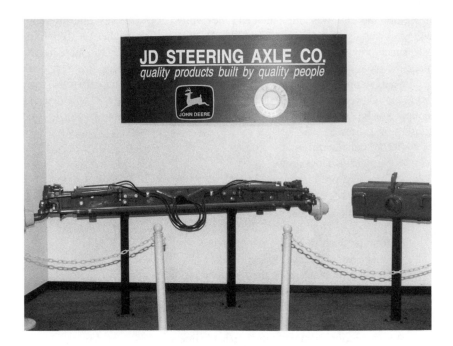

Figure 21-20. *The pride of the steering axle cell team is visible in its representation of itself as a self-directed "company" within John Deere, made possible by a joint effort between the company and the UAW.*

John Deere Harvester Works dedicated a full-time team supported by part-time members to begin the planning effort. The team's first responsibility was to review all the possible candidates for manufacture inside the V-Building complex. One of these major components was the rear-axle assembly. Available space was not an issue for this component. Additionally, machines for the cell were readily available and the savings potential was high. Once the rear-axle assembly had been identified as a pilot project, the team verified basic criteria for the cell, including:

- Machine move costs,
- Floor space required, and
- Extent of appropriate vertical integration.

The conceptual information and detailed plans were turned over to the rear-axle assembly team made up of engineers, supervisors, operators, and buyers. This team then detailed the cost reductions along with process and method changes and the capital requirements, holding quality as the overall driver.

The cost to implement the cell amounted to 44% of the total annual manufacturing cost savings, with a projected return on investment of six months.

Results

Before the cell was implemented, the typical lot size was greater than a two-week supply of rear-axle assemblies. With the much smaller lot sizes in the cell, space previously occupied by parts could be occupied by the machine that made the parts. Machines were moved closer together, resulting in a very dense cell layout, so operators could easily hand parts to their customers. In total, space requirements were reduced by 25%. As illustrated in *Figure 21-21*, the rear-axle manufacturing arrangement before cellularization was a long-traveled, time-consuming process:

- Before cellularization:
  —Number of flow routes: 30;
  —Total network distance: 225,000 feet;
  —Total yearly mileage: 2,800 miles.
- After cellularization:
  —Number of flow routes: 14;
  —Total network distance: 78,000 feet;
  —Total yearly mileage: 970 miles.

Specifically, the benefits of the rear-axle module include:

- 55% inventory reduction;
- 65% reduction in the distance travelled by rear-axle parts;
- 55% reduction in material handling;
- 80% reduction in manufacturing lead time.

These results were gained almost immediately upon startup of the rear-axle cell. There were additional tangible and intangible benefits as well. For example, one of the intangible aspects of the cell is the simplicity brought to the

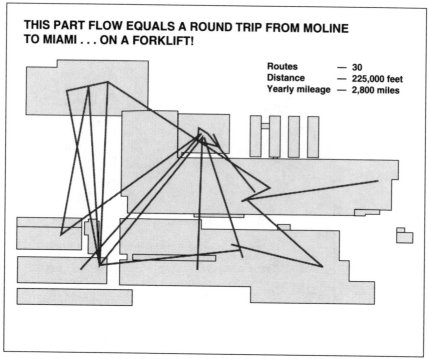

THIS PART FLOW EQUALS A ROUND TRIP FROM MOLINE TO MIAMI . . . ON A FORKLIFT!

Routes	— 30
Distance	— 225,000 feet
Yearly mileage	— 2,800 miles

Figure 21-21. *Rear-axle part flow before cellularization.*

manufacturing setting. Under traditional methods, a randomly selected part was routed through 38 different inventory control transactions as it progressed through the old flow routes; the cell reduced this to three transactions. More tangible were the quality improvements; in the first six months of tracking, no rear-axle rejects occurred in the assembly operation. (See *Figure 21-22.*)

Conclusions

One of the more significant results of the project was the change in thinking. Manufacturing people began to believe that product cells could be much more beneficial than group technology cells producing families of parts. The vertical integration concept permitted many people from different disciplines to come together to generate benefits that were not achievable in a conventionally oriented factory. The success of the rear-axle cell has resulted in many more product cells being planned and implemented at John Deere Harvester Works.

According to the manager of combine manufacturing, change itself is the most difficult challenge; spending ample time trying to educate and convert the organization is essential. His advice:

- Communicate with all areas.
- Be sure to involve the areas that are affected most.
- Include both salaried and hourly employees.
- Top-down commitment is necessary.

377

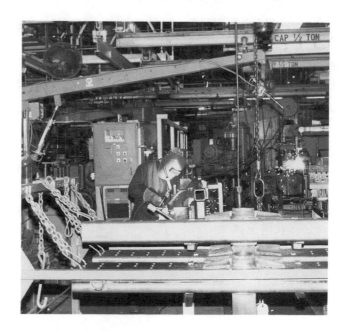

Figure 21-22. *By combining both manual and robotic welding within the cell, Deere was able to reach the ultimate balance between people and machines in an extremely dense layout.*

INDEX

Temporary equipment, 237
Testing, 184, 289
Throughput, 104, 109, 130, 151, 231, 241, 244, 246, 292, 298, 300
Throughput times, 70, 72, 129, 217, 243, 245, 306, 326, 338
Throughput velocity, 104
Time-based management, 3
Timeliness of measurements, 312, 313
Timing schedule, 285
Tolerances, 254, 274, 294, 320
Tool
  changing, 256, 265
  control, 317
  delivery, 322
  management, 219, 256, 252
  management specifications, 257
  management systems, 269
  probing, 219
  setting, 110
  wear, 267
Tooling, 233, 240, 251, 365, 366
Total preventive maintenance program, 295
Tracking, 297-301
Traditional performance measurements, 309
Training, 45-47, 61, 189, 232, 240, 261, 313-314
Training manuals, 261
Transport, 293
Turf protection, 55
Turnkey
  conversions, 333
  equipment specifications, 259
  inclusions, 274
  machines, 254
  package, 255
  programs, 323
  proposal, 255
  specifications, 257
  systems, 330
  vendor, 256

U

Uninterrupted production/schedules, 237
Unions, 38

Unit load principles, 151, 159
Unproductive teams, 57
Upstream work, 346
Uptime, 254, 272, 274
Utility requirements, 285

V

Value-adding operations, 150, 195, 208
Value-adding time, 246, 247
Velocity, 116, 224-225, 243, 245-246
Velocity of orders, 237
Vendor
  arranged visits, 272
  certification, 192, 194
  evaluation matrix, 263-269
  identification, 251-253
  list, 263
  partnering, 330
  performance, 274
  proposals, 270-272
  selection, 273
Verification tools, 297
Vision sensors, 173

W

War room, 58, 62, 63, 64, 225
Waste, 183, 186, 192, 237, 239
Water jets, 139, 164
Weight factors, 115, 120-121, 265, 270
Workcenters, 85, 289
Workholding devices, 256
Work-in-process, 24, 298, 315, 341, 352, 372
Work-in-process inventory, 30, 116, 247
Work plan development, 64-65

Z

Zero defects, 187, 296
Zero-risk option, 31